TAKING SIDES

Clashing Views on Controversial

Environmental Issues

NINTH EDITION

Clashing Views on Controversial

Environmental Issues

NINTH EDITION

Selected, Edited, and with Introductions by

Theodore D. Goldfarb
State University of New York at Stony Brook

McGraw-Hill/Dushkin
A Division of The McGraw-Hill Companies

This book is dedicated to my children and grandchildren as well as all other children for whom the successful resolution of these issues is of great urgency.

Photo Acknowledgment
Cover image: © 2001 by PhotoDisc, Inc.

Cover Art Acknowledgment
Charles Vitelli

Copyright © 2001 by McGraw-Hill/Dushkin,
A Division of The McGraw-Hill Companies, Inc., Guilford, Connecticut 06437

Copyright law prohibits the reproduction, storage, or transmission in any form by any means of any portion of this publication without the express written permission of McGraw-Hill/Dushkin and of the copyright holder (if different) of the part of the publication to be reproduced. The Guidelines for Classroom Copying endorsed by Congress explicitly state that unauthorized copying may not be used to create, to replace, or to substitute for anthologies, compilations, or collective works.

Taking Sides ® is a registered trademark of McGraw-Hill/Dushkin

Manufactured in the United States of America

Ninth Edition

123456789BAHBAH4321

Library of Congress Cataloging-in-Publication Data
Main entry under title:
Taking sides: clashing views on controversial environmental issues/selected, edited, and with introductions by Theodore D. Goldfarb.—9th ed.
Includes bibliographical references and index.
1. Environmental policy. 2. Environmental protection. I. Goldfarb, Theodore D., *comp.*
363.7

0-07-243097-4
1091-8825

Printed on Recycled Paper

Preface

$\mathbf{F}$or the past 24 years I have been teaching an environmental chemistry course, and my experience has been that the critical and complex relationship we have with our environment is of vital and growing concern to students, regardless of their majors. Consequently, for this ninth edition, I again sought to shape issues and to select readings that do not require a technical background or prerequisite courses in order to be understood. In addition to the sciences, this volume would be appropriate for such disciplines as philosophy, law, sociology, political science, economics, and allied health—any course where environmental topics are addressed.

Faculty are divided about whether or not it is appropriate to use a classroom to advocate a particular position on a controversial issue. Some believe that the proper role of a teacher is to maintain neutrality in order to present the material in as objective a manner as possible. Others, like myself, find that students rarely fail to recognize their instructors' points of view. Rather than reveal which side I am on through subtle hints, I prefer to be forthright about it, while doing my best to encourage students to develop their own positions, and I do not penalize them if they disagree with my views. No matter whether the goal is to attempt an objective presentation or to encourage advocacy, it is necessary to present both sides of any argument. To be a successful proponent of any position, it is essential to understand your opponents' arguments. The format of this text, with 38 essays arranged in pro and con pairs on 19 environmental controversies, is designed with these objectives in mind.

In the *introduction* to each issue, I present the historical context of the controversy and some of the key questions that divide the disputants. The *postscript* that follows each pair of essays includes comments offered to provoke thought about aspects of the issue that are suitable for classroom discussion. A careful reading of my remarks may reveal the positions I favor, but the essays themselves and the *suggestions for further reading* in each postscript should provide the student with the information needed to construct and support an independent perspective. Also, the *On the Internet* page that accompanies each part opener provides Internet site addresses (URLs) that should prove useful as starting points for further research.

Changes to this edition This ninth edition has been extensively revised and updated. There are two completely new issues: *Is the Precautionary Principle a Sound Basis for Environmental Policy?* (Issue 4) and *Is Sustainable Development Compatible With Human Welfare?* (Issue 16). For one of the issues retained from the previous edition, the issue question has been significantly modified to focus the debate more sharply and to bring it up-to-date: *Should Environmental Policy Attempt to Cure Environmental Racism?* (Issue 6). For one of these issues both selections have been replaced: *Are Aggressive International Efforts Needed*

i

to Slow Global Warming? (Issue 18). For the other modified issues, either the Yes selection or the No selection has been replaced. The Yes selection has been replaced for Issue 1: *Should a Price Be Put on the Goods and Services Provided by the World's Ecosystems?* and Issue 19: *Are Major Changes Needed to Avert a Global Environmental Crisis?* The No selection has been replaced for Issue 10: *Is Biotechnology an Environmentally Sound Way to Increase Food Production?* In all, 11 of the 38 selections in this ninth edition are new.

A word to the instructor   An *Instructor's Manual With Test Questions* (multiple-choice and essay) is available through the publisher for the instructor using *Taking Sides* in the classroom. Also available is a general guidebook, *Using Taking Sides in the Classroom,* which has general suggestions for adapting the pro-con approach in any classroom setting. An online version of *Using Taking Sides in the Classroom* and a correspondence service for Taking Sides adopters can be found at http://www.dushkin.com/usingts/.

Taking Sides: Clashing Views on Controversial Environmental Issues is only one title in the Taking Sides series. If you are interested in seeing the table of contents for any of the other titles, please visit the Taking Sides Web site at http://www.dushkin.com/takingsides/.

Acknowledgments   I received many helpful comments and suggestions from friends and readers across the United States and Canada. Their suggestions have markedly enhanced the quality of this edition and are reflected in the new issues and the updated selections.

Special thanks go to those who responded to the questionnaire with specific suggestions for the ninth edition:

Ruth Barczewski
Kaskaskia College

Duane Bartak
University of Northern Iowa

John Bumpus
University of Northern Iowa

R. Laurence Davis
University of New Haven

David Leonard Downie
Columbia University

George Dugan
Middlesex Community College

Thomas Duncan
Nichols College

Ray Faber
St. Mary's University

Sandra Frankmann
University of Southern Colorado

William Hallahan
Nazareth College

Nan Ho
Las Positas College

Nayyer Hussain
Tougaloo College

Dale Lightfoot
Oklahoma State University

Charles Maier
Wayne State University

John Martin
University of Maine

Jane McElroy
University of Wisconsin

David McKee
Texas A&M University

Donald Spano
University of Southern Colorado

Timothy O'Keefe
California State Polytechnic University

Carol Stanton
Southern California University

Martin L. Saradjian
Endicott College

Richard Sylves
University of Delaware

Finally, I am grateful to Theodore Knight, list manager for the Taking Sides program, for his assistance.

Theodore D. Goldfarb
State University of New York at Stony Brook

Contents In Brief

Contents

Worldwatch Institute senior researcher Janet N. Abramovitz argues that failing to assign an appropriate economic value to the "free" services provided by the ecosystem encourages the misuse and destruction of the systems that provide these services. Environmental ethicist and philosopher Mark Sagoff agrees that it is important to recognize the great value of nature's services, but he rejects efforts to attach a price to them as futile attempts to legitimize the standard cost-benefit policy framework, which he believes undermines the struggle to protect the natural environment.

Nature writer Rick Bass defends the need for true wildlands, rather than managed ecosystems, if we are to preserve our ecological heritage and the cultural treasures that it inspires. William Tucker, a writer and social critic, asserts that wilderness areas are elitist preserves designed to keep people out.

David Langhorst, executive board member of the Idaho Wildlife Federation, asserts that the Endangered Species Act has saved hundreds of plant and animal species that were in serious decline and that reauthorization of the act is in the public interest. Mark L. Plummer, an environmental economist, argues that the act's goal of bringing listed species to full recovery is not achievable.

Environmental attorneys Patti Goldman and J. Martin Wagner argue that the precautionary principle is essential to protecting public health and the environment; it must not be abandoned in the pursuit of international policy. Science correspondent Ronald Bailey counters that the precautionary principle slows development and is a luxury affordable only by "those who live in societies already replete with technology."

Bruce Yandle, a professor of economics and legal studies, argues that technological development is transforming the world into a Garden of Eden. He maintains that environmental regulation is an unnecessary, misguided effort that threatens private property rights. *Business Week*'s Washington correspondent Doug Harbrecht contends that it is absurd to have to pay owners of private property for obeying environmental regulations.

Health planning sociologist Jan Marie Fritz discusses the national and international manifestations of environmental racism and the global imperative in the search for environmental justice. Writer and social analyst David Friedman denies the evidence of environmental racism. He argues that the environmental justice movement is a government-sanctioned political ploy that will hurt urban minorities by driving away industrial jobs.

The International Food Information Council, a food industry–supported education organization, asserts that biotechnology can safely modify crops in ways that will help feed the increasing world population while reducing the resulting toll on the environment. Worldwatch Institute researcher Brian Halweil argues that the genetic modification of crops threatens to produce pesticide-resistant insect pests and herbicide-resistant weeds, will victimize poor farmers, and is unlikely to feed the world.

Zoologist Theo Colborn, journalist Dianne Dumanoski, and John Peterson Myers, director of the W. Alton Jones Foundation, present evidence suggesting that low environmental levels of hormone-mimicking chemicals may threaten the health of humans and other animals. Toxicologist Stephen H. Safe argues that the suggestion that industrial estrogenic compounds contribute to increased cancer incidence and reproductive problems in humans is not plausible.

Carol M. Browner, administrator for the Environmental Protection Agency (EPA), summarizes the evidence and arguments that were the basis for the EPA's proposal for more stringent standards for ozone and particulates. Daniel B. Menzel, a professor of environmental medicine and a researcher on air pollution toxicology, argues that adequate research has not been done to demonstrate that the new standards will result in the additional public health benefits that would justify the difficulty and expense associated with their implementation.

DuPont corporate counsel Bernard J. Reilly argues that the Superfund legislation has led to unfair standards and waste cleanup cost delegation. *Audubon* contributing editor Ted Williams warns against turning Superfund into a public welfare program for polluters.

Environmental Defense Fund scientist Richard A. Denison and economic analyst John F. Ruston rebut a series of myths that they say have been promoted by industrial opponents in an effort to undermine the environmentally valuable and successful recycling movement. Engineering and economics researchers Chris Hendrickson, Lester Lave, and Francis McMichael assert that ambitious recycling programs are often too costly and are of dubious environmental value.

Science writer Luther J. Carter and nuclear engineer Thomas H. Pigford argue that establishing clear goals that would culminate in a safe, permanent nuclear waste repository at Yucca Mountain is a realistic and sensible strategy. Risk assessment expert D. Warner North discusses the many formidable technical and political problems with the current strategy to site a permanent nuclear waste facility at Yucca Mountain and argues that a new, more open and flexible paradigm is needed to deal with disposal of radioactive materials.

Chris Bright argues that human impacts on the environment are so extensive that we face an era of catastrophic surprises unless we learn to think of the world as a complex system and behave accordingly. The late professor of economics and business administration Julian L. Simon predicts that over the long term, the brainpower of more people coupled with the market forces of a free economy will lead to improved standards of living and a healthier environment.

Introduction

The Environmental Movement

Theodore D. Goldfarb

Environmental Consciousness

The twentieth century saw immense changes in the conditions of human life and in the environment that surrounds and supports us all. According to historian J. R. McNeill, in *Something New Under the Sun: An Environmental History of the Twentieth-Century World* (W. W. Norton, 2000), the environmental impacts that resulted from the interactions of burgeoning population, technological development, shifts in energy use, politics, and economics in that period are unprecedented in both degree and kind. However, the worst impact may be the acceptance as "normal" of a very temporary situation that "is an extreme deviation from any of the durable, more 'normal,' states of the world over the span of human history, indeed over the span of earth history." We are thus not prepared for the inevitable, and perhaps, drastic, changes ahead.

Environmental factors cannot be denied their role in human affairs. Nor can human affairs be denied their place in any effort to understand environmental change. As McNeill says, "Both history and ecology are, as fields of knowledge go, supremely integrative. They merely need to integrate with each other." The environmental movement, which grew during the twentieth century in response to increasing awareness of human impacts, is a step in that direction.

In June 1992 Rio de Janeiro was the site of the United Nations Conference on Environment and Development (UNCED), popularly billed as the Earth Summit. UNCED, which was the follow-up to a much more modest United Nations conference held 20 years earlier, consisted of two massive, global conferences—one of official government delegations and the other of a diverse array of nongovernmental organizations (NGOs)—as well as a separate "Earth Parliament" comprised of 800 delegates of indigenous peoples. The most far-reaching outcome of UNCED was a 600-page agreement called Agenda 21, which sets guidelines for how, under UN leadership, the governments and businesses of the world should attempt to achieve economic growth while maintaining environmental quality. Two years prior to the Earth Summit, on April 22, 1990, 200 million people in 140 countries around the world participated in a variety of activities to celebrate Earth Day. It was also a follow-up to an event that took place two decades earlier, the first Earth Day (celebrated only in the United

States), which many social historians credit with spawning the ongoing global environmental movement.

Comparing the enormous increase in size, complexity, range of issues, and diversity of participation in either UNCED or Earth Day 1990 with its predecessor event reveals the explosive growth in political, scientific and technical, regulatory, financial, industrial, and educational activity related to an expanding list of environmental problems that has developed in the intervening years. Industrial development has reached a level at which pollutants threaten not only local environments but also the global ecosystems that control the Earth's climate and the ozone shield that filters out potentially lethal solar radiation. The elevation of environmental concern to a prominent position on the international political agenda persuaded commentators on Earth Day 1990 events to speculate that the world was entering "the decade—or even era—of the environment." The initial attention given to UNCED and the ongoing activities it spawned at first appeared to confirm this prediction. However, as the world enters the new millenium with increasing concern about future worldwide economic prosperity, there is growing resistance from the international industrial community to the imposition of further environmental regulations and restrictions. In June 1997 a five-year review at a special session of the UN General Assembly revealed little progress in implementing the Earth Summit agreements. The unprecedented and surprising progress made in controlling the release of pollutants that destroy stratospheric ozone has not resulted in similar, rapid progress in reducing the emission of "greenhouse gases" that threaten global climatic stability. In December 1997 an international forum was held in Kyoto, Japan, to consider such actions, but it produced only a modest protocol that even the most optimistic assessments judged to be no more than a first step. Yet even this agreement has encountered strong congressional opposition in the United States, and its implementation seems unlikely.

The History of U.S. Environmentalism

The current interest in environmental issues in the United States has its historical roots in the conservation movement of the late nineteenth and early twentieth centuries. This earlier, more limited, recognition of the need for environmental preservation was a response to the destruction wrought by uncontrolled industrial exploitation of natural resources in the post–Civil War period. Clear-cutting forests, in addition to producing large devastated areas, resulted in secondary disasters. Bark and branches left in the cutover areas fueled several major midwestern forest fires. Severe floods were caused by the loss of trees that previously had helped to reduce surface water runoff. The Sierra Club and the Audubon Society, the two oldest environmental organizations still active today, were founded around the turn of the century and helped to organize public opposition to the destructive practice of uncontrolled natural resource exploitation. Mining, grazing, and lumbering were brought under government control by such landmark legislation as the Forest Reserve Act of 1891 and the Forest Management Act of 1897. Schools of forestry were established at several of the land grant colleges to help develop the scientific expertise needed for the wise management of forest resources.

The present environmental movement can be traced back to 1962, when Rachel Carson's book *Silent Spring* appeared. The book's emotional warning about the inherent dangers in the excessive use of pesticides ignited the imagination of an enormous and disparate audience who had become uneasy about the proliferation of new synthetic chemicals in agriculture and industry. The atmospheric testing of nuclear weapons had resulted in widespread public concern about the effects of nuclear radiation. City dwellers were beginning to recognize the connection between the increasing prevalence of smoky, irritating air and the daily ritual of urban commuter traffic jams. The responses to Carson's book included not only a multitude of scientific and popular debates about the issues she had raised but also a groundswell of public support for increased controls over all forms of pollution.

The rapid rise in the United States of public concern about environmental issues is apparent from the results of opinion polls. Similar surveys taken in 1965 and 1970 showed an increase from 17 to 53 percent in the number of respondents who rated "reducing pollution of air and water" as one of the three problems they would like the government to pay more attention to. By 1984 pollster Louis Harris was reporting to Congress that 69 percent of the public favored making the Clean Air Act more stringent. A CBS News/*New York Times* survey revealed that 74 percent of respondents in 1990 (up from 45 percent in 1981) supported protecting the environment *regardless of the cost.*

The growth of environmental consciousness in the United States swelled the ranks of the older voluntary organizations, such as the national Wildlife Federation, the Sierra Club, the Isaac Walton League, and the Audubon Society, and has led to the establishment of more than 200 new national and regional associations and 3,000 local ones. Such national and international groups as the Environmental Defense Fund, Friends of the Earth, the National Resources Defense Council, Environmental Action, the League of Conservation Voters, and Zero Population Growth have become proficient at lobbying for legislation, influencing elections, and litigating in the courts.

Environmental literature has also grown exponentially since the appearance of *Silent Spring.* Many popular magazines, technical journals, and organizational newsletters devoted to environmental issues have been introduced, as well as hundreds of books, some of which, like Paul Ehrlich's *The Population Bomb* (1968) and Barry Commoner's *The Closing Circle* (1972), have become best-sellers.

Clashing Views From Conflicting Values

As with all social issues, those on opposite sides of environmental disputes have conflicting personal values. On some level, almost everyone would admit to being concerned about threats to the environment. However, enormous differences exist in individual perceptions about the seriousness of some environmental threats, their origins, their relative importance, and what to do about them. In most instances, very different conclusions, drawn from the same basic scientific evidence, can be expressed on these issues.

What are these different value systems that produce such heated debate? Some are obvious: An executive of a chemical company has a vested interest in placing greater value on the financial security of the company's stockholders than on the possible environmental effects of the company's operation. He or she is likely to interpret the potential health effects of what comes out of the plant's smokestacks or sewer pipes differently than would a resident of the surrounding community. These different interpretations need not involve any conscious dishonesty on anyone's part. There is likely to be sufficient scientific uncertainty about the pathological and ecological consequences of the company's effluents to enable both sides to reach very different conclusions from the available "facts."

Less obvious are the value differences among scientists that can divide them in an environmental dispute. Unfortunately, when questions are raised about the effects of personal value systems on scientific judgments, the twin myths of scientific objectivity and scientific neutrality get in the way. Neither the scientific community nor the general population appear to understand that scientists are very much influenced by subjective, value-laden considerations and will frequently evaluate data in a manner that supports their own interests. For example, a scientist employed by a pesticide manufacturer may be less likely than a similarly trained scientist working for an environmental organization to take data that show that one of the company's products is a low-level carcinogen in mice and interpret those data to mean that the product therefore poses a threat to human health.

Even self-proclaimed environmentalists frequently argue over environmental issues. Hunters, while supporting the prohibition of lumbering and mining on their favorite hunting grounds, strongly oppose the designation of these regions as wilderness areas because that would result in their being prohibited from using their vehicles to bring home their bounty. Also opposed to wilderness designation are foresters, who believe that forest lands should be scientifically managed rather than left alone to evolve naturally.

Political ideology can also have a profound effect on environmental attitudes. Those critical of the prevailing socioeconomic system are likely to attribute environmental problems to the industrial development supported by that system. Others are likelier to blame environmental degradation on more universal factors, such as population growth.

Changes in prevailing social attitudes influence public response to environmental issues. The American pioneers were likely to perceive their natural surroundings as being dominated by hostile forces that needed to be conquered or overcome. The notion that humans should conquer nature has only slowly been replaced by the alternative view of living in harmony with the natural environment, but the growing popularity of the environmental movement evinces the public's acceptance of this goal.

Protecting the Environment

There has always been strong resistance to regulatory restraints on industrial and economic activity in the United States. The most ardent supporters of America's capitalist economy argue that pollution and other environmental effects have certain costs and that regulation will take place automatically through the marketplace. Despite mounting evidence that the social costs of polluted air and water are usually external to the economic mechanisms affecting prices and profits, prior to the 1960s, Congress imposed very few restrictions on the types of technology and products industry could use or produce.

As noted above, the turn-of-the-century conservation movement did result in legislation restricting the exploitation of lumber and minerals on federal lands. Similarly, in response to public outrage over numerous incidents of death and illness from adulterated foods, Congress established the Food and Drug Administration (FDA) in 1906.

Regulatory Legislation

The environmental movement of the 1960s and 1970s produced a profound and controversial change in the political climate concerning regulatory legislation. Concerns such as the proliferation of new synthetic chemicals in industry and agriculture, the increased use of hundreds of inadequately tested additives in foods, and the effects of automotive emissions were pressed on Congress by increasingly influential environmental organizations. Beginning with the Food Additives Amendment of 1958, which required FDA approval of all new chemicals used in the processing and marketing of foods, a series of federal and state legislative and administrative actions resulted in the creation of numerous regulations and standards aimed at reducing and reversing environmental degradation.

Congress responded to the environmental movement with the National Environmental Policy Act of 1969. This act pronounced a national policy requiring an ecological impact assessment for any major federal action. The legislation called for the establishment of a three-member Council on Environmental Quality to initiate studies, make recommendations, and prepare an annual Environmental Quality Report. It also requires all agencies of the federal government to prepare a detailed environmental impact statement (EIS) for any major project or proposed legislation in which they are involved. Despite some initial attempts to evade this requirement, court suits by environmental groups have forced compliance, and now, new facilities like electrical power plants, interstate highways, dams, harbors, and interstate pipelines can proceed only after preparation and review of an EIS.

Another major step in increasing federal antipollution efforts was the establishment in 1970 of the Environmental Protection Agency (EPA). Many programs previously administered by a variety of agencies, such as the departments of the Interior, Agriculture, Health, Education, and Welfare, were transferred to this new, central, independent agency. The EPA was granted authority to do research, propose new legislation, and implement and enforce laws concerning air and water pollution, pesticide use, radiation exposure, toxic substances, solid waste, and noise abatement. The year 1970 also marked the establishment of the Occupational Safety and Health Administration (OSHA), the result of a long struggle by organized labor and independent occupational health organizations to focus attention on the special problems of the workplace.

The first major legislation to propose the establishment of national standards for pollution control was the Air Quality Act of 1967. The Clean Air Act of 1970 specified that ambient air quality standards were to be achieved by July 1, 1975 (a goal that was not met and remains elusive), and that automotive hydrocarbon, carbon monoxide, and nitrogen oxide emissions were to be reduced by 90 percent within five years—a deadline that was repeatedly extended. Specific standards to limit the pollution content of effluent wastewater were prescribed in the Water Pollution Control Act of 1970. The Safe Drinking Water Act of 1974 authorized the EPA to establish federal drinking water standards, applicable to all public water supplies. The Occupational Safety and Health Act of 1970 allowed OSHA to establish strict standards for exposure to harmful substances in the workplace. The Environmental Pesticide Control Act of 1972 gave the EPA authority to regulate pesticide use and to control the sale of pesticides in interstate commerce. In 1976 the EPA was authorized to establish specific standards for the disposal of hazardous industrial wastes under the Resource Conservation and Recovery Act—but it was not until 1980 that the procedures for implementing this legislative mandate were announced. Finally, in 1976, the Toxic Substance Control Act became law, providing the basis for the regulation of public exposure to toxic materials not covered by any other legislation.

All of this environmental legislation in such a short time span produced a predictable reaction from industrial spokespeople and free-market economists. By the late 1970s attacks on what critics referred to as overregulation appeared with increasing frequency in the media. Antipollution legislation was criticized as a significant contributor to inflation and a serious impediment to continued industrial development.

One of the principal themes of Ronald Reagan's first presidential campaign was a pledge to get regulators off the backs of entrepreneurs. He interpreted his landslide victory in 1980 to mean that the public supported a sharp reversal of the federal government's role as regulator in all areas, including the environment. Two of Reagan's key appointees were Interior Secretary James Watt and EPA Administrator Ann Gorsuch Burford, both of whom set about to reverse the momentum of their agencies with respect to the regulation of pollution and environmental degradation. It soon became apparent that Reagan and his advisers had misread public attitudes. Sharp staffing and budget cuts at the EPA and OSHA produced a counterattack by environmental organizations whose membership rolls had continued to swell. Mounting public criticism of the neglect of environmental concerns by the Reagan administration was compounded by allegations of misconduct and criminal activity against environmental officials, including Ms. Burford, who was forced to resign. President Reagan attempted to mend fences with environmentalists by recalling William Ruckelshaus, the popular, first EPA administrator, to again head the agency.

But throughout Reagan's presidency, few new environmental initiatives were carried out.

Despite campaign promises to return to vigorous efforts to curb pollution, President George Bush received poor grades for the overall environmental policies of his administration. However, he can be credited with providing the support that resulted in the enactment of the long-stalled 1990 Clean Air Act amendments. Despite some criticisms concerning compromises with the automobile and fossil fuel industries, most environmentalists were pleased with many aspects of the new law, particularly its provisions designed to decrease the threat of acid rain. This early optimism was soon negated by what many perceived to be weak efforts to implement and enforce this legislation. Bush has also been faulted for his failure to implement an environmentally sound energy policy and his refusal to support other industrial nations' proposed initiatives to slow global warming and deforestation.

Once again a new president, Bill Clinton, was elected in 1992 on a platform that pledged to reverse the environmental neglect of his predecessors. This pledge was reinforced by the fact that his choice for vice president, Al Gore, had gained a reputation as an environmental activist. The administration failed to make much headway in fulfilling its campaign promises during its first two years in office, despite the appointment of committed environmentalist Carol Browner to head the EPA. Initially encouraged by the selection of environmental advocate Bruce Babbitt as secretary of the interior, environmentalists were soon disheartened by his failure to successfully press for restrictions on the ecological damage that results from the commercial exploitation of public lands. Since the 1994 elections, the U.S. Congress has been dominated by legislators who once again echo Reagan's promise to reduce the burden of environmental restrictions on industry and commerce. In their successful 1996 reelection campaign, Clinton and Gore again promised vigorous promotion of an environmental protection agenda, but little progress has been made. Efforts to reauthorize and update such important legislation as the Endangered Species Act, the Resource Conservation and Recovery Act, and the Comprehensive Environmental Response, Compensation, and Liability Act (Superfund) have been stalled since 1993. Having failed to override presidential vetoes of legislation designed to weaken environmental protection, the antienvironmental members of Congress switched to the tactic of trying to attach riders and other such legislation to important, end-of-the-year budget appropriations bills. A strong, organized response from the environmental community has thus far succeeded in having most of these riders removed. Among the riders that survived and were signed into law in October 1998 is one that could double logging in California's ancient forest ecosystem and another that blocks the upgrading of automobile fuel efficiency standards.

Global Developments

Although initially lagging behind the United States in environmental regulation, many other developed industrial countries have been moving rapidly over the past two decades to catch up. In a few Western European countries where "green parties" have become influential participants in the political process, and in Japan, certain pollutant emission standards are now more stringent than their U.S. counterparts. A uniform system of environmental regulations and controls is prominent among the controversial issues being planned and implemented by the nations of the European Economic Community.

Following the dissolution of the Soviet Union, it became clear that the former Soviet bloc countries had enforced few environmental restraints during their postwar industrialization. As a result, these countries—as well as China—presently suffer the consequences of the most severely degraded environments in the world. In many cases economic priorities preclude the prospect of a rapid reversal of this unhealthy situation.

Although the feeding and clothing of their growing populations continue to be the dominant concerns of developing countries, they too are paying increasing attention to environmental protection. Suggestions that they forgo the use of industrial technologies that have resulted in environmental degradation in developed countries are often viewed as an additional obstacle to the goal of raising their standard of living.

During the past decade, attention has shifted from a focus on local pollution to concern about global environmental degradation. Studies of the potential effects of several gaseous atmospheric pollutants on the Earth's climate and its protective ozone layer have made it apparent that human activity has reached a level that can result in major impacts on the planetary ecosystems. A series of major international conferences of political as well as scientific leaders have been held with the goal of seeking solutions to threatening worldwide environmental problems. The "North-South" disputes that limited the agreements reached at the Rio Earth Summit were about how to promote future industrial development to avert or minimize the threats to the world's ecosystems, while satisfying the frequently conflicting socioeconomic needs of the developed "North" and developing "South" nations.

Current Environmental Issues

Most analyses of the ongoing environmental movement conclude that it has been unsuccessful in stemming the tide of ecosystem degradation while acknowledging that the world would now be in much worse shape without the educational and regulatory response that the movement has generated to date. Among the proposals for the future is for regulatory agencies to adopt a more holistic approach to environmental protection, rather than continue their attempts to impose separate controls on what are actually interconnected problems. Another idea, supported by business leaders and even some environmental organizations, is the use of market-based strategies, such as pollution taxes or the trading of pollution rights (see Issue 8), which some feel are potentially more effective than regulatory emission standards. However, many environmentalists are enraged by the idea of selling the right to pollute, which they consider immoral.

An entirely new paradigm for protecting the environment is embodied in the concept "sustainable development," whose advocates propose replacing our entire system of energy production, transportation, and industrial technology with systems that are designed from the start to produce minimal cumulative environmental degradation. An excellent introduction to this concept is included in the 1987 World Commission on the Environment report *Our Common Future* (often referred to as the *Brundtland Report* after its principal author, commission chairperson Gro Harlem Brundtland). This idea is related to the theme of prominent environmentalist Barry Commoner's book *Making Peace With the Planet* (Pantheon Books, 1990). In *Making Peace,* Commoner argues that attempts to merely limit pollution that is produced by existing inappropriate technologies, in the face of continuing global development, are doomed to failure.

An important recent development is the growing movement for environmental justice (see Issue 6). Largely absent from the group of activists who promoted the early agenda of the environmental movement, representatives of minority groups and the economically disadvantaged are organizing around the contention that they have been made to bear far more than their fair share of the effects of pollution. In response to this development, President Clinton has committed the EPA to making environmental equity an important component of future policy decisions.

Much concern in the environmental community has resulted from the emergence of what has been popularly labeled the "environmental backlash movement." With considerable funding and support from regulated industries, organizations with environmental-sounding names like Wise Use/Property Rights, the Council on Energy Awareness, and the Information Council on the Environment have rallied together to oppose environmental regulations. Some of these groups go so far as to claim that environmental problems like ozone depletion, global warming, and acid rain do not exist. One strategy of these groups that has had some success is fighting restrictions on land development on the basis of the constitutional prohibition against the "taking" of private property, which they argue applies to virtually any governmental action (see Issue 5).

Another complaint of many environmentalists is the increasing popularity of a tactic by which self-proclaimed "green" corporations mislead the public by falsely portraying their developmental strategies as environmentally sound. An example of this public-relations ploy (which has been labeled "greenwashing") that received much criticism was the extent to which the organizers of the 1990 Earth Day events allowed wealthy industrial sponsors to control the agenda and to promote their self-serving propaganda.

In response to the antienvironmental backlash, a militant wing has sprung up within the movement. Building on the confrontational tactics used by the highly successful Greenpeace organization, a radical group calling itself Earth First! staged a campaign in 1990 that they called Redwood Summer. During this campaign members chained themselves to trees and embedded metal spikes in the tree trunks to prevent the cutting of redwoods in the ancient forests of northern California. In advocating a tactic referred to as "monkey wrenching," Earth First! and other new, radical groups have condoned the destruction of earth-moving tractors and other equipment that is used for development projects that threaten the habitats of endangered animals. Recently, a small fringe group calling itself the Earth Liberation Front has upped the ante even further by burning ski lifts and buildings in Vail, Colorado, in response to a planned expansion into a National Forest area that is a habitat for lynx.

In a more moderate, mainstream response to the backlash, environmental activists are proposing that the movement return to its grassroots origins. The growing chorus of advocates of this strategy assert that in recent years the cause of environmental protection has lost its activist, populist base and is now mostly in the hands of the leaders of the major environmental organizations. These leaders, it is argued, have interests and lifestyles that are closer to those of the executives of the major polluting corporations than to those of the general public, and the policies advocated by these leaders too often involve unwise compromises that fall short of what is needed to promote a change toward sustainable development.

Ecology and Environmental Studies

Efforts to protect the environment from the far-reaching effects of human activity require a detailed understanding of the intricate web of interconnected cycles that constitute our natural surroundings. The recent blossoming of ecology and environmental studies into respectable fields of scientific study has provided the basis for such an understanding. Traditional fields of scientific endeavor such as geology, chemistry, and physics are too narrowly focused to successfully describe a complex ecosystem. Thus, it is not surprising that chemists who helped to promote the use of DDT and other pesticides failed to predict the harmful effects that accumulation of these substances in biological food chains had on birds and marine life.

Ecology and environmental studies involve a holistic study of the relationships among living organisms and their environment. It is clearly an ambitious undertaking, and ecologists are only beginning to advance our ability to predict the effects of human intrusions into natural ecosystems.

It has been suggested that our failure to recognize the potentially harmful effects of our activities is related to the way we lead our lives. Industrial development has produced lifestyles that separate most of us from direct contact with the natural systems upon which we depend for sustenance. We buy our food in supermarkets and get our water from a kitchen faucet. We tend to take the availability of these essentials for granted until something threatens the supply.

Some Thoughts on Armed Conflict and International Cooperation

It has long been recognized that a major nuclear war would produce devastating environmental consequences. In *The Fate of the Earth* (Alfred A. Knopf, 1982), Jonathan Schell provides a chilling analysis of the likely effects of radioactive fallout, including destruction of the ozone layer and radioactive contamination of the food chain. In 1983 a group of eminent scientists initiated a controversial debate by predicting that a "nuclear winter" that could threaten the continued existence of human civilization might result from even a limited nuclear conflict.

Perhaps, as some political analysts suggest, the realignment of power following the demise of the Soviet Union has reduced the threat of nuclear war. Unfortunately, we have recently learned from the Persian Gulf War that modern, *conventional*, nonnuclear war can also produce catastrophic ecological damage. The intentional release of huge quantities of petroleum into the Persian Gulf and the ignition of the vast Kuwaiti oil fields produced severe water and air pollution problems whose long-term effects are still being assessed. Several analysts have suggested that environmental factors will figure prominently as both causes and effects of future armed conflicts. Whether or not this proves to be the case, it is beyond doubt that solutions to the growing list of threats to global and regional ecosystems will require unprecedented efforts toward international cooperation.

International Organization for Standardization

The International Organization for Standard (ISO) is a nongovernmental organization that promotes the development of standardization and related activities with a view to facilitating the international exchange of goods and services.

http://www.iso.ch/infoe/intro.htm

National Endangered Species Act Reform Coalition

The National Endangered Species Act Reform Coalition is a broad-based coalition of roughly 200 member organizations, representing millions of individuals across the United States, that is dedicated to bringing balance back to the Endangered Species Act.

http://www.nesarc.org

The U.S. Environmental Protection Agency

The mission of the U.S. Environmental Protection Agency (EPA) is to protect human health and to safeguard the natural environment—air, water, and land —upon which life depends. The EPA's home page is a major resource for information on a wide variety of current environmental issues.

http://www.epa.gov

The Natural Resources Defense Council

This is the home page of the Natural Resources Defense Council, one of the most active environmental research and advocacy organizations. It includes links to fact sheets and reports on many environmental issues.

http://www.nrdc.org/index.html

The National Wildlife Federation

The National Wildlife Federation is dedicated to wilderness preservation and the protection of endangered species. The organization's home page contains links to information about these issues.

http://www.nwf.org

PART 1

General Philosophical and Political Issues

*P*eople who regard themselves as environmentalists can be found on both sides of all the issues in this section. But the participants in these debates nonetheless strongly disagree—due to differences in personal values, political beliefs, and what they perceive as their own self-interests— on how best to prevent environmental degeneration.

Understanding the general issues raised in this initial section is useful preparation for examining the more specific controversies that follow in later sections.

- Should a Price Be Put on the Goods and Services Provided by the World's Ecosystems?

- Does Wilderness Have Intrinsic Value?

- Is the U.S. Endangered Species Act Fundamentally Sound?

- Is the Precautionary Principle a Sound Basis for Environmental Policy?

- Does Environmental Regulation Unnecessarily Limit Private Property Rights?

- Should Environmental Policy Attempt to Cure Environmental Racism?

- Is Limiting Population Growth a Key Factor in Protecting the Global Environment?

- Will Pollution Rights Trading Effectively Control Environmental Problems?

ISSUE 1

Should a Price Be Put on the Goods and Services Provided by the World's Ecosystems?

YES: Janet N. Abramovitz, from "Putting a Value on Nature's 'Free' Services," *World Watch* (January/February 1998)

NO: Mark Sagoff, from "Can We Put a Price on Nature's Services?" *Report from the Institute for Philosophy and Public Policy* (Summer 1997)

ISSUE SUMMARY

YES: Worldwatch Institute senior researcher Janet N. Abramovitz argues that failing to assign an appropriate economic value to the "free" services provided by the ecosystem encourages the misuse and destruction of the systems that provide these services.

NO: Environmental ethicist and philosopher Mark Sagoff agrees that it is important to recognize the great value of nature's services, but he rejects efforts to attach a price to them as futile attempts to legitimize the standard cost-benefit policy framework, which he believes undermines the struggle to protect the natural environment.

The importance of halting the continued degradation of the world's ecosystems, which grows more urgent as worldwide environmental consciousness continues to expand, has been confounded by disputes about how best to accomplish this goal. Believers in strong central planning and governmental control insist on top-down regulation. Advocates of unfettered, industrial development promote reliance on market-based incentives coupled with self-regulation. Most industrial polluters continue to resist all efforts to regulate and restrict their activities.

All but the most single-minded supporters of a free-market economy recognize that the economic marketplace alone cannot be expected to produce consistent, environmentally sound industrial development decisions. This is because it would be a violation of their fiduciary responsibilities for the board of directors of a corporation to voluntarily decide to take expensive steps to prevent its activities from contributing to the degradation of land, air, or water when there are no costs associated with failing to do so. Thus, a corporation engaged in strip-mining cannot be expected to voluntarily restore the site of its activities to its original ecological health or aesthetic beauty, nor can a coal-burning electrical utility be expected to install costly smokestack scrubbers to remove the sulfur dioxide that contributes to acid precipitation 100 miles downwind.

The recognition that resource limitations dictate the extent to which it is possible to remediate environmental problems also links environmentalism to economics. Thus, agencies that are charged with serving as guardians of the environment must prioritize their goals and pursue the most cost-effective strategies in order to appropriately allocate their limited funds.

These are among the issues that have given rise to the relatively new field of environmental economics. Many of the policies that have been recommended by practitioners of this field have generated controversy. The use of cost-benefit comparisons to prioritize environmental goals or to test the acceptability of proposed environmental regulations has been opposed by those who question the objectivity, fairness, and appropriateness of such analyses. Recently, proposals have been made to replace such economic measures of a nation's productivity as the gross national product (GNP)—which fails to take into account the depletion of nonrenewable resources and actually includes credits for such environmental disasters as major oil spills by incorporating the economic value generated by the clean-up activities—with alternative assessment schemes that deduct the loss of resource capital and the cost of damaged ecosystems. These proposals have failed to win support from establishment economists and many environmentalists due to ideological and methodological disagreements.

Environmental economists have recently focused attention on the fact that the goods and services that nature provides in support of the economy—such as the cycling of nutrients for the production of renewable resources (like fish and forest products), the pollination of flowering plants, and the regulation of the climate—are free. As a result, they argue, little attention is given to preventing activities that may compromise nature's ability to sustain human life. In response to this concern, a group of economists, ecologists, and other scientists headed by Robert Costanza made the first comprehensive effort to assess the dollar value of ecosystem services. By reporting their findings in the British science journal *Nature*—that the median estimated value for the entire biosphere is $33 trillion per year—they hoped to draw attention to the importance of environmental protection. This report has received much media attention and has generated a flurry of responses.

In the following selections, Janet N. Abramovitz describes how most of the value in the world economy is provided, free of cost, by healthy ecosystems. She suggests that attaching economic value to these services may encourage their long-term protection. Mark Sagoff not only derides the methods employed by Costanza et al. but he also raises serious ideological and practical objections to the concept of ecovaluing.

Janet N. Abramovitz  **YES**

Putting a Value on Nature's "Free" Services

During the last half of 1997, massive fires swept through the forests of Sumatra, Borneo, and Irian Jaya, which together form a stretch of the Indonesian archipelago as wide as all of Europe. By November, almost 2 million hectares had burned, leaving the region shrouded in haze and more than 20 million of its people breathing hazardous air. Tens of thousands of people had been treated for respiratory ailments. Hundreds had died from illness, accidents and starvation. The fires, though by then out of control, had been set deliberately and systematically—not by small farmers, and not by El Niño, but by commercial outfits operating with implicit government approval. Strange as this immolation of some of the world's most valuable natural assets may seem, it was not unique. The same year, a large part of the Amazon Basin in Brazil was blanketed by smoke for similar reasons. The fires in the Amazon have been set annually, but in 1997 they destroyed over 50 percent more forest than the year before, which in turn had recorded five times as many fires (some 19,115 fires during a single six-week period) as in 1995.

For the timber and plantation barons of Indonesia, as for the cattle ranchers and frontier farmers of Amazonia, setting fires to clear forests has become standard practice. To them, the natural rainforests are an obstruction that must be sold or burned to make way for their profitable pulp and palm oil plantations. Yet, these are the same forests that for many others serve as both homes and livelihoods. For the hundreds of millions who live in Indonesia and in the neighboring nations of Malaysia, Singapore, Brunei, southern Thailand and the Philippines, it is becoming painfully apparent that without healthy forests, it is difficult to remain healthy people.

As this issue of WORLD WATCH went to press [December 1997], the fires in Southeast Asia were still generating enough smoke to be visible from space. Some relief was expected with the arrival of the seasonal rains, but those rains were past due—in part because of an unusually strong El Niño effect. Along with the trees, the region's large underground peat deposits have caught on fire, and such fires are perniciously difficult to put out; they can continue smoldering for years.

From Janet N. Abramovitz, "Putting a Value on Nature's 'Free' Services," *World Watch* (January/February 1998). Copyright © 1998 by The Worldwatch Institute. Reprinted by permission of *World Watch*.

When the smoke does finally clear, Southeast Asia—and the world—will attempt to tally the costs. There are the costs of impaired health and sometimes death, from both lung diseases and accidents caused by poor visibility. There is the productivity that was lost as factories, schools, roads, docks, and airports were shut down (over 1,000 flights in and out of Malaysia were cancelled in September alone); there are the crop yields that fell as haze kept the region in day-long twilight, and the harvests of forest products that were wiped out. Timber (some of the most valuable species in the world) and wildlife (some of the most endangered in the world) are still being consumed by flames. Over three-fourths of the world's remaining wild orangutans live on the fire-ravaged provinces of Sumatra and Kalimantan. Some of them, caught fleeing the flames, have become part of the illegal trade. Because of their location, the Indonesian fires, like those in the Amazon, have dealt a heavy blow to the biodiversity of the earth as a whole.

As the smoke billowed dramatically from Southeast Asia, a much less visible—but similarly costly—ecological loss was taking place in a very different kind of location. While the Indonesian haze was being photographed from satellites, this other loss might not be noticed by a person standing within an arm's length of the evidence—yet, in its implications for the human future, it is a close cousin of the Asian catastrophe. In the United States, more than 50 percent of all honeybee colonies have disappeared in the last 50 years, with half of that loss occurring in just the last 5 years. Similar losses have been observed in Europe. Thirteen of the 19 native bumblebee species in the United Kingdom are now extinct. These bees are just two of the many kinds of pollinators and their decline is costing farmers, fruit growers, and beekeepers hundreds of millions of dollars in losses each year.

What the ravaged Indonesian forests and disappearing bees have in common is that they are both examples of "free services" that are provided by nature and consumed by the human economy—services that have immense economic value, but that go largely unrecognized and uncounted until they have been lost. Many of those services are indispensable to the people who exploit them, yet are not counted as real benefits, or as a part of GNP.

Though widely taken for granted, the "free" services provided by the natural world form the invisible foundation that supports all societies and economies. We rely on the oceans to provide abundant fish, on forests for wood and new medicines, on insects and other creatures to pollinate our crops, on birds and frogs to keep pests in check, and on forests and rivers to supply clean water. We take it for granted that when we need timber we can cut trees, or that when we need water we can find a spring or drill a well. We assume that clean air will blow the smog out of our cities, that the climate will be stable and predictable, and that the mounting quantity of waste we generate will continue to disappear, if we can just get it out of sight. Nature's services have always been there, free for the taking, and our expectations—and economies—are based on the premise that they always will be. A timber magnate or farmer may have to pay a price for the land, but assumes that what happens naturally on the land—the growing of trees, or pollinating of crops by wild bees, or filtering of fresh water—usually happens for free. We are like young children who think that food comes from the refrigerator, and who do not yet understand that what now seems free is not.

Ironically, by undervaluing natural services, economies unwittingly provide incentives to misuse and destroy the very systems that produce those services; rather than protecting their assets, they squander them. Nature, in turn becomes increasingly less able to supply the prolific range services that the earth's expanding population and economy demand. . . . It is no exaggeration to suggest that the continued erosion of natural systems threatens not only the continuing viability of today's human enterprise, but ultimately the prospects for our continued existence.

Underpinning the steady stream of services nature provides to us, there is a more fundamental service these systems provide—a kind of self-regulating process by which ecosystems and the biosphere are kept relatively stable and resilient. The ability to withstand disturbances like fires, floods, diseases, and droughts, and to rebound from the shocks these events inflict, is essential to keeping the life-support system operating. As systems are simplified by monoculture or cut up by roads, and the webs that link systems become disconnected, they become more brittle and vulnerable to catastrophic, irreversible decline. We are being confronted by ample evidence, now—from the breakdown of the ozone layer to the increasingly severity of fires, floods and droughts, to the diminished productivity of fruit and seed sets in wild and agricultural plants —that the biosphere is becoming less resilient.

Unfortunately, much of the human economy is based on practices that convert natural systems into something simpler, either for ease of management (it's easier to harvest straight rows of trees that are all the same age than to harvest carefully from complex forests) or to maximize the production of a desired commodity (like corn). But simplified systems lack the resilience that allows them to survive short-term shocks such as outbreaks of diseases or pests, or forest fires, or even longer-term stresses such as that of global warming. One reason is that the conditions within these simplified systems are not hospitable to all of the numerous organisms and processes needed to keep such systems running. A tree plantation or fish farm may provide some of the products we need, but it cannot supply the array of services that natural diverse systems do—and must do—in order to survive over a range of conditions. To keep our own economies sustainable, then, we need to use natural systems in ways that capitalize on, rather than destroy, their regenerative capacity. For humans to be healthy and resilient, nature must be too.

Resiliency is destroyed by fragmentation, as well as by simplification. Fires in healthy rainforests are very rare. By nature, they are too wet to burn. But as they are opened up and fragmented by roads and logging and pasture, they become drier and more prone to fire. When fire strikes forests that are not adapted to fire (as is the case in the rainforests of both Brazil and Indonesia), it is exceptionally destructive and tends to kill a majority of the trees. The fires in Southeast Asia's peat swamp rainforests bring further disruption, by releasing long-sequestered carbon into the atmosphere.

The fires in Indonesia are not being started by poor slash-and-burn peasants, but by "slash-and-burn industrialists"—owners of rubber, palm oil, rice, and timber plantations who have been taking advantage of a dry year to clear as much natural forest as they can. Though it issued a recent law forbidding the burning, the government in Indonesia is in fact pushing for higher production levels from these export sectors. In both the rainforests and the peat swamps, it has given the plantation owners large concessions to encourage continued "conversion" to one-crop commodities. And the government continues to push costly agricultural settlements into peat forests ill-suited to rice. After the fires became a serious regional problem (and international embarrassment), the government revoked the permits of 29 companies, but such actions were too little, too late.

The current fires are not the first to ravage parts of Southeast Asia; extensive logging in Indonesia and Malaysia led to a major conflagration in 1983 that burned over 3 million hectares and wiped out $5 billion worth of standing timber in Indonesia alone. After 1983, fires that had once been rare became a common occurrence. The 1997 fire will likely turn out to have been the most costly yet. Unless policies change, the fires will be reignited this year.

What Forests Do

Around the world, the degradation, fragmentation, and simplification—or "conversion"—of ecosystems is progressing rapidly. Today, only 1 to 5 percent of the original forest cover of the United States and Europe remains. One-third of Asia's forest has been lost since 1960, and half of what remains is threatened by the same industrial forest activities responsible for the Indonesian fires. In the Amazon, 13 percent of the natural cover has already been cleared, mostly for cattle pasture. In many countries, including some of the largest, more than half of the land has been converted from natural habitat to other uses that are less resilient. In countries that stayed relatively undisturbed until the 1980s, significant portions of remaining ecosystems have been lost in the last decade. These trends have been accelerating everywhere. As the natural ecosystems disappear, so do many of the goods and services they provide.

That may seem to contradict the premise that people want those goods and services and would not deliberately destroy them. But there's a logical explanation: governments and business owners typically perceive that the way they can make the most profit from an ecosystem is to maximize its production of a single commodity, such as timber from a forest. For the community (or society) as a whole, however, that is often the least profitable or sustainable use. The economic values of other uses, and the number of people who benefit, added up, can be enormous. A forest, if not cut down to make space for a one-commodity plantation, can produce a rich variety of nontimber forest products (NTFPs) on one hand, while providing essential watershed protection and climate regulation, on the other. These uses not only have more immediate economic value but can also be sustained over a longer term and benefit more people.

In 1992, alternative management strategies were reviewed for the mangrove forests of Bintuni Bay in Indonesia. When nontimber uses such as fish, locally used products, and erosion control were included in the calculations, the researchers found that the most economically profitable strategy was to keep the forest standing with only a modest amount of timber cutting—yielding $4,800 per hectare. If the forest was managed only for timber-cutting, it would yield only $3,600 per hectare. Over the longer term, it was calculated that keeping the forest intact would ensure continued local uses of the area worth $10 million a year (providing 70 percent of local income) and protect fisheries worth $25 million a year—values that would be lost if the forest were cut.

The variety and value of goods produced and collected from forests, and their importance to local livelihoods and national economies, is an economic reality worldwide. For instance, rattan—a vine that grows naturally in tropical forests—is widely used to make furniture. Global trade in rattan is worth $2.7 billion in exports each year, and in Asia it employs a half-million people. In Thailand, the value of rattan exports is equal to 80 percent of the legal timber exports. In India, such "minor" products account for three fourths of the net export earnings from forest produce, and provide more than half of the formal employment in the forestry sector. And in Indonesia, hundreds of thousands of people make their livelihoods collecting and processing NTFPs for export, a trade worth at least $25 million a year. Many of these forests were destroyed in the fire.

Even so, non-timber commodities are only part of what is lost when a forest is converted to a one-commodity industry. There is a nexus between the two catastrophes of the Indonesian fires and the North American and European bee declines, for example, since forests provide habitat for bees and other pollinators. They also provide habitat for birds that control disease-carrying and agricultural pests. Their canopies break the force of the winds and reduce rainfall's impact on the ground, which lessens soil erosion. Their roots hold soil in place, further stemming erosion. In purely monetary terms, a forest's capacity to protect a watershed alone can exceed the value of its timber. Forests also act as effective water-pumping and recycling machinery, helping to stabilize local climate. And, through photosynthesis, they generate enough of the planet's oxygen, while absorbing and storing so much of its carbon (in living trees and plants), that they are essential to the stability of climate worldwide.

Beyond these general functions, there are services that are specific to particular kinds of forests. Mangrove forests and coastal wetlands, notably, play critical roles in linking land and sea. They buffer coasts from storms and erosion, cycle nutrients, serve as nurseries for coastal and marine fisheries, and supply critical resources to local communities. For flood control alone, the value of mangroves has been calculated at $300,000 per kilometer of coastline in Malaysia—the cost of the rock walls that would be needed to replace them. Protecting coasts from storms will be especially important as climate change makes storms more violent and unpredictable. One force driving the accelerated loss of these mangroves in the last two decades has been the explosive growth of intensive commercial aquaculture, especially for shrimp export. Another has been the excess diversion of inland rivers and streams, which reduces downstream flow and allows the coastal waters to become too salty to support the coastal forests.

The planet's water moves in a continuous cycle, falling as precipitation and moving slowly across the landscape to streams and rivers and ultimately to the sea, being absorbed and recycled by plants along the way. Yet, human actions have changed even that most fundamental force of nature by removing natural plant cover, draining swamps and wetlands, separating rivers from their floodplains, and paving over land. The slow natural movement of water across the landscape is also vital for refilling nature's underground reservoirs, or aquifers, from which we draw much of our water. In many places, water now races across the landscape much too quickly, causing flooding and droughts, while failing to adequately recharge aquifers.

The value of a forested watershed comes from its capacity to absorb and cleanse water, recycle excess nutrients, hold soil in place, and prevent flooding. When plant cover is removed or disturbed, water and wind not only race across the land, but carry valuable topsoil with them. According to David Pimentel, an agricultural ecologist at Cornell University, exposed soil is eroded at several thousand times the natural rate. Under normal conditions, each hectare of land loses somewhere between 0.004 and 0.05 tons of soil to erosion each year—far less than what is replaced by natural soil building processes. On lands that have been logged or converted to crops and grazing, however, erosion typically takes away 17 tons in a year in the United States or Europe, and 30 to 40 tons in Asia, Africa, or South America. On severely degraded land, the hemorrhage can rise to 100 tons in a year. The eroded soil carries nutrients, sediments, and chemicals valuable to the system it leaves, but often harmful to the ultimate destination.

One way to estimate the economic value of an ostensibly free service like that of a forested watershed is to estimate what it would cost society if that service had to be replaced. New York City, for example, has always relied on the natural filtering capacity of its rural watersheds to cleanse the water that serves 10 million people each day. In 1996, experts estimated that it would cost $7 billion to build water treatment facilities adequate to meet the city's future needs. Instead, the city chose a strategy that will cost it only one-tenth that amount: simply helping upstream counties to protect the watersheds around its drinking water reservoirs.

Even an estimate like that tends to greatly understate the real value, however, because it covers the replacement cost of only one of the many services the ecosystem provides. A watershed, for example, also contributes to the regulation of the local climate. After forest cover is removed, an area can become hotter and drier, because water is no longer cycled and recycled by plants (it has been estimated that a single rainforest tree pumps 2.5 million gallons of water into the atmosphere during its lifetime.) Ancient Greece and turn-of-the-century Ethiopia, for example, were moister, wooded regions before extensive deforestation, cultivation, and the soil erosion that followed transformed them into the hot, rocky countries they are today. The global spread of desertification offers brutal evidence of the toll of lost ecosystem services.

The cumulative effects of local land use changes have global implications. One of the planet's first ecosystem services was the production of oxygen over billions of years of photosynthetic activity, which allowed oxygen-breathing organisms—such as ourselves—to evolve. Humans have begun to unbalance the global climate regulation system, however, by generating too much carbon dioxide and reducing the capacity of ecosystems to absorb it. Burning forests and peat deposits only makes the problem worse. The fires in Asia sent about as much carbon into the atmosphere last year as did all of the factories, power plants, and vehicles in the United Kingdom. For carbon sequestration alone, economists have been able to estimate the value of intact forests at anywhere from several hundred to several thousand dollars per hectare. As the climate changes the value of being able to regulate local and global climates will only increase.

What Bees Do

If we are often blind to the value of the free products we take from nature, it is even easier to overlook the value of those products we don't harvest directly—but without which our economies could not function. Among these less conspicuous assets are the innumerable creatures that keep potentially harmful organisms in check, build and maintain soils, and decompose dead matter so it can be used to build new life, as well as those that pollinate crops. These various birds, insects, worms, and microorganisms demonstrate that small things can have hugely disproportionate value. Unfortunately, their services are in increasingly short supply because pesticides, pollutants, disease, hunting, and habitat fragmentation or destruction have drastically reduced their numbers and ability to function. As Stephen Buchmann and Gary Paul Nabhan put it in a recent book on pollinators, "nature's most productive workers [are] slowly being put out of business."

Pollinators, for example, are of enormous value to agriculture and the functioning of natural ecosystems. Without them, plants cannot produce the seeds that ensure their survival—and ours. Unlike animals, plants cannot roam around looking for mates. To accomplish sexual reproduction and ensure genetic mixing, plants have evolved strategies for moving genetic material from one plant to the next, sometimes over great distances. Some rely on wind or water to carry pollen to a receptive female, and some can self-pollinate. The most highly evolved are those that use flowers, scents, oils, pollens, and nectars to attract and reward animals to do the job. In fact, more than 90 percent of the world's quarter-million flowering plant species are animal-pollinated. When animals pick up the flower's reward, they also pick up its pollen on various body parts—faces, legs, torsos. Laden with sticky yellow cargo, they can appear comical as they veer through the air—but their evolutionary adaptations are uncannily potent.

Developing a mutually beneficial relationship with a pollinator is a highly effective way for a plant to ensure reproductive success, especially when individuals are isolated from each other. Spending energy producing nectars and extra pollen is a small price to pay to guarantee reproduction. Performing this matchmaking service are between 120,000 and 200,000 animal species, including bees, beetles, butterflies, moths, ants, and flies, along with more than 1,000 species of vertebrates such as birds, bats, possums, lemurs, and even geckos. New evidence shows that many more of these pollinator species than previously believed are threatened with extinction.

Eighty percent of the world's 1,330 cultivated crop species (including fruits, vegetables, beans and legumes, coffee and tea, cocoa, and spices) are pollinated by wild and semi-wild pollinators. One-third of U.S. agricultural output is from insect-pollinated plants (the remainder is from wind-pollinated grain plants such as wheat, rice, and corn). In dollars, honeybee pollination services are 60 to 100 times more valuable than the honey they produce. The value of wild blueberry bees is so great, with each bee pollinating 15 to 19 liters (about 40 pints) of blueberries in its life, that they are viewed by farmers as "flying $50 bills."

Without pollinator services, crops would yield less, and wild plants would produce few seeds—with large economic and ecological consequences. In Europe, the contribution of honey bee pollination to agriculture was estimated to be worth $100 billion in 1989. In the Piedmont region of Italy, poor pollination of apple and apricot orchards cost growers $124 million in 1996. The most pervasive threats to pollinators include habitat fragmentation and disturbance, loss of nesting and over-wintering sites, intense exposure of pollinators to pesticides and of nectar plants to herbicides, breakdown of "nectar corridors" that provide food sources to pollinators during migration, new diseases, competition from exotic species, and excessive hunting. The rapid spread of two parasitic mites in the United States and Europe has wiped out substantial numbers of honeybee colonies. A "forgotten pollinators" campaign was recently launched by the Arizona Sonoran Desert Museum and others, to raise awareness of the importance and plight of these service providers.

Ironically, many modern agricultural practices actually limit the productivity of crops by reducing pollination. According to one estimate, for example, the high levels of pesticides used on cotton reduce annual yields by 20 percent (worth $400 million) in the United States alone by killing bees and other insect pollinators. One-fifth of all honeybee losses involve pesticide exposure, and honeybee poisonings may cost agriculture hundreds of millions of dollars each year. Wild pollinators are particularly vulnerable to chemical poisoning because their colonies cannot be picked up and moved in advance of spraying the way domesticated hives can. Herbicides can kill the plants that pollinators need to sustain themselves during the "off-season" when they are not at work pollinating crops. Plowing to the edges of fields to maximize planting area can reduce yields by disturbing pollinator nesting sites. Just one hectare of unplowed land, for example, provides nesting habitat for enough wild alkali bees to pollinate 100 hectares of alfalfa.

Domesticated honeybees cannot be expected to fill the gap left when wild pollinators are lost. Of the world's major crops, only 15 percent are pollinated by domesticated and feral honeybees, while at least 80 percent are serviced by wild pollinators. Honeybees do not "fit" every type of flower that needs pollination. And because honeybees visit so many different plant species, they are not very "efficient"—that is, there is no guarantee that the pollen will be carried to a potential mate of the same species and not deposited on a different species.

Many plants have developed interdependencies with particular species of pollinators. In peninsular Malaysia, the bat *Eonycteris spelea* is thought to be the exclusive pollinator of the durian, a large spiny fruit that is highly valued in Southeast Asia. The bats' primary food supply is a coastal mangrove that flowers continuously throughout the year. The bats routinely fly tens of kilometers from their roost sites to the mangrove stands, pollinating durian trees along the way. However, mangrove stands in Malaysia and elsewhere are under siege, as are the inland forests. Without both, the bats are unlikely to survive.

Pollinators that migrate long distances, such as bats, monarch butterflies, and hummingbirds, need to follow routes that offer a reliable supply of nectar-providing plants for the full journey. Today, however, such nectar corridors are being stretched increasingly thin and are breaking. When the travelers cannot rest and "refuel" every day, they may not survive the journey.

The migratory route followed by long-nosed bats from their summer breeding colonies in the desert regions of the U.S. Southwest to winter roosts in central Mexico illustrates the problems faced by many service providers. To fuel trips of up to 150 kilometers a night, these bats rely on the sequential flowering of at least 16 plant species—particularly century plants and columnar cacti. Along much of the migratory route, the nectar corridor is being fragmented. On both U.S. and Mexican rangelands, ranchers are converting native vegetation into exotic pasture grasses for grazing cattle. In the Mexican state of Sonora, an estimated 376,000 hectares have been stripped of nectar source plants. In parts of the Sierra Madre, the bat-pollinators are threatened by competition from human bootleggers, who have been over-harvesting century plants to make the alcoholic beverage mescal. And the latest threat comes from dynamiting and burning of bat roosts by Mexican ranchers attempting to eliminate vampire bats that feed on cattle and spread livestock diseases. The World Conservation Union estimates that worldwide, 26 percent of bat species are threatened with extinction.

Many of the disturbances that have harmed pollinators are also hurting creatures that provide other beneficial services, such as biological control of pests and disease. Much of the wild and semi-wild habitat inhabited by beneficial predators such as birds has been wiped out. The "pest control services" that nature provides are incalculable, and do not have the fundamental flaws of chemical pesticides (which kill beneficial insects along with the pests and harm people). Individual bat colonies in Texas can eat 250 tons of insects each night. Without birds, leaf-eating insects are more abundant and can slow the growth of trees or damage crops. Biologists Paul and Anne Ehrlich speculate that without birds, insects would have become so dominant that humans might never have been able to achieve the agricultural revolution that set the stage for the rise of civilization.

It is not too late to provide essential protections to the providers of such essential services—by using no-till farming to reduce soil erosion and allow nature's underground economy to flourish, by cutting back on the use of toxic agricultural chemicals, and by protecting migratory routes and nectar corridors to ensure the survival of wild pollinators and pest control agents.

Buffer areas of native vegetation and trees can have numerous beneficial effects. They can serve as havens for resident and migratory insects and animals that pollinate crops and control pests. They can also help to reduce wind erosion, and to absorb nutrient pollution that leaks from agricultural fields. Such zones have been eliminated from many agricultural areas that are modernized to accommodate new equipment or larger field sizes. The "sacred groves" in South Asian and African villages—natural areas intentionally left undeveloped —still provide such havens. Where such buffers have been removed, they can be reestablished; they can be added not only around farmers' fields, but along highways and river banks, links between parks, and in people's back yards.

People can also encourage pollinators by providing nesting sites, such as hollow logs, or by ensuring that pollinators have the native plants they need during the "off-season" when they are not working on the agricultural crops. Changing some prevalent cultural or industrial practices, too, can help. There is the practice, for example, of growing tidy rows of cocoa trees. These may make for a handsome plantation. But midges, the only known pollinator of cultivated cacao (the source of chocolate), prefer an abundance of leaf litter and trees in a more natural array. Plantations that encourage midges can have ten times the yield of those that don't.

Scientists have begun to ratchet up their study of wild pollinators and to domesticate more of them. The bumblebee, for example, was domesticated ten years ago and is now a pollinator of valuable greenhouse grown crops.

The Other Service Economy

Natural services have been so undervalued because, for so long, we have viewed the natural world as an inexhaustible resource and sink. Human impact has been seen as insignificant or beneficial. The tools used to gauge the economic health and progress of a nation have tended to reinforce and encourage these attitudes. The gross domestic product (GDP), for example, supposedly measures the value of the goods and services produced in a nation. But the most valuable goods and services—the ones provided by nature, on which all else rests—are measured poorly or not at all.... The unhealthy dynamic is compounded by the fact that activities that pollute or deplete natural capital are counted as contributions to economic wellbeing. As ecologist Norman Myers puts it, "Our tools of economic analysis are far from able to apprehend, let alone comprehend, the entire range of values implicit in forests."

When economies and societies use misleading signals about what is valuable, people are encouraged to make decisions that run counter to their own long-range interests—and those of society and future generations. Economic calculations grossly underestimate the current and future value of nature. While a fraction of nature's goods are counted when they enter the marketplace, many of them are not. And nature's services—the life-support systems—are not counted at all. When the goods are considered free and therefore valued at zero, the market sends signals that they are only economically valuable when converted into something else. For example, the profit from deforesting land is counted as a plus on a nation's ledger sheet, because the trees have been converted to saleable lumber or pulp, but the depletions of the timber stock, watershed, and fisheries are not subtracted.

Last year, an international team of researchers led by Robert Costanza of the University of Maryland's Institute for Ecological Economics, published a landmark study on the importance of nature's services in supporting human economies. The study provides, for the first time, a quantification of the current economic value of the world's ecosystem services and natural capital. The researchers synthesized the findings of over 100 studies to compute the average per hectare value for each of the 17 services that world's ecosystems provide. They concluded that the current economic value of the world's ecosystem services is in the neighborhood of $33 trillion per year, exceeding the global GNP of $25 trillion.

Placing a monetary value on nature in this way has been criticized by those who believe that it commoditizes and cheapens nature's infinite value. But in practice, we all regularly assign value to nature through the choices we make. The problem is that in normal practice, many of us don't assign such value to nature until it is converted to something man-made—forests to timber, or swimming fish to a restaurant meal. With a zero value, it's easy to see why nature has almost always been the loser in standard economic equations. As the authors of the Costanza study note, " . . . the decisions we make about ecosystems imply valuations (although not necessarily expressed in monetary terms). We can choose to make these valuations explicit or not . . . but as long as we are forced to make choices, we are going through the process of valuation." The study is also raising a powerful new challenge to those traditional economists who are accustomed to keeping environmental costs and benefits "external" to their calculations.

While some skeptics will doubtless argue that the global valuation reported by Costanza and his colleagues overestimates the current value of nature's services, if anything it is actually a very conservative estimate. As the authors point out, values for some biomes (such as mountains, arctic tundra, deserts, urban parks) were not included. Further, they note that as ecosystem services become scarcer, their economic value will only increase.

Clearly, failure to value nature's services is not the only reason why these services are misused. Too often, illogical and inequitable resource use continues —even in the face of evidence that it is ecologically, economically, and socially unsustainable—because powerful interests are able to shape policies by legal or illegal means. Frequently, some individuals or entities get the financial benefits from a resource while the losses are distributed across society. Economists call this "socializing costs." Stated simply, the people who get the benefits are not the ones who pay the costs. Thus, there is little economic incentive for those exploiting a resource to use it judiciously or in a manner that maximizes public good. Where laws are lax or are ignored, and where people do not have an opportunity for meaningful participation in decision-making, such abuses will continue.

The liquidation of 90 percent of the Philippines' forest during the 1970s and 1980s under the Ferdinand Marcos dictatorship, for example, made a few hundred families over $42 billion richer. But 18 million forest dwellers became much poorer. The nation as a whole went from being the world's second largest log exporter to a net importer. Likewise, in Indonesia today, the "benefits" from burning the forest will enrich a relatively few well-connected individuals and companies but tens of millions of others are bearing the costs. Even in wealthy nations, such as Canada, the forest industry wields heavy influence over how the forests are managed, and for whose benefit.

We have already seen that the loss of ecosystem services can have severe economic, social, and ecological costs even though we can only measure a fraction of them. The loss of timber and lives in the Indonesian fires, and the lower production of fruits and vegetables from inadequate pollination, are but the tip of the iceberg. The other consequences for nature are often unforeseen and unpredictable. The loss of individual species and habitat, and the degradation and simplification of ecosystems, impair nature's ability to provide the services we need. Many of these changes are irreversible, and much of what is lost is simply irreplaceable.

By reducing the number of species and the size and integrity of ecosystems, we are also reducing nature's capacity to evolve and create new life. Almost half of the forests that once covered the Earth are now gone, and much of what remains is in fragmented patches. In just a few centuries we have gone from living off nature's interest to spending down the capital that has accumulated over millions of years of evolution. At the same time we are diminishing the capacity of nature to create new capital. Humans are only one part of the evolutionary product. Yet we have taken on a major role in shaping its future production course and potential. We are pulling out the threads of nature's safety net even as we depend on it to support the world's expanding human population and economy.

In that expanding economy, consumers now need to recognize that it is possible to reduce and reverse the destructive impact of our activities by consuming less and by placing fewer demands on those services we have so mistakenly regarded as free. We can, for example, reduce the high levels of waste and overconsumption of timber and paper. We can also increase the efficiency of water and energy use. In agricultural fields we can leave hedgerows and unplowed areas that serve as nesting and feeding sites for pollinators. We can sharply reduce reliance on agricultural chemicals, and improve the timing of their application to avoid killing pollinators.

Maintaining nature's services requires looking beyond the needs of the present generation, with the goal of ensuring sustainability for many generations to come. We have no honest choice but to act under the assumption that future generations will need at least the same level of nature's services as we have today. We can neither practically nor ethically decide what future generations will need and what they can survive without.

Mark Sagoff

 **NO**

Can We Put a Price on Nature's Services?

In 1962, the Drifters, a popular rock 'n' roll group, sang:

> At night the stars put on a show for free,
>
> And darling, you can share it all with me...
>
> Up on the roof...

Nature provides many goods and services which we, like the Drifters, enjoy for free. But, as Thomas Paine said about liberty, "What we obtain too cheap, we esteem too lightly." Recently, a group of ecological economists led by Robert Costanza of the University of Maryland has argued that if the importance of nature's free benefits could be adequately quantified in economic terms, policy decisions could "better reflect the value of ecosystem services and natural capital." Drawing upon earlier studies that have "aimed at estimating the value of a wide variety" of ecosystem goods and services—from waste assimilation and the renewal of soil fertility to climate stabilization and the tempering of floods and droughts—the research team has estimated the "current economic value" of the entire biosphere at between 16 and 54 trillion dollars per year. Its "average" value, according to Costanza and colleagues, is about $33 trillion per year.

So tremendous an estimate—especially when presented in a lead article in the British science journal *Nature*—was bound to attract public attention. In feature stories with titles like "How Much Is Nature Worth? For You, $33 Trillion" and "What Has Mother Nature Done for You Lately?" dozens of newspapers and magazines, including the *New York Times, Newsweek,* and *U.S. News and World Report,* covered the Costanza study. "What is the natural environment worth in cold cash?" asked a story in the *San Francisco Chronicle.* "No one knows for sure, but a team of economists and scientists figures $33 trillion, more or less, for such 'free' goods and services as water, air, crop pollination, fish, pollution control and splendid scenery.... For comparison, the gross national product of all the world's countries put together is around $18 trillion."

From Mark Sagoff, "Can We Put a Price on Nature's Services?" *Report from the Institute for Philosophy and Public Policy,* vol. 17, no. 3 (Summer 1997). Copyright © 1997 by The Institute for Philosophy and Public Policy, University of Maryland at College Park. Reprinted by permission.

Costanza and colleagues acknowledge that their estimates are fraught with uncertainties; their study, they say, provides only "a first approximation of the relative magnitude of global ecosystem services." Their caution is understandable. No one can doubt that "ecological systems... contribute to human welfare, both directly and indirectly," or that the world's economies depend on the "ecological life-support systems" that nature provides. "Once explained, the importance of ecosystem services is typically quickly appreciated," writes Gretchen Daily, an ecologist at Stanford University who has edited a collection of papers on this theme. And yet, as Daily goes on to say, "the actual assigning of value to ecosystem services may arouse great suspicion." This is because "valuation involves resolving fundamental philosophical issues"—about the role of economic values in the policy process, and about the relation between economic value and human welfare. Methodological problems also haunt any attempt to impute prices to the services of nature.

Even so, Daily concludes that "nothing could matter more" than attaching economic values to ecological services. "The way our decisions are made today is based almost entirely on economic values," she told the *Chronicle*. "We have to completely rethink how we deal with the environment, and we should put a price on it." This essay will review critically the attempt to set prices for the benefits that nature provides "for free."

Calculating Values

Environmentalists have long noted that many of nature's gifts, such as the show the stars put on at night, are "public goods"; in other words, they are not traded in commercial markets, no one can be excluded from using them, and one person's use does not limit another's, at least up to some congestion point. "Because ecosystem services are not fully 'captured' in commercial markets or adequately quantified in terms comparable with economic services and manufactured capital, they are often given too little weight in policy decisions," Costanza and colleagues argue. Public goods notoriously have no market prices. If prices could be imputed to ecosystem services, wouldn't these prices help us better to appreciate their worth?

The many studies which Costanza and colleagues have assembled use a great variety of methods by which to impute economic value to ecosystem services. Some employ experimental techniques, including "contingent valuation," to estimate the aesthetic or "nonuse" value of natural settings. The large majority of studies, however, estimate either the value of ecosystem outputs, such as fish, fiber, and food, or the costs of replicating ecosystem services. Costanza and colleagues use estimates of these two kinds—output values and replacement costs—to account for most of the $33 trillion price tag they impute to ecosystem services. As we will see, however, neither kind of estimate can serve as a basis for measuring the economic value of ecosystem services, even though those services are essential to human well-being.

Costanza and colleagues gathered data reporting the market value of the outputs of the world's fisheries, forests, and farms. To calculate the contribution of ecosystem services to the oceans' fisheries, for example, the Costanza team multiplied the world's fish catch in kilograms by an average market price per kilogram. In other words, they used "price times quantity as a proxy for the economic value of the service." They apparently reasoned that since there would be no fish harvests if not for ecosystem services, the economic value of these services should include the value of those harvests. They then used data about the value of the total harvest to calculate what they identify as "the 'incremental' or 'marginal' value of ecosystem services." To arrive at a "marginal" or "unit" price in fisheries, they divided the overall value of fish harvests by the number of hectares of ocean to reach an estimated ecosystem service contribution of $15 per hectare.

The researchers used a slightly different approach to measure the value of ecosystem services in forestry and agriculture. Timber values "were estimated from global value of production, adjusted for average harvest cost... assumed to be 20% of revenues." In this instance, the researchers used "the net rent (or producer surplus)"—that is, proceeds to producers minus their costs—to estimate the overall value of ecosystem services. They used rents to farmers—that is, the value of crops less production costs—to compute the value of ecosystem services to agriculture. To obtain a "per unit" value for ecosystem services in forestry and agriculture, the researchers divided the resulting timber values and crop values by the number of hectares of forests and farmland.

We can see one problem in reasoning from the value of an output to that of an input if we assume that several different ecosystem services, such as climate stabilization or nutrient cycling, are each essential to production in fisheries, forests, and farms. If so, each of these services would possess individually a value commensurate with the output of fishing, forestry, or agriculture. This same difficulty arises with respect to inputs other than ecosystem services that may also be essential to production. Ships are indispensable for fisheries, saws for forestry, and tractors for farming. If we were to use the Costanza team's approach to estimate the value of these inputs, we would infer a price for each of them by dividing the number used into the value of the proceeds or profits of the industry. In the aggregate, ships would be worth just as much as ecosystem services to fishing, saws to forestry, and tractors to farming. Labor, being essential, would also have the same price as ecosystem services collectively and individually in all of these industries.

It is understandable that Costanza and colleagues would want the economic value of ecosystem services to reflect the values of the industries to which they are essential. It is a mistake to assume, however, that if x is essential to the production of y, the price of x can be inferred from that of y. Rather, prices for inputs—or "factors"—of production are determined by market forces, that is, by supply and demand. The marginal economic value of a ship, for example, equals the amount it fetches in a market in which shipwrights compete for buyers on the basis of quality and price. Ships are essential to the fishing industry, to be sure, but this does not suggest that the price of ships can be inferred from the price of fish.

If the costs of providing ecosystem services are zero—if Mother Nature supplies them free—then the prices (and, in that sense, the economic value) of these services must approach zero as well. This is true because competition among suppliers for buyers tends to drive prices down to costs. If we fantasize that Mother Nature is a monopoly provider of ecosystem services, she may charge whatever the market will bear, gouging consumers for all they are willing to pay and extracting whatever profits producers might otherwise obtain. This seems to be the situation that Costanza and colleagues envision. Monopoly prices, however, do not represent fair marginal value. Prices, to be meaningful at all, must arise in competitive markets. If Nature sought to operate as a monopoly, the government would rightfully either set the price of an ecosystem service at a small percentage above costs (as it does with utilities) or break up Ma Nature into competing units (as it did Ma Bell).

Substitution and Replication

Costanza and colleagues also use the costs of creating technological substitutes for ecosystem services as a basis for inferring their incremental or marginal value. In an accompanying article in *Nature,* Stuart Pimm illustrates this process by explaining how the researchers determined that the nutrient-cycling services of the world's oceans are worth $17 trillion:

> If the oceans were not there, re-creating their nutrient cycling would require removing the nutrients from the land's runoff and returning them. The estimate of this service's $17 trillion value is arrived at by multiplying the cost of removing phosphorus and nitrogen from a liter of waste water by the 40,000 cubic kilometers of water that flow from the land each year.

Similarly, Costanza and colleagues estimate what it would cost to re-create, with levees and other structures, natural flood control and storm protection ($1.8 trillion); to replicate artificially the pollination of plants ($1.8 trillion); to provide technological substitutes for natural waste treatment and breakdown of toxins ($2.7 trillion); and to replace the outdoor recreation and "esthetic, artistic, educational, spiritual, and/or scientific" benefits people find in natural places ($3.83 trillion).

When economists speak about substitution, they do not generally refer to alternative and sometimes more costly methods of providing some good or service. Rather, they refer to consumer indifference between alternatives at given prices. For example, the economic value of a beefsteak will be determined in part by the price at which consumers will switch to some other item on the menu instead. In the absence of cattle, it would be very expensive to produce beef. This fact suggests nothing, however, about the goods or services people would substitute for beef when beef prices increase.

Plainly, it would cost a great deal to replicate technologically the experience the Drifters enjoy up on the roof, where "it's peaceful as can be" and "the air is fresh and sweet." One cannot meaningfully impute an economic value to the rooftop experience, however, by determining how much it would cost to replicate it technologically—for example, by building an air-conditioned planetarium. Rather, to get at the economic value of the rooftop experience, one would ask the Drifters at what price they would choose a different activity—

to venture under the boardwalk down by the sea, for example, or to spend Saturday night at the movies. No matter how much it might cost to replicate an ecosystem service technologically, that amount does not tell us the economic value of the service.

To impute economic value, one would have to determine the price at which people would cease to demand that service and spend their money on some other source of satisfaction instead. A $17 trillion dollar price tag on the oceans' work of "pure ablution round earth's human shores" (as the poet John Keats described it) would price these services out of the market. According to Pimm, people would not think these services are worth that kind of money: "In the short term, many would not notice (and perhaps not care) what happens to the elements as they flow into the ocean." New York City agreed to pay $660 million over several years to enhance the Catskill area as an aquifer, but rather than pay $4 billion for a technological alternative—a treatment plant— people might have moved from the city. Thus, the economic value of ecosystem services has no clear relation to the costs of replicating them technologically.

The Basis of Decision Making

The issue of determining economic prices for ecological services, Costanza and colleagues write, is inseparable from the choices and decisions we have to make about ecological systems. These authors continue:

> Some argue that valuation of ecosystems is either impossible or unwise, that we cannot place a value on such "intangibles" as human life, environmental aesthetics, or long-term ecological benefits. But, in fact, we do so every day. When we set construction standards for highways, bridges and the like, we value human life (acknowledged or not) because spending more money on construction would save lives.

Many people share the suspicion that public policy is often based on implicit valuations that have never been articulated or defended. Part of the appeal of the Costanza study lies in its insistence that these matters of valuation be confronted directly.

Contrary to what Costanza and colleagues suggest, however, risk regulation does not necessarily imply an implicit economic valuation of "intangibles" such as human life. Decisions in this area more typically respond to public attitudes, statutory guidance, and relevant legal history. This is why our society protects human life and the environment much more stringently in some moral contexts than in others. We do not seek to save lives up to some predetermined economic value. Rather, we control risk more or less strictly on a number of moral grounds—for example, insofar as risks are involuntary or coerced, connected to dreaded events such as cancer, associated with industry and the workplace, unfamiliar, unnatural, and so on.

To be sure, one can infer an imputed or implicit economic value for human life from any of thousands of governmental regulations. Mandatory seat-belt laws cost $69 per year of life saved, while laws requiring uranium fuel-cycle facilities to purchase radionuclide emission-control technology cost an estimated $34 billion for every year of life saved. Safety controls involving chloroform at paper mills weigh in at $99 billion cost/life-year. For any number you pick between $20 (motorcycle helmet requirements) and $20 billion (benzene emission control at rubber-tire manufacturing plants), there is a governmental program from which that number can be inferred as the value of a statistical year of life.

Every situation—every regulatory decision—responds to different ethical, economic, political, historical, and other conditions. A national speed limit of 55 miles per hour on highways and interstates would save a statistical year of life at a cost of only $6,600, but it is politically unpopular. Strict enforcement of such a speed limit, even more unpopular, would save an additional life/year at a cost of $16,000. Still more unpopular are random motor-vehicle inspections—but these could save lives at even less expense. Can we infer that people value their lives at only a few thousand dollars? No; it is simply that people fear and resent some risks less than others, and least of all those risks they control themselves. These moral factors affect private and public decisions about risk. To impute a value-per-life/year to any regulation or policy, such as highway construction, is to create an epiphenomenon, a statistical abstraction, or descriptive convention, but not to identify a value judgment that necessarily affected that program.

Value in Exchange and Value in Use

The interest among academics and others in "green accounting," of which the Costanza study is a model, seeks to serve an important political purpose, namely, to restrain the commercial juggernaut that is destroying the health, beauty, and integrity of the natural world. The *Nature* article urges us to recognize the benefits ecosystems provide for free, in the hope that this will prompt us to defend these systems from relentless exploitation and destruction.

Whatever the study's political uses, however, it is difficult to see what it would mean for the researchers to "get the prices right." In an actual market, budgetary constraints and consumers' willingness to pay limit the number of dollars at which goods and services change hands. Prices represent bargains struck between willing buyers and sellers, for example, between Romeo and the honest apothecary. Prices do not represent the contribution a good or service makes to human welfare, if welfare is measured in any other way—or has any other sense—than market exchange.

Economists often describe their discipline as a positive rather than as a normative science. (This is the source of the maxim, quoted by Stuart Pimm, that economists "know the price of everything and the value of nothing.") As a positive science, economics concerns itself with value in exchange, which is to say, the prices at which goods and services change hands in competitive markets. Economic science does not traditionally claim that these prices indicate or reveal anything about the contribution goods or services make to human welfare in any substantive sense. Economic science uses the term "welfare" or "utility" in a purely formal way, to designate whatever it is that price is supposed to measure. This may have no relation to the true sources of human well-being or flourishing.

The classic articulation of this view is in Adam Smith's *Wealth of Nations*:

> The word VALUE, it is to be observed, has two different meanings, and sometimes expresses the utility of some particular object, and sometimes the power of purchasing other goods which the possession of that object conveys. The one may be called "value in use"; the other, "value in exchange." The things which have the greatest value in use have frequently little or no value in exchange; and on the contrary, those which have the greatest value in exchange have frequently little or no value in use. Nothing is more useful than water: but it will purchase scarce any thing; scarce any thing can be had in exchange for it. A diamond, on the contrary, has scarce any value in use; but a very great quantity of other goods may frequently be had in exchange for it.
>
> In order to investigate the principles which regulate the exchangeable value of commodities, I shall endeavor to shew... what is the real measure of this exchangeable value; or, wherein consists the real price of all commodities.

Cigarettes illustrate the difference between value in exchange and in use. The price of cigarettes reflects the costs of production, competition among suppliers, and levels of demand. The price has no relation to human well-being as society judges it. As a society, we have reached a judgment that cigarettes have a negative welfare value—a deleterious effect on actual human well-being. The more consumers are willing to pay to smoke, the worse off they are, according to doctors and other respected social authorities. Cigarettes, therefore, have a positive exchange value but a negative value in use.

Suppose a well-meaning team of economists, seeing that cigarettes are bad for health, wished to correct the price of cigarettes to make it better reflect their negative welfare effect. If these economists thought—as the Costanza team apparently does—that prices should rise with contribution to welfare, they would recommend lowering the price of tobacco. In fact, society may set tobacco prices higher, not to reflect its great contribution to human welfare but to discourage its use.

To achieve social goals and values, including human well-being, we may adjust the prices of goods and services—for example, by taxing tobacco products. This kind of price-fixing, although often justifiable in terms of human welfare, does not reflect "true" or "correct" market value. To bring prices into line with societal goals, values, and judgments is not to embrace but to reject market exchange as a criterion or basis for social valuation. It is to recognize that market or exchange value bears no necessary connection to value in use.

Growth vs. Development

In important and insightful earlier essays, Robert Costanza and other ecological economists have criticized GNP [gross national product] as a measure of human welfare and economic growth as a goal of public policy. Improvements in the quality of human life, these analysts have argued, are not to be confused with increases in the size of the economy. Costanza has written elsewhere that economic growth "cannot be sustainable indefinitely on a finite planet." Economic development, in contrast, "which is an improvement in the quality of life ... may be sustainable."

In these writings, Costanza and others insist on a distinction similar to the one that Adam Smith draws between "value in exchange" and "value in use." Economic growth is measured in terms of value in exchange; it is the rate of increase of GNP, which is to say, the total market value of all goods and services produced or consumed as measured in current prices. Development has to do with value in use—true human flourishing, including happiness and contentment—and is measured in terms of indices of human welfare such as nutrition, education, and longevity. In these earlier writings, Costanza and other ecological economists did not try to "correct" the exchange or economic value of goods and services to make them better reflect their "true" contribution to human well-being. For example, they would not have imputed a higher exchange value to water and a lower one to diamonds to make the prices of these goods more commensurate with their importance to human survival.

The *Nature* article, in contrast, seeks to "correct" market prices "to better reflect the value of ecosystem services and natural capital." The article concludes that world "GNP would be very different in both magnitude and composition if it adequately incorporated ecosystem services." In seeking to "get the prices right," however, Costanza and colleagues discard the earlier insight that measures of economic value, which arise from the play of market forces, have no clear relation to human welfare or well-being in any substantive sense. The effort to "correct" the prices of ecological services and natural capital confuses value in exchange with value in use, or, in contemporary terms, measures of economic growth with indices of human development.

The Drifters recognized the importance of nature to their well-being even though its services are free. Like the Drifters, Costanza and colleagues understand the abiding importance of nature's services to the quality of our lives. Our dependence on ecosystem services cannot be overstated, and our efforts to sustain them can never be too great. To try to "get the prices right" as a way to protect nature, however, is to lend support to economic measures of welfare, such as economic growth and GNP, ecological economists rightly reject. The effort Costanza and colleagues undertake to "estimate the 'incremental' or 'marginal' value of ecosystem services" should be seen as an aberration within the program of ecological economics. It can succeed only in lowering the credibility of that discipline while increasing the legitimacy of the standard cost-benefit policy framework most likely to defeat attempts to protect the natural environment.

POSTSCRIPT

Should a Price Be Put on the Goods and Services Provided by the World's Ecosystems?

Abramovitz argues that undervaluing natural services encourages the squandering of these vital resources. She does not specify how it would be possible to make the beneficiaries of nutrient recycling or of the biodiversity protected by forest preservation pay the large price that ecovaluing would establish for these presently free services. One problem is the notion that a consumer's "willingness to pay" the marginal cost of an incremental unit of an ecosystem service is a reflection of the social value of that service. However, this seems to require an unrealistic degree of public wisdom about the requirements of ecological sustainability. Although Sagoff rejects the possibility and the efficacy of trying to establish dollar prices for the services provided by nature, he does not propose an alternative means of persuading the beneficiaries of these services to refrain from taking them for granted and from pursuing developmental strategies that undermine them.

A detailed critique of the work by Costanza et al. by environmental economics professor David Pearce appears in the March 1998 issue of *Environment*. Pearce supports Costanza's goal but argues that the methodology is seriously flawed. The same journal issue contains a pointed reply to Pearce's analysis by Costanza and his coauthors. Gretchen C. Dailey et al., in "The Value of Nature and the Nature of Value," *Science* (July 21, 2000), discuss valuation as an essential step in all decision making and note that efforts "to capture the value of ecosystem assets . . . can lead to profoundly favorable effects."

For an introduction to the field of environmental economics written for a general audience, see *Environmental Economics* by R. Kerry Turner, David Pearce, and Ian Bateman (Johns Hopkins University Press, 1993). One current school of thought that has won near-unanimous praise from environmentalists puts ecosystems first and rejects the standard notion that the marketplace requires unlimited growth. Such green perspectives are described and praised in Jeff Gersh's article "Bigger, Badder—But Not Better," *The Amicus Journal* (Winter 1999).

Readers who find Sagoff's arguments persuasive are referred to his essay "At the Shrine of Our Lady of Fatima *or* Why Political Questions Are Not All Economic," *Arizona Law Review* (vol. 23, 1981), in which Sagoff turns his critical eye to the uses of cost-benefit analyses.

ISSUE 2

Does Wilderness Have Intrinsic Value?

YES: Rick Bass, from "On Wilderness and Wallace Stegner," *The Amicus Journal* (Spring 1997)

NO: William Tucker, from "Is Nature Too Good for Us?" *Harper's Magazine* (March 1982)

ISSUE SUMMARY

YES: Nature writer Rick Bass defends the need for true wildlands, rather than managed ecosystems, if we are to preserve our ecological heritage and the cultural treasures that it inspires.

NO: William Tucker, a writer and social critic, asserts that wilderness areas are elitist preserves designed to keep people out.

The environmental destruction that resulted from the exploitation of natural resources for private profit during the founding of the United States and its early decades gave birth after the Civil War to the progressive conservation movement. Naturalists such as John Muir (1839–1914) and forester and politician Gifford Pinchot (1865–1946) worked to gain the support of powerful people who recognized the need for resource management. Political leaders such as Theodore Roosevelt (1858–1912) promoted legislation during the last quarter of the nineteenth century that led to the establishment of Yellowstone, Yosemite, and Mount Rainier national parks and the Adirondack Forest Preserve. This period also witnessed the founding of the Sierra Club and the Audubon Society, whose influential, upper-class members worked to promote the conservationist ethic.

Two conflicting positions on resource management emerged. Preservationists, like Muir, argued for the establishment of wilderness areas that would be off-limits to industrial or commercial development. Conservationists, like Roosevelt and Pinchot, supported the concept of "multiple use" of public lands, which permitted limited development and resource consumption to continue. The latter position prevailed and, under the Forest Management Act of 1897, mining, grazing, and lumbering were permitted on U.S. forest lands and were regulated through permits issued by the U.S. Forestry Division.

The first "primitive areas," where all development was prohibited, were designated in the 1920s. Aldo Leopold and Robert Marshall, two officers in the Forest Service, helped establish 70 such areas by administrative fiat. Leopold and Marshall did this in response to their own concerns about the failure of some of the National Forest Service's management practices. Many preservationists were heartened by this development, and the Wilderness Society was organized in 1935 to press for the preservation of additional undeveloped land.

It became increasingly apparent during the 1940s and 1950s that the administrative mechanism whereby land was designated as either available for development or off-limits was vulnerable. Because of pressure from commercial interests (lumber, mining, and so on), an increasing number of what were then called wilderness areas were lost through reclassification. This set the stage for an eight-year-long campaign that ended in 1964 with the passage of the Federal Wilderness Act. But this was by no means the end of the struggle. The process of implementing this legislation and determining which areas to set aside has been long and tortuous and will probably continue into the next century.

There are more clear-cut differences between values espoused by the opposing factions in the battle over wilderness preservation than in many other environmental conflicts. On one side are the naturalists who see undeveloped "wild" land as a precious resource, where people can go to seek solace and solitude—provided they do not leave their mark. On the opposite extreme are the entrepreneurs whose principal concern is the profit that can be made from utilizing the resources on these lands.

It has become apparent that industrial pollutants move through the air and water and find their way into every nook and cranny of the ecosphere. The notion of totally protecting any area of the Earth from contamination is an ideal that cannot be fully realized. This knowledge has encouraged those who advocate replacing the practice of wilderness designation with the strategy of "scientific ecosystem management," whereby developmental activities are restricted rather than completely prohibited in areas that are habitats to many species.

Many environmentalists and ecologists doubt the efficacy of ecosystem management. They continue to advocate the maintenance of wilderness areas as the only effective means of preserving biodiversity. This position is supported in the following selection by Rick Bass. In his essay, Bass argues that we need to preserve wild areas not only for ecological reasons but also to protect the natural heritage that nurtured such great literary giants as Henry David Thoreau, Ralph Waldo Emerson, and John Muir. In the second selection, William Tucker, who is critical of environmentalism, views the wilderness movement as elitist and the idea of excluding most human activity from wilderness areas as a consequence of a misguided, romantic, ecological ethic.

Rick Bass  **YES**

On Wilderness and Wallace Stegner

I keep waiting each day to make friends with the Forest Service—not with the individuals, but with the agency itself. The agency harbors, as a rotting log harbors nutrients and hope-for-the-future, some of the country's best and most passionate hydrologists, entomologists, range managers, recreation specialists, ornithologists, wilderness specialists, big-game biologists. But the gears and levers of the agency are still pulled and fitted in Washington, in an agency run by a Congress that in turn is run not by the people but by the corporations that funded their election campaigns.

We all know this. The simplicity of it makes us want to shriek. Its inevitability—the brute force, the economic biomass behind this process—also makes us want to shriek. Artists in the West continue to struggle daily with the question of how best to combat the madness of this loss: whether to lay down works of beauty—classical art, in the form of song, sculpture, stories, paintings, poems—or to lay down works of essay and activism; whether or not to speak directly to the politics of this loss, as the health of our communities and of our wild heritage continues to be taken from us. It is a taking, a theft, that is funded by our own dollars, as if we were some hideous, wounded wolverine caught in a trap, eating its own entrails.

I am speaking about wilderness, of course, or about the lack of wilderness—not just in Montana's Yaak Valley, where I live, but all across the Rocky Mountain West: the failure to protect as wilderness anything beyond rock and ice.

There is much talk now by some in the Forest Service about "ecosystem management" as a new and somehow a better way to draw profit from the land, but there are those of us who will tell you, and who believe, that this is only a new pretext for building more roads into the last roadless areas. A Forest Service Chief once issued a memo directing his regional foresters that if they were denied entrance to a roadless area for a timber sale that would violate the law, they should not try to "make up" that "lost" volume by moving the sale to a roaded area, but should instead try to substitute entrance into a different roadless area.

I have not yet heard ecosystem management talk about conservation biology or wilderness cores. It has thus far skirted this issue so completely that

From Rick Bass, "On Wilderness and Wallace Stegner," *The Amicus Journal* (Spring 1997). Copyright © 1997 by *The Amicus Journal*. Reprinted by permission of *The Amicus Journal*, a quarterly publication of The Natural Resources Defense Council, 40 West 20th Street, New York, NY 10011. http://www.nrdc.org/amicus. NRDC membership dues or nonmember subscriptions: $10 annually.

I believe its true heart has been revealed: that it is a ploy to further fragment wild places, rather than begin healing and weaving them back together.

≈⟨⊙⟩≈

Central to the science of conservation biology is the need for cornerstone or foundation reserves, undisturbed cores of diversity—forests, or any other ecosystems, of such radiant health and strength that they not only exist strong and free in the world by themselves, but pass on their genetic and spiritual vigor to things and places beyond their perimeters.

You don't want to try to figure out how to go into those places and dissect them. You want to move in the opposite direction—as the Forest Service has yet to do, in the Yaak and many other areas. You want to devise ways to protect these places—to turn away from them and walk in the opposite direction.

You want to preserve them, not extinguish them.

≈⟨⊙⟩≈

For me, the question of harm and injustice and what's wrong with these initial visions of ecosystem management, and the question of what's right and diverse and healthy about a whole, untouched wilderness, resonates most clearly and painfully in the specific example of the Yaak Valley. The Yaak rests up against the Idaho and British Columbia borders. It is a pipeline, a thin straw, drawing genetic diversity down out of Canada and into the rest of the West. It is also utterly unprotected for the future. The valley is almost 500,000 acres in size, and within it exist over 150,000 acres of roadless cores, still connected or nearly connected in an archipelago of vibrant health—and yet not a single acre of wilderness is protected for future generations.

The Yaak lies in a seam, a crevice, between the rainforest ecosystems of the Pacific Northwest and the jagged mountainscapes of the northern Rockies. The richness of these two systems combines to create a richness that is even greater, and that is palpable. When you sleep in the Yaak for the first time, you have dreams you never had before; as a writer, you think of stories you never previously imagined; as a painter, you see shadows and colors not earlier noticed; as a hunter, you see and feel more acutely the different movements, different relationships to each other, of the animals in the forest. As a scientist, you think of connections you never made before. The double-richness of the landscape of the Yaak is like the mysterious, tempting, rich and troubling territories of the heart in the areas between art and activism.

Woodland caribou use this country occasionally, as do, with great frequency, moose, elk, mule deer and whitetails, mountain goats and bighorn sheep along the Kootenai River. It is, to the best of my knowledge, a zone of unprecedented speciation and uniqueness in the West, a secret gift of life. On any given mountain you can find three species of grouse. There are groupings of vegetation as yet undescribed, and some even still unknown—the stuff of literature, dreams, and mystery.

The Yaak is a predators' showcase, home to a snarling and scrapping, reclusive combination of tooth and claw: wolves, wolverines, lynx, black bears, grizzlies, bobcats, martens, fishers, coyotes, mountain lions, hawks, owls, golden eagles, bald eagles. It is a valley of giants—five-hundred-year-old cedar trees and tamaracks; great blue herons, sturgeon, bull trout weighing twenty and twenty-five pounds. I have lived and camped all over the country, and the Yaak is the most savage and delicate place I've ever seen. It is a vital organ of the West. Yet it continues to be ignored for wilderness protection.

❧

For a Western artist, to speak about the wilderness system of the West is to speak indirectly about the work of Wallace Stegner; it is to speak about the vision of wilderness that he put forth in works such as the Pulitzer Prize-winning novel *Angle of Repose* and the essay collection *The Sound of Mountain Water,* among so many others. As a team of oxen pulls in a double-yoke, he used his talents as an artist and as an activist, all his life, to help give us what we have now: what we have as a community of artists and what we have as a community of those who love the landscape of the West.

Is it too easy a metaphor to discuss Stegner's work as a core of community health, similar to a core of wilderness health? In a healthy forest, vertical and horizontal matrices of diversity—by species, age, structure, and every other factor—are interwoven. If you can gauge the life of an artist by such a measure, Stegner's was the healthiest I know. In every dimension—as a writer of novels, short stories, essays; as an activist for wilderness; as a teacher, a father, and a husband—he was exemplary in the truest sense of that word. He was example, bedrock, and touchstone for the rest of the country around him.

I keep finding myself trying to figure out how he was able to maintain this vertical and horizontal strength so forcefully throughout his working career —while publishing in seven decades, roughly one third of the United States' history. But I believe that, in the end, the answer to this question is really no mystery at all. It's like wondering how a forest that has such big trees can also have such rich soil, or how a forest that has such a diversity of bird life can also have so many different mammals.

In the essay, "The Law of Nature and the Dream of Man," Stegner wrote: "How to write a story, though ignorant or baffled? You take something that is important to you, something you have brooded about. You try to see it as clearly as you can, and to fix it in a transferable equivalent. All you want in the finished print is the clean statement of the lens, which is yourself, on the subject that has been absorbing your attention. Sure, it's autobiography. Sure, it's fiction. Either way if you have done it right, it's true."

❧

Ecosystem management is not yet true. It will not succeed without vital cores —anchor points—of wilderness in each ecosystem. Only a few islands of health currently exist in the West, and even the health of those is suspect. And the fact that agency discussions of ecosystem management continue to avoid acknowledging that there are relationships we can never understand convinces me that timber managers are speaking only of managing those few factors they think they *can* understand and perhaps get a handle on: more fiber production from one or two species of tree, over the short run; maybe, over the short run, more summertime forage for big game.

We can never manage or control the balance of, say, seed-eating versus insect-eating birds within a region. It is due to factors perhaps within that region, but perhaps beyond—tropical deforestation, global warming, worldwide ecosystem fragmentation. The faces of different forest types, especially in a land as diverse as the Yaak, are still in wild flux—especially compared to our knowledge, or lack of it. Even a 500-year-old larch forest is in relative flux, part of an earth-desired, rock-and-soil-desired cycle of progression and regression—a pulse—that is specific to that particular spot on the earth, and yet connected to all others. Core samples in the bogs in the interior of an old tamarack forest will reveal the ashes of sagebrush and juniper from only a few thousand years ago. In the wilderness, the forest continues to tilt, to change, under its own rules, with all the wonderful accompanying (and invisible) genetic alterations in species and speciation, of the trees themselves and of everything above and beneath and around them—birds-plants-mammals-insects-fungus; the forest changes through the centuries and millennia like the shadows of clouds drifting across a mountain.

Of course there are places where we need to attempt, with respect, to do our awkward best. Ecosystem management acknowledges this. But again: ecosystem management does *not* yet acknowledge the necessity of protecting significant wilderness cores in each and every watershed—not shifting these wildernesses around, like moving old folks from one rest home to another on a Forest Service shuttle bus, but committing the cores, the anchor points, to nature for the duration of humankind's time on earth.

The number that is expected with regard to tithes, both spiritual and biological, is 10 percent. I propose that in fragile or ravaged landscapes such as the Yaak, 15 or 20 percent is entirely more appropriate—and that in some landscapes, 100 percent is appropriate—in an attempt to initiate the healing process, to re-establish health and balance and cycles. A solution in the Yaak, a place wildly out of balance (the bug-killed lodgepole was ignored as timber in the 1980s, and two thirds of the harvest comprised instead green larch and fir) is still within [r]each. Wilderness designation of at least the last roadless areas in the Yaak would still leave almost 350,000 acres for the hard-core, high-volume timber yearnings of Congress and the Forest Service, and for our own consuming hungers. Let the ecosystem managers then tie in their activities to the wild cores or anchor points, rather than riding over and erasing these last fixed points of reason and last fixed points of data.

The idea behind ecosystem management is that we humans can enter a forest, or a desert or a meadow, and with our scientific tools and studies divine where to cut and where to burn or even build so as to imitate the actions of nature.

But is chaos theory applicable to insect and fire patterns in lodgepole stands, and soil changes, and forest succession through the centuries' cycles? Should we walk along every streambank following every fire, whether natural or prescribed, and attempt to manage or evaluate, as nature does, whether each and every burned snag should be left standing for one of the forty-seven species of cavity nesters that use the Yaak, or pushed over at a 45- to 90-degree angle into the stream to help trap ash and other sediment runoff? Or pushed over so as to land parallel to the slope, to help hold ash and soil in place on-site? Or pushed over so as to land upslope, to rot in the soil and produce a seedbed for ceanothus or kinnikinnick? But wouldn't that then help the seed-eating birds instead of the insect-eaters? And wouldn't the insects get out of hand, then? And then after the insects took over the world, wouldn't that mean more dead trees, hence still more fires? Maybe the fires would fry some of the insects, but then wouldn't it just start all over again? Maybe we need to re-think this. Maybe we need to do another study.

Even today—as recently as 1996—you will find Forest Service officials making such statements as, "It's comparatively easy for foresters to emulate nature's large severe fires . . . by clearcutting large areas and burning the slash." The truth is, we haven't figured it all out, and I don't think we ever will—not to the extent that we can outmanage the wilderness cores that inspire and nurture an ecosystem's health. We can't even make up our minds about whether to burn slash or leave it on the ground, whether to try to aerate the compacted soil of clearcuts or leave it alone to recover in the next millennium on its own.

We're only just beginning to figure out site-specific light management for overstory openings: the balance of photosynthesis versus UV shielding required by different seedlings. What about the understory, and what about the mechanics of soil? Does anyone really think we can manage dirt—two million, or four million, or twelve million acres of dirt? Wilderness cores are not only sources of vital health. They are buffers against our trials and errors. Wilderness cores forgive us our trespasses into other areas.

The light touches that Stegner could wield with his pen, in his art, cannot of course be wielded by humans upon the land. We are too small and the land is far larger than an 8½-by-11-inch sheet of paper, and infinitely deeper than the little three-pound electrical impulse-generator we know as the human brain, marvelous as that organ is. The earth is an enormous, unimaginably complex brain, and we ought to let pieces of it, places of it, function under the grace and power of its own miracle. The Yaak is only one instance of our present failure to do this.

I like to believe that Stegner was aware of the healthy influence exerted by the artistic cores, the anchors, he created through his work: not just the interconnectedness of his work to that of his many students—like migration corridors for diversity—but the sanctuaries of his individual books. It was in Stegner's era that we evolved, in the manner of a forest approaching its fullest health and complexity as it matures, our fullest tradition of nature writing. In this country's first century and a half, a few individuals carried most of the load of ecological literacy and the obligation to disperse it. The shifts of duty among these writers followed a somewhat linear model, similar to that of the beginnings of a forest: new seedlings of a few species concentrating on vertical growth. Only a handful of names led the way through this period—Thoreau, Emerson, Muir, Austin, Leopold, Carson, Stegner among them, and relatively few others.

Stegner's work acted as a core, an incubator and radiant source, of health and diversity in the literature of nature. Due in large part to his teaching and writing and his example, a critical threshold of literary health was reached. There are now hundreds, even thousands, of nature writers, blossoming from Stegner's era as if from a nurse log. In literature, if not out on the land, there is now a community of health.

≈⊙≈

If the last roadless, wild cores of the West are lost—entered, whether by ecosystem management or clearcutting; further fragmented, rather than re-connected —will this, over time, cause the artistic works that sprang up out of love of earth, love of country, to lose part or all of their power—to become like ghosts, tales of things-gone-by, like empty insect-husks in the autumn?

As much as I love the works of Stegner and other writers whose work is based in the roadless wilderness and in the healthy country that lies on its perimeter, I cannot argue that the power of those works is not at risk. They are too intimately and fully connected, not just to the spirit of these places, but also to the physical elements, the presence, of these places. Those who have visited these sanctuaries, and even those who only hope to visit them, can feel their existence. There is a blood of vitality that still flows from the land to its literature (and perhaps from the literature back to the land—perhaps the dirt desires stories, as it desires life). The land and the literature are still connected. Harm the land further and a case can be made that it will diminish our literature, both that which has already been gifted to us and that which is still to come.

≈⊙≈

If ecosystem management continues to avoid committing to the protection of these last undesignated wilderness cores, it is nothing more than another blueprint for extinction, and extinguishment. We might just as well enter these last roadless areas now, and we might just as well gather up all of Stegner's books and get it over with: rip the pages out of them, or hire lesser writers to manage them—to re-write, re-shape, re-imagine, and re-create them.

New York literary folks were not always able to understand Stegner's work, nor were the extractive industry corporations and chamber of commerce flash-in-the-pan boosters always overly fond of it. If they could have outlawed or fragmented him, I think they would have. If they could have ignored him, they would have. If they could have clearcut his work, or even if they could have ecosystem-managed it, they would have.

But they couldn't. Art, like nature, desires life. His books and his life have everything to teach us about wilderness and ecosystem management, and we all need to go back and re-read them, and then re-read them again, and keep re-reading them. And we need to protect the Yaak and our other last wild and roadless places: to guard them as fiercely as we would our libraries or any other heritage, against intruders either foreign or domestic.

We need to keep using and saying the word *wilderness*—not replacing it, through time, with lesser phrases—with diluted, vanishing, and finally invisible non-words such as "ecosystem management." We owe it to Stegner and we owe it to ourselves and we owe it to those who will be following after us.

I want to believe in ecosystem management and I keep waiting to be friends with the Forest Service again, but in the meantime, there is still no protected wilderness in the Yaak Valley, nor is the lack of it being discussed enough.

NO ↩

William Tucker

Is Nature Too Good for Us?

Probably nothing has been more central to the environmental movement than the concept of wilderness. "In wildness is the preservation of the world," wrote Thoreau, and environmental writers and speakers have intoned his message repeatedly. Wilderness, in the environmental pantheon, represents a particular kind of sanctuary in which all true values—that is, all nonhuman values —are reposited. Wildernesses are often described as "temples," "churches," and "sacred ground"—refuges for the proposed "new religion" based on environmental consciousness. Carrying the religious metaphor to the extreme, one of the most famous essays of the environmental era holds the Judeo-Christian religion responsible for "ecological crisis."

The wilderness issue also has a political edge. Since 1964, long-standing preservation groups like the Wilderness Society and the Sierra Club have been pressuring conservation agencies like the National Forest Service and the Bureau of Land Management to put large tracts of their holdings into permanent "wilderness designations," countering the "multiple use" concept that was one of the cornerstones of the Conservation Era of the early 1900s.

Preservation and conservation groups have been at odds since the end of the last century, and the rift between them has been a major controversy of environmentalism. The leaders of the Conservation Movement—most notably Theodore Roosevelt, Gifford Pinchot, and John Wesley Powell—called for rational, efficient development of land and other natural resources: multiple use, or reconciling competing uses of land, and also "highest use," or forfeiting more immediate profits from land development for more lasting gains. Preservationists, on the other hand, the followers of California woodsman John Muir, have advocated protecting land in its natural state, setting aside tracts and keeping them inviolate. "Wilderness area" battles have become one of the hottest political issues of the day, especially in western states—the current "Sagebrush Revolt" comes to mind—where large quantities of potentially commercially usable land are at stake.

The term "wilderness" generally connotes mountains, trees, clear streams, rushing waterfalls, grasslands, or parched deserts, but the concept has been institutionalized and has a careful legal definition as well. The one given by the 1964 Wilderness Act, and that most environmentalists favor, is that wilderness

From William Tucker, "Is Nature Too Good for Us?" *Harper's Magazine* (March 1982). Adapted from William Tucker, *Progress and Privilege: America in the Age of Environmentalism* (Doubleday, 1982). Copyright © 1982 by William Tucker. Reprinted by permission.

is an area "where man is a visitor but does not remain." People do not "leave footprints there," wilderness exponents often say. Wildernesses are, most importantly, areas in which *evidence of human activity is excluded;* they need not have any particular scenic, aesthetic, or recreational value. The values, as environmentalists usually say, are "ecological"—which means, roughly translated, that natural systems are allowed to operate as free from human interference as possible.

The concept of excluding human activity is not to be taken lightly. One of the major issues in wilderness areas has been whether or not federal agencies should fight forest fires. The general decision has been that they should not, except in cases where other lands are threatened. The federal agencies also do not fight the fires with motorized vehicles, which are prohibited in wilderness areas except in extreme emergencies. Thus in recent years both the National Forest Service and the National Park Service have taken to letting forest fires burn unchecked, to the frequent alarm of tourists. The defense is that many forests require periodic leveling by fire in order to make room for new growth. There are some pine trees, for instance, whose cones will break open and scatter their seeds only when burned. This theoretical justification has won some converts, but very few in the timber companies, which bridle at watching millions of board-feet go up in smoke when their own "harvesting" of mature forests has the same effect in clearing the way for new growth and does less damage to forest soils.

The effort to set aside permanent wilderness areas on federal lands began with the National Forest Service in the 1920s. The first permanent reservation was in the Gila National Forest in New Mexico. It was set aside by a young Forest Service officer named Aldo Leopold, who was later to write *A Sand County Almanac,* which has become one of the bibles of the wilderness movement. Robert Marshall, another Forest Service officer, continued the program, and by the 1950s nearly 14 million of the National Forest System's 186 million acres had been administratively designated wilderness preserves.

Leopold and Marshall had been disillusioned by one of the first great efforts at "game management" under the National Forest Service, carried out in the Kaibab Plateau, just north of the Grand Canyon. As early as 1906 federal officials began a program of "predator control" to increase the deer population in the area. Mountain lions, wolves, coyotes, and bobcats were systematically hunted and trapped by game officials. By 1920, the program appeared to be spectacularly successful. The deer population, formerly numbering 4,000, had grown to almost 100,000. But it was realized too late that it was the range's limited food resources that would threaten the deer's existence. During two severe winters, in 1924-26, 60 percent of the herd died, and by 1939 the population had shrunk to only 10,000. Deer populations (unlike human populations) were found to have no way of putting limits on their own reproduction. The case is still cited as the classic example of the "boom and bust" disequilibrium that comes from thoughtless intervention in an ecological system.

The idea of setting aside as wilderness areas larger and larger segments of federally controlled lands began to gain more support from the old preservationists' growing realizations, during the 1950s, that they had not won the battle during the Conservation Era, and that the national forests were not parks that would be protected forever from commercial activity.

Pinchot's plan for practicing "conservation" in the western forests was to encourage a partnership between the government and large industry. In order to discourage overcutting and destructive competition, he formulated a plan that would promote conservation activities among the larger timber companies while placing large segments of the western forests under federal control. It was a classic case of "market restriction," carried out by the joint efforts of larger businesses and government. Only the larger companies, Pinchot reasoned, could generate the profits that would allow them to cut their forest holdings *slowly* so that the trees would have time to grow back. In order to ensure these profit margins, the National Forest Service would hold most of its timber lands out of the market for some time. This would hold up the price of timber and prevent a rampage through the forests by smaller companies trying to beat small profit margins by cutting everything in sight. Then, in later years, the federal lands would gradually be worked into the "sustained yield" cycles, and timber rights put up for sale. It was when the national forests finally came up for cutting in the 1950s that the old preservation groups began to react.

The battle was fought in Congress. The 1960 Multiple Use and Sustained Yield Act tried to reaffirm the principles of the Conservation Movement. But the wilderness groups had their day in 1964 with the passing of the Wilderness Act. The law required all the federal land-management agencies—the National Forest Service, the National Park Service, and the Fish and Wildlife Service—to review all their holdings, keeping in mind that "wilderness" now constituted a valid alternative in the "multiple use" concept—even though the concept of wilderness is essentially a rejection of the idea of multiple use. The Forest Service, with 190 million acres, and the Park Service and Fish and Wildlife Service, each with about 35 million acres, were all given twenty years to start designating wilderness areas. At the time, only 14.5 million acres of National Forest System land were in wilderness designations.

The results have been mixed. The wilderness concept appears valid if it is recognized for what it is—an attempt to create what are essentially "ecological museums" in scenic and biologically significant areas of these lands. But "wilderness," in the hands of environmentalists, has become an all-purpose tool for stopping economic activity as well. This is particularly crucial now because of the many mineral and energy resources available on western lands that environmentalists are trying to push through as wilderness designations. The original legislation specified that lands were to be surveyed for valuable mineral resources before they were put into wilderness preservation. Yet with so much land being reviewed at once, these inventories have been sketchy at best. And once land is locked up as wilderness, it becomes illegal even to explore it for mineral or energy resources.

Thus the situation in western states—where the federal government still owns 68 percent of the land, counting Alaska—has in recent years become a race between mining companies trying to prospect under severely restricted conditions, and environmental groups trying to lock the doors to resource development for good. This kind of permanent preservation—the antithesis of conservation—will probably have enormous effects on our future international trade in energy and mineral resources.

At stake in both the national forests and the Bureau of Land Management holdings are what are called the "roadless areas." Environmentalists call these lands "de facto wilderness," and say that because they have not yet been explored or developed for resources they should not be explored and developed in the future. The Forest Service began its Roadless Area Resources Evaluation (RARE) in 1972, while the Bureau of Land Management began four years later in 1976, after Congress brought its 174 million acres under jurisdiction of the 1964 act. The Forest Service is studying 62 million roadless acres, while the BLM is reviewing 24 million.

In 1974 the Forest Service recommended that 15 million of the 50 million acres then under study be designated as permanent wilderness. Environmental groups, which wanted much more set aside, immediately challenged the decision in court. Naturally, they had no trouble finding flaws in a study intended to cover such a huge amount of land, and in 1977 the Carter administration decided to start over with a "RARE II" study, completed in 1979. This has also been challenged by a consortium of environmental groups that includes the Sierra Club, the Wilderness Society, the National Wildlife Federation, and the Natural Resources Defense Council. The RARE II report also recommended putting about 15 million acres in permanent wilderness, with 36 million released for development and 11 million held for further study. The Bureau of Land Management is not scheduled to complete the study of its 24 million acres until 1991.

The effects of this campaign against resource development have been powerful. From 1972 to 1980, the price of a Douglas fir in Oregon increased 500 percent, largely due to the delays in timber sales from the national forests because of the battles over wilderness areas. Over the decade, timber production from the national forests declined slightly, putting far more pressure on the timber industry's own lands. The nation has now become an importer of logs, despite the vast resources on federal lands. In 1979, environmentalists succeeded in pressuring Congress into setting aside 750,000 acres in Idaho as the Sawtooth Wilderness and National Recreational Area. A resource survey, which was not completed until *after* the congressional action, showed that the area contained an estimated billion dollars' worth of molybdenum, zinc, silver, and gold. The same tract also contained a potential source of cobalt, an important mineral for which we are now dependent on foreign sources for 97 percent of what we use.

Perhaps most fiercely contested are the energy supplies believed to be lying under the geological strata running through Colorado, Wyoming, and Montana just east of the Rockies, called the Overthrust Belt. Much of this land is still administered by the Bureau of Land Management for multiple usage. But with the prospect of energy development, environmental groups have been rushing to try to have these high-plains areas designated as wilderness areas as well (cattle grazing is still allowed in wilderness tracts). On those lands permanently withdrawn from commercial use, mineral exploration will be allowed to continue until 1983. Any mines begun by then can continue on a very restricted basis. But the exploration in "roadless areas" is severely limited, in that in most cases there can be no roads constructed (and no use of off-road vehicles) while exploration is going on. Environmentalists have argued that wells can still be drilled and test mines explored using helicopters. But any such exploration is likely to be extraordinarily expensive and ineffective. Wilderness restrictions are now being drawn so tightly that people on the site are not allowed to leave their excrement in the area.

Impossible Paradises

What is the purpose of all this? The standard environmental argument is that we have to "preserve these last few wild places before they all disappear." Yet it is obvious that something more is at stake. What is being purveyed is a view of the world in which human activity is defined as "bad" and natural conditions are defined as "good." What is being preserved is evidently much more than "ecosystems." What is being preserved is an *image* of wilderness as a semisacred place beyond humanity's intrusion.

It is instructive to consider how environmentalists themselves define the wilderness. David Brower, former director of the Sierra Club, wrote in his introduction to Paul Ehrlich's *The Population Bomb* (1968):

> Whatever resources the wilderness still held would not sustain (man) in his old habits of growing and reaching without limits. Wilderness could, however, provide answers for questions he had not yet learned how to ask. He could predict that the day of creation was not over, that there would be wiser men, and they would thank him for leaving the source of those answers. Wilderness would remain part of his geography of hope, as Wallace Stegner put it, and could, merely because wilderness endured on the planet, prevent man's world from becoming a cage.

The wilderness, he suggested, is a source of peace and freedom. Yet setting wilderness aside for the purposes of solitude doesn't always work very well. Environmentalists have discovered this over and over again, much to their chagrin. Every time a new "untouched paradise" is discovered, the first thing everyone wants to do is visit it. By their united enthusiasm to find these "sanctuaries," people bring the "cage" of society with them. Very quickly it becomes necessary to erect bars to keep people *out*—which is exactly what most of the "wilderness" legislation has been all about.

In 1964, for example, the Sierra Club published a book on the relatively "undiscovered" paradise of Kauai, the second most westerly island in the Hawaiian chain. It wasn't long before the island had been overrun with tourists. When *Time* magazine ran a feature on Kauai in 1979, one unhappy island resident wrote in to convey this telling sentiment: "We're hoping the shortages of jet fuel will stay around and keep people away from here." The age of environmentalism has also been marked by the near overrunning of popular national parks like Yosemite (which now has a full-time jail), intense pressure on woodland recreational areas, full bookings two and three years in advance for raft trips through the Grand Canyon, and dozens of other spectacles of people crowding into isolated areas to get away from it all. Environmentalists are often critical of these inundations, but they must recognize that they have at least contributed to them.

I am not arguing against wild things, scenic beauty, pristine landscapes, and scenic preservation. What I am questioning is the argument that wilderness is a value against which every other human activity must be judged, and that human beings are somehow unworthy of the landscape. The wilderness has been equated with freedom, but there are many different ideas about what constitutes freedom. In the Middle Ages, the saying was that "city air makes a man free," meaning that the harsh social burdens of medieval feudalism vanished once a person escaped into the heady anonymity of a metropolitan community. When city planner Jane Jacobs, author of *The Death and Life of Great American Cities,* was asked by an interviewer if "overpopulation" and "crowding into large cities" weren't making social prisoners of us all, her simple reply was: "Have you ever lived in a small town?"

It may seem unfair to itemize the personal idiosyncrasies of people who feel comfortable only in wilderness, but it must be remembered that the environmental movement has been shaped by many people who literally spent years of their lives living in isolation. John Muir, the founder of the National Parks movement and the Sierra Club, spent almost ten years living alone in the Sierra Mountains while learning to be a trail guide. David Brower, who headed the Sierra Club for over a decade and later broke with it to found the Friends of the Earth, also spent years as a mountaineer. Gary Snyder, the poet laureate of the environmental movement, has lived much of his life in wilderness isolation and has also spent several years in a Zen monastery. All these people far outdid Thoreau in their desire to get a little perspective on the world. There is nothing reprehensible in this, and the literature and philosophy that merge from such experiences are often admirable. But it seems questionable to me that the ethic that comes out of this wilderness isolation—and the sense of ownership of natural landscapes that inevitably follows—can serve as the basis for a useful national philosophy.

That Frontier Spirit

The American frontier is generally agreed to have closed down physically in 1890, the year the last Indian Territory of Oklahoma was opened for the settlement. After that, the Conservation Movement arose quickly to protect the remaining resources and wilderness from heedless stripping and development. Along with this came a significant psychological change in the national character, as the "frontier spirit" diminished and social issues attracted greater attention. The Progressive Movement, the Social Gospel among religious groups, Populism, and Conservation all arose in quick succession immediately after the "closing of the frontier." It seems fair to say that it was only after the frontier had been settled and the sense of endless possibilities that came with open spaces had been constricted in the national consciousness that the country started "growing up."

Does this mean the new environmental consciousness has arisen because we are once again "running out of space"? I doubt it. Anyone taking an airplane across almost any part of the country is inevitably struck by how much greenery and open territory remain, and how little room our towns and cities really occupy. The amount of standing forest in the country, for example, has not diminished appreciably over the last fifty years, and is 75 percent of what it was in 1620. In addition, as environmentalists constantly remind us, trees are "renewable resources." If they continue to be handled intelligently, the forests will always grow back. As farming has moved out to the Great Plains of the Middle West, many eastern areas that were once farmed have reverted back to trees. Though mining operations can permanently scar hillsides and plains, they are usually very limited in scope (and as often as not, it is the roads leading to these mines that environmentalists find most objectionable).

It seems to be that the wilderness ethic has actually represented an attempt psychologically to reopen the American frontier. We have been desperate to maintain belief in unlimited, uncharted vistas within our borders, a preoccupation that has eclipsed the permanent shrinking of the rest of the world outside. Why else would it be so necessary to preserve such huge tracts of "roadless territory" simply because they are now roadless, regardless of their scenic, recreational, or aesthetic values? The environmental movement, among other things, has been a rather backward-looking effort to recapture America's lost innocence.

The central figure in this effort has been the backpacker. The backpacker is a young, unprepossessing person (inevitably white and upper middle class) who journeys into the wilderness as a passive observer. He or she brings his or her own food, treads softly, leaves no litter, and has no need to make use of any of the resources at hand. Backpackers bring all the necessary accouterments of civilization with them. All their needs have been met by the society from which they seek temporary release. The backpacker is freed from the need to support itself in order to enjoy the aesthetic and spiritual values that are made available by this temporary *removal* from the demands of nature. Many dangers—raging rivers or precipitous cliffs, for instance—become sought-out adventures.

Yet once the backpacker runs out of supplies and starts using resources around him—cutting trees for firewood, putting up a shelter against the rain—he is violating some aspect of the federal Wilderness Act. For example, one of the issues fought in the national forests revolves around tying one's horse to a tree. Purists claim the practice should be forbidden, since it may leave a trodden ring around the tree. They say horses should be hobbled and allowed to graze instead. In recent years, the National Forest Service has come under pressure from environmental groups to enforce this restriction.

Wildernesses, then, are essentially parks for the upper middle class. They are vacation reserves for people who want to rough it—with the assurance that few other people will have the time, energy, or means to follow them into the solitude. This is dramatically highlighted in one Sierra Club book that shows a picture of a professorial sort of individual backpacking off into the woods. The ironic caption is a quote from Julius Viancour, an official of the Western Council of Lumber and Sawmill Workers: "The inaccessible wilderness and primitive areas are off limits to most laboring people. We must have access...." The implication for Sierra Club readers is: "What do these beer-drinking, gun-toting, working people want to do in *our* woods?"

This class-oriented vision of wilderness as an upper-middle-class preserve is further illustrated by the fact that most of the opposition to wilderness designations comes not from industry but from owners of off-road vehicles. In most northern rural areas, snowmobiles are now regarded as the greatest invention since the automobile, and people are ready to fight rather than stay cooped up all winter in their houses. It seems ludicrous to them that snowmobiles (which can't be said even to endanger the ground) should be restricted from vast tracts of land so that the occasional city visitor can have solitude while hiking past on snowshoes.

The recent Boundary Waters Canoe Area controversy in northern Minnesota is an excellent example of the conflict. When the tract was first designated as wilderness in 1964, Congress included a special provision that allowed motorboats into the entire area. By the mid-1970s, outboards and inboards were roaming all over the wilderness, and environmental groups began asking that certain portions of the million-acre preserve be set aside exclusively for canoes. Local residents protested vigorously, arguing that fishing expeditions, via motorboats, contributed to their own recreation. Nevertheless, Congress eventually excluded motorboats from 670,000 acres to the north.

A more even split would seem fairer. It should certainly be possible to accommodate both forms of recreation in the area, and there is as much to be said for canoeing in solitude as there is for making rapid expeditions by powerboat. The natural landscape is not likely to suffer very much from either form of recreation. It is not absolute "ecological" values that are really at stake, but simply different tastes in recreation.

Not Entirely Nature

At bottom, then, the mystique of the wilderness has been little more than a revival of Rousseau's Romanticism about the "state of nature." The notion that "only in wilderness are human beings truly free," a credo of environmentalists, is merely a variation on Rousseau's dictum that "man is born free, and everywhere he is in chains." According to Rousseau, only society could enslave people, and only in the "state of nature" was the "noble savage"—the preoccupation of so many early explorers—a fulfilled human being.

The "noble savage" and other indigenous peoples, however, have been carefully excised from the environmentalists' vision. Where environmental efforts have encountered primitive peoples, these indigenous residents have often proved one of the biggest problems. One of the most bitter issues in Alaska is the efforts by environmentalist groups to restrict Indians in their hunting practices.

At the same time, few modern wilderness enthusiasts could imagine, for example, the experience of the nineteenth-century artist J. Ross Browne, who wrote in *Harper's New Monthly Magazine* after visiting the Arizona territories in 1864:

> Sketching in Arizona is ... rather a ticklish pursuit. ... I never before traveled through a country in which I was compelled to pursue the fine arts with a revolver strapped around my body, a double-barreled shot-gun lying across my knees, and half a dozen soldiers armed with Sharpe's carbines keeping guard in the distance. Even with all the safeguards ... I am free to admit that on occasions of this kind I frequently looked behind to see how the country appeared in its rear aspect. An artist with an arrow in his back may be a very picturesque object ... but I would rather draw him on paper than sit for the portrait myself.

Wilderness today means the land *after* the Indians have been cleared away but *before* the settlers have arrived. It represents an attempt to hold that particular moment forever frozen in time, that moment when the visionary American settler looked out on the land and imagined it as an empty paradise, waiting to be molded to our vision.

In the absence of the noble savage, the environmentalist substitutes himself. The wilderness, while free of human dangers, becomes a kind of basic-training ground for upper-middle-class values. Hence the rise of "survival" groups, where college kids are taken out into the woods for a week or two and let loose to prove their survival instincts. No risks are spared on these expeditions. Several people have died on them, and a string of lawsuits has already been launched by parents and survivors who didn't realize how seriously these survival courses were being taken.

The ultimate aim of these efforts is to test upper-middle-class values against the natural environment. "Survival" candidates cannot hunt, kill, or use much of the natural resources available. The true test is whether their zero-degree sleeping bags and dried-food kits prove equal to the hazards of the tasks. What happens is not necessarily related to nature. One could as easily test survival skills by turning a person loose without money or means in New York City for three days.

I do not mean to imply that these efforts do not require enormous amounts of courage and daring—"survival skills." I am only suggesting that what the backpacker or survival hiker encounters is not entirely "nature," and that the effort to go "back to nature" is one that is carefully circumscribed by the most intensely civilized artifacts. Irving Babbitt, the early twentieth-century critic of Rousseau's Romanticism, is particularly vigorous in his dissent from the idea of civilized people going "back to nature." This type, he says, is actually "the least primitive of all beings":

> We have seen that the special form of unreality encouraged by the aesthetic romanticism of Rousseau is the dream of the simple life, the return to a nature that never existed, and that this dream made its special appeal to an age that was suffering from an excess of artificiality and conventionalism.

Babbitt notes shrewdly that our concept of the "state of nature" is actually one of the most sophisticated productions of civilization. Most primitive peoples, who live much closer to the soil than we do, are repelled by wilderness. The American colonists, when they first encountered the unspoiled landscape, saw nothing but a horrible desert, filled with savages.

What we really encounter when we talk about "wilderness," then, is one of the highest products of civilization. It is a reserve set up to keep people *out*, rather than a "state of nature" in which the inhabitants are "truly free." The only thing that makes people "free" in such a reservation is that they can leave so much behind when they enter. Those who try to stay too long find out how spurious this "freedom" is. After spending a year in a cabin in the north Canadian woods, Elizabeth Arthur wrote in *Island Sojourn:* "I never felt so completely tied to *objects,* resources, and the tools to shape them with."

What we are witnessing in the environmental movement's obsession with purified wilderness is what has often been called the "pastoral impulse." The image of nature as unspoiled, unspotted wilderness where we can go to learn the lessons of ecology is both a product of a complex, technological society and an escape from it. It is this undeniable paradox that forms the real problem of setting up "wildernesses." Only when we have created a society that gives us the leisure to appreciate it can we go out and experience what we imagine to be untrammeled nature. Yet if we lock up too much of our land in these reserves, we are cutting into our resources and endangering the very leisure that allows us to enjoy nature.

The answer is, of course, that we cannot simply let nature "take over" and assume that because we have kept roads and people out of huge tracts of land, then we have absolved ourselves of a national guilt. The concept of stewardship means taking responsibility, not simply letting nature take its course. Where tracts can be set aside from commercialism at no great cost, they should be. Where primitive hiking and recreation areas are appealing, they should be maintained. But if we think we are somehow appeasing the gods by *not* developing resources where they exist, then we are being very shortsighted. Conservation, not preservation, is once again the best guiding principle.

The cult of wilderness leads inevitably in the direction of religion. Once again, Irving Babbitt anticipated this fully.

> When pushed to a certain point the nature cult always tends toward sham spirituality.... Those to whom I may seem to be treating the nature cult with undue severity should remember that I am treating it only in its pseudo-religious aspect.... My quarrel is only with the asthete who assumes an apocalyptic pose and gives forth as a profound philosophy what is at best only a holiday or weekend view of existence....

It is often said that environmentalism could or should serve as the basis of a new religious consciousness, or a religious "reawakening." This religious trend is usually given an Oriental aura. E. F. Schumacher has a chapter on Buddhist economics in his classic *Small Is Beautiful.* Primitive animisms are also frequently cited as attitudes toward nature that are more "environmentally sound." One book on the environment states baldly that "the American Indian lived in almost perfect harmony with nature." Anthropologist Marvin Harris has even put forth the novel view that primitive man is an environmentalist, and that many cultural habits are unconscious efforts to reduce the population and conserve the environment. He says that the Hindu prohibition against eating cows and the Jewish tradition of not eating pork were both efforts to avoid the ecological destruction that would come with raising these grazing animals intensively. The implication in these arguments is usually that science and modern technology have somehow dulled our instinctive "environmental" impulses, and that Western "non-spiritual" technology puts us out of harmony with the "balance of nature."

Perhaps the most daring challenge to the environmental soundness of current religious tradition came early in the environmental movement, in a much quoted paper by Lynn White, professor of the history of science at UCLA. Writing in *Science* magazine in 1967, White traced "the historical roots of our ecological crisis" directly to the Western Judeo-Christian tradition in which "man and nature are two things, and man is master." "By destroying pagan animism," he wrote, "Christianity made it possible to exploit nature in a mood of indifference to the feelings of natural objects." He continued:

> Especially in its Western form, Christianity is the most anthropocentric religion the world has seen.... Christianity, in absolute contrast to ancient paganism and Asia's religions (except, perhaps, Zoroastrianism), not only established a dualism of man and nature but also insisted that it is God's will that man exploit nature for his proper ends.... In antiquity every tree, every spring, every stream, every hill had its own *genius loci*, its guardian spirit.... Before one cut a tree, mined a mountain, or dammed a brook, it was important to placate the spirit in charge of that particular situation, and keep it placated.

But the question here is not whether the Judeo-Christian tradition is worth saving in and of itself. It would be more than disappointing if we canceled the accomplishments of Judeo-Christian thought only to find that our treatment of nature had not changed a bit.

There can be no question that White is onto a favorite environmental theme here. What he calls the "Judeo-Christian tradition" is what other writers often term "Western civilization." It is easy to go through environmental books and find long outbursts about the evils that "civilization and progress" have brought us. The long list of Western achievements and advances, the scientific men of genius, are brought to task for creating our "environmental crisis." Sometimes the condemnation is of our brains, pure and simple. Here, for example, is the opening statement from a book about pesticides, written by the late Robert van den Bosch, an outstanding environmental advocate:

> Our problem is that we are too smart for our own good, and for that matter, the good of the biosphere. The basic problem is that our brain enables us to evaluate, plan, and execute. Thus, while all other creatures are programmed by nature and subject to her whims, we have our own gray computer to motivate, for good or evil, our chemical engine.... Among living species, we are the only one possessed of arrogance, deliberate stupidity, greed, hate,

jealousy, treachery, and the impulse to revenge, all of which may erupt spontaneously or be turned on at will.

At this rate, it can be seen that we don't even need religion to lead us astray. We are doomed from the start because we are not creatures of *instinct*, programmed from the start "by nature."

This type of primitivism has been a very strong, stable undercurrent in the environmental movement. It runs from the kind of fatalistic gibberish quoted above to the Romanticism that names primitive tribes "instinctive environmentalists," from the pessimistic predictions that human beings cannot learn to control their own numbers to the notion that only by remaining innocent children of nature, untouched by progress, can the rural populations of the world hope to feed themselves. At bottom, as many commentators have pointed out, environmentalism is reminiscent of the German Romanticism of the nineteenth century, which sought to shed Christian (and Roman) traditions and revive the Teutonic gods because they were "more in touch with nature."

But are progress, reason, Western civilization, science, and the cerebral cortex really at the root of the "environmental crisis"? Perhaps the best answer comes from an environmentalist himself, Dr. Rene Dubos, a world-renowned microbiologist, author of several prize-winning books on conservation and a founding member of the Natural Resources Defense Council. Dr. Dubos takes exception to the notion that Western Christianity has produced a uniquely exploitative attitude toward nature:

> Erosion of the land, destruction of animal and plant species, excessive exploitation of natural resources, and ecological disasters are not peculiar to the Judeo-Christian tradition and to scientific technology. At all times, and all over the world, man's thoughtless interventions into nature have had a variety of disastrous consequences or at least have changed profoundly the complexity of nature.

Dr. Dubos has catalogued the non-Western or non-Christian cultures that have done environmental damage. Plato observed, for instance, that the hills in Greece had been heedlessly stripped of wood, and erosion had been the result; the ancient Egyptians and Assyrians exterminated large numbers of wild animal species; Indian hunters presumably caused the extinction of many large paleolithic species in North America; Buddhist monks building temples in Asia contributed largely to deforestation. Dubos notes:

> All over the globe and at all times . . . men have pillaged nature and disturbed the ecological equilibrium . . . nor did they have a real choice of alternatives. If men are more destructive now . . . it is because they have at their command more powerful means of destruction, not because they have been influenced by the Bible. In fact, the Judeo-Christian peoples were probably the first to develop on a large scale a pervasive concern for land management and an ethic of nature.

The concern that Dr. Dubos cites is the same one we have rescued out of the perception of environmentalism as a movement based on aristocratic conservatism. That is the legitimate doctrine of *stewardship* of the land. In order to take this responsibility, however, we must recognize the part we play in nature

—that "the land is ours." It will not do simply to worship nature, to create a cult of wilderness in which humanity is an eternal intruder and where human activity can only destroy.

"True conservation," writes Dubos, "means not only protecting nature against human misbehavior but also developing human activities which favor a creative, harmonious relationship between man and nature." This is a legitimate goal for the environmental movement.

POSTSCRIPT

Does Wilderness Have Intrinsic Value?

Bass's rhapsodic descriptions of the beauty and wonder of the Yaak Valley and other wild areas that have inspired many great nature writers would surely qualify him as a member of what Tucker refers to as the "cult of wilderness," but his preservationist motives include such practical ecological goals as preserving biodiversity, which Tucker does not discuss.

Despite the increasing popularity of backpacking, Tucker is correct in maintaining that it is still primarily a diversion of the economically privileged. Indeed, a lack of financial resources and leisure time prevents the majority of U.S. citizens from taking advantage of the tax-supported parks that multiple-use conservationists such as Tucker support, as well as from enjoying a small fraction of the acreage that has been set aside as protected wilderness.

The controversies over oil development in Alaska's Arctic National Wildlife Refuge and mineral exploitation in Utah's vast red rock region are currently the two most bitterly contested struggles concerning U.S. wilderness areas. Although the huge 1989 oil spill in Prince William Sound by the supertanker *Exxon Valdez* dealt a temporary setback to proponents of oil exploration in the Alaskan wilderness, the development lobby has continued its efforts. President Bill Clinton's executive action to protect part of Utah's remaining wilderness by creating the Grand Staircase—Escalante National Monument— is still being challenged by congressional opponents. For information about the Alaskan wilderness controversy, see the article by Douglas Kuzmiak in *The Geographical Magazine* (April 1994) and the report *The Arctic National Wildlife Refuge* by M. Lynne Corn, Lawrence C. Kumins, and Pamela Baldwin, available from the Committee for the National Institute for the Environment in Washington, D.C. Daniel Glick presents a strong argument for preserving Utah's wilderness region in the Winter 1995 issue of *Wilderness*. The results of a recent survey that expands by 3 million acres the area of Utah that is eligible for inclusion in the National Wilderness Preservation System are reported by T. H. Watkins in the November/December 1998 issue of *Sierra*. See also his book *The Redrock Chronicles: Saving Wild Utah* (Johns Hopkins University Press, 2000).

Two moving and thought-provoking essays commemorating the 30th anniversary of the U.S. Wilderness Act are "An Enduring Wilderness," by Bruce Hamilton, in the September/October 1994 issue of *Sierra* and "Toward Wild Heartlands," by John Daniel, *Audubon* (September/October 1994). For a comprehensive collection of essays on the subject, see *Voices for the Wilderness* edited by William Schwartz (Ballantine Books, 1969). For a survey of the wilderness system created by the 1964 act, see John G. Mitchell and Peter Essick, "Wilderness: America's Land Apart," *National Geographic* (November 1998).

ISSUE 3

Is the U.S. Endangered Species Act Fundamentally Sound?

YES: David Langhorst, from "Is the Endangered Species Act Fundamentally Sound? Pro," *Congressional Digest* (March 1996)

NO: Mark L. Plummer, from "Is the Endangered Species Act Fundamentally Sound? Con," *Congressional Digest* (March 1996)

ISSUE SUMMARY

YES: David Langhorst, executive board member of the Idaho Wildlife Federation, asserts that the Endangered Species Act has saved hundreds of plant and animal species that were in serious decline, in a manner sensitive to economic concerns, and that reauthorization of the act is in the public interest.

NO: Mark L. Plummer, an environmental economist and fellow of the Discovery Institute, denies that species extinction is a catastrophe. He argues that the act's goal of bringing listed species to full recovery is not achievable.

Extinction of biological species is not necessarily a phenomenon initiated by human activity. Although the specific role of extinction in the process of evolution is still being researched and debated, it is generally accepted that the demise of any biological species is inevitable. Opponents of special efforts to protect endangered species invariably point this out. They also suggest that the role of *Homo sapiens* in causing extinction should not be distinguished from that of any other species.

This position is contrary to some well-established facts. Unlike other creatures that have inhabited the Earth, human beings are the first to possess the technological ability to cause wholesale extermination of species, genera, or even entire families of living creatures. This process is accelerating. Between 1600 and 1900, humans hunted about 75 known species of mammals and birds to extinction. Wildlife management efforts initiated during the twentieth century have been unsuccessful in stemming the tide, as indicated by the fact that the rate of extinction of species of mammals and birds has jumped to approximately one per year.

In 1973 the Endangered Species Act was adopted, and an international treaty was negotiated, in an effort to combat this worldwide problem. This act united a variety of industrial and business interests with the commercial hunters and trappers who traditionally objected to efforts to restrict their activities. Opposition developed because the act prohibited construction projects that threatened to cause the extinction of any species. The most celebrated example of a confrontation brought about under this act was an effort by environmentalists to halt construction of the Tellico Dam in Loudon County, Tennessee, on the Little Tennessee River, because it threatened a small fish called the snail darter. Much publicity has also been given to the lumber industry's opposition to efforts to protect the spotted owl, whose habitat is in valued timber areas in the U.S. Northwest. Opponents of the act cite such controversies as evidence that the nation's economic well-being is undermined by species preservation efforts. The act's supporters counter by claiming that the number of serious conflicts between development and species protection have been very few and that despite the uncompromising language of the law, no major project has been prevented because of its enforcement.

Scientists fear that the vitality of our ecology may be undermined by the reduction of biodiversity resulting from genetic resources lost through species extinction. The ability of species to evolve and adapt to environmental change depends on a vast pool of genetic material. The present principal human threat to species survival is habitat destruction, rather than hunting or trapping. This links the issues of wilderness preservation and endangered species protection.

The authorization of the Endangered Species Act expired on October 1, 1992; the appropriation of funds for the act's enforcement, however, has been repeatedly extended. The battle over reauthorization continues to be one of the key unresolved environmental questions before the U.S. Congress. In 1995 a moratorium was imposed on the listing of new endangered or threatened species. In 1996 President Bill Clinton vetoed a bill that would have extended the moratorium for another year. A bill that would substitute habitat conservation plans for the protection of individual species is backed by the Clinton administration and has bipartisan support in the Senate, although it failed to come to the floor for a vote in 1998. In the House, this bill is opposed by representatives who, like most major environmental organizations, believe that it does not offer enough protection to endangered species as well as by conservatives who object to the bill's failure to compensate property owners who lose money because of habitat protection restrictions. The debate continues in 2000, with reauthorization bills before both the House and the Senate.

The following arguments are from testimony before a congressional committee hearing on the Endangered Species Act. David Langhorst presents the position of the National Wildlife Federation, which is that the act is an effective and vital tool for the preservation of biodiversity. He argues strongly against using cost/benefit analysis to decide which species to protect. Mark L. Plummer argues that the goal of the act, which is to bring endangered species to recovery, is not achievable and results in species' remaining on the list for an indefinite time. He maintains that the costs of wildlife protection should be balanced against other social goals.

David Langhorst  **YES**

Is the Endangered Species Act Fundamentally Sound? Pro

I am here to testify on behalf of the National Wildlife Federation [NWF], the Nation's largest conservation education organization. I serve as Executive Board member of the Idaho Wildlife Federation, one of NWF's 45 affiliated conservation organizations throughout the United States.

The ESA [Endangered Species Act] has produced a remarkable string of successes. In its 23-year history, it has stabilized or improved the conditions of hundreds of plant and animal species that had been in serious decline. In my own work, I have seen large numbers of concerned citizens work with the ESA to help bring about the recovery of the gray wolf in the Northern Rockies ecosystem. By educating communities about the importance of the wolf to the health of the ecosystem and using the ESA's flexible provisions, we are successfully restoring this wonderful animal to the wild in a manner sensitive to local economic interests.

The gray wolf recovery effort is a model of how diverse groups of local citizens can work together to achieve results using the ESA. However, as a result of delaying tactics by narrow ranching interests, wolf recovery is taking too many years and is generating inordinate costs to the Federal taxpayer. Meanwhile, during the period of the wolf recovery effort, the recovery of numerous other listed species is being neglected.

Certain regulated industry groups are now advocating that the ESA's goal of protecting and recovering all of the Nation's imperiled plant and animal species be abandoned and that the fate of each species be left to the discretion of the Secretaries of Interior and Commerce. Such an abandonment of the ESA's goal would be unwise for at least two reasons. First, conserving the fullest extent of our natural heritage provides enormous benefits to people, benefits that greatly exceed the costs of protection measures. Second, the alternative —separately deciding the fate of each species using a cost/benefit analysis—is simply unnecessary, unworkable, and would be extremely wasteful considering the numerous ESA procedures already in place to ensure that economic consequences are considered before the law is implemented.

From David Langhorst, "Is the Endangered Species Act Fundamentally Sound? Pro," in "Saving Endangered Species: Wildlife Conservation and Property Rights," *Congressional Digest,* vol. 75, no. 3 (March 1996). Copyright © 1996 by The Congressional Digest Corp., Washington, DC, (202) 333-7332. Reprinted by permission.

Congress established the goal of protecting and recovering all imperiled species when it first enacted the ESA in 1973. This ambitious goal was not chosen carelessly, but was arrived at after Congress determined that the rapid loss of biodiversity in the U.S. and abroad posed a direct threat to the well-being of the American people. When the law was reauthorized in 1978, 1982, and 1988, Congress reaffirmed that recovering all threatened and endangered species was essential.

The scientific evidence that motivated previous Congresses to set the goal of recovering all species has only strengthened in recent years. Today there is no dispute in the scientific community that human activity has brought about a loss of biodiversity not witnessed since the cataclysmic changes ending the dinosaur era 65 million years ago. Edmund O. Wilson, the eminent Harvard biologist, estimates that the current extinction rate in the tropical rain forests is somewhere between 1,000 to 10,000 times the rate that would exist without human disturbances of the environment. According to the recent study of the ESA by the National Academy of Sciences, the "current accelerated extinction rate is largely human-caused and is likely to increase rather than decrease in the near future."

This rapid loss of biodiversity is occurring not just in the tropical rain forests. In the nearly 400 years since the Pilgrims arrived to settle in North America, about 500 extinctions of plant and animal species and subspecies have occurred—a rate of extinction already much greater than the natural rate. According to recent calculations by Peter Hoch of the Missouri Botanical Garden, over the next five to 10 years, another 4,000 species in the U.S. alone could become extinct. This evidence of increased extinctions provides sad testimony to the need for improving the ESA rather than scaling back its fundamental goal.

Species are essential components of natural life-support systems that provide medicines, food, and other essential materials, regulate local climates and watersheds, and satisfy basic cultural, aesthetic, and spiritual needs. Below are six examples of how endangered species protections help people.

New medicines to respond to the health crises of tomorrow Wild plant and animal species are an essential part of the $79 billion annual U.S. pharmaceutical industry. One-fourth of all prescriptions dispensed in the U.S. contain active ingredients extracted from plants. Many other drugs that are now synthesized such as aspirin, were first discovered in the wild.

Researchers continue to discover new potential applications of wild plants and animals for life-saving or life-enhancing drugs. In fact, many pharmaceutical companies screen wild organisms for their medicinal potential. Yet, to date, less than 10 percent of known plant species have been screened for their medicinal values, and only 1 percent have been intensively investigated. Thus, species protections are essential to ensure that the full panoply of wild plants and animals remains available for study and future use.

Wild plant species that safeguard our food supply The human population depends upon only 20 plant species, out of over 80,000 edible plant species, to supply 90 percent of its food. These plants are the product of centuries of genetic cross-breeding among various strains of wild plants. Continual cross-breeding enables these plant species to withstand ever-evolving new diseases, pests, and changes in climatic and soil conditions. According to a recent study, the constant infusion of genes from wild plant species adds approximately $1 billion per year to U.S. agricultural production.

If abundant wild plant species were unavailable to U.S. agriculture companies for crossbreeding, entire crops would be vulnerable to pests and disease, with potentially devastating repercussions for U.S. farmers, consumers, and the economy.

Renewable resources for a sustainable future   At existing levels of consumption, nonrenewable resources such as petroleum will inevitably become increasingly costly and scarce in the coming decades. To prepare the U.S. for the global economy's certain transition toward renewable resources, Congress must ensure the health of the U.S. biological resource base. Fish, wildlife, and plant species could potentially supply the ingredients for the products that drive the U.S. economy of the 21st century. The substance that holds mussels to rocks through stormy seas, for example, may hold clues for a better glue to use in applications from shipbuilding to dentistry.

Early warning of ecosystem decline   Scientists have long known that the loss of any one species is a strong warning sign that the ecosystem that supported the species may be in decline. A recent study reported that loss of species could directly curtail the vital services that ecosystems provide to people. A subsequent study suggests that destruction of habitat could lead to the selective extinction of an ecosystem's "best competitors," causing a more substantial loss of ecosystem functions than otherwise would be expected.

Negative impacts in wild species often portend negative impacts for human health and quality of life. For example, some animal species are critical indicators of the harm that heavy chemicals can cause in our environment.

Ecosystems: life-support systems for people   Our society has become so alienated from nature that sometimes we forget that we rely on ecosystems for our survival. Ecosystems carry out essentially natural processes, such as those that purify our water and air, create our soil, protect against floods and erosion, and determine our climate.

For example, the Chesapeake Bay, the Nation's largest estuary, not only supports 2,700 plant and animal species, but also plays a major role in regulating environmental quality for humans. Rapid development around the Bay has sent countless tons of sediment downstream, landlocking communities that were once important ports. The construction of seawalls and breakwaters in some areas has led to rapid beach erosion in others.

Ecosystems: industries and jobs depend on them   Healthy ecosystems enable multi-billion dollar, job-intensive industries to survive. Examples of industries that are dependent on the health of ecosystems are: tourism, commercial fishing, recreational fishing, hunting, and wildlife watching.

When ecosystems are degraded, the result is economic distress. Destruction of salmon runs on the Columbia and Snake river systems in the Pacific Northwest led to the near collapse of that region's multi-billion dollar commercial and sport fishing industries.

Anti-ESA advocates propose to replace the goal of saving all species with a cost/benefit analysis to determine whether to save each species. Such cost/benefit analysis would likely produce an extinction of hundreds of endangered species due to human disturbances of habitat.

In the absence of any legal obligation to recover species, the Secretary of Interior could ultimately succumb to political pressures and choose meager objectives for any species that dare to get in the way of industry or development. For most species, any objective short of full recovery would effectively perpetuate the continued slide toward extinction.

Even if the cost/benefit analysis could somehow be insulated from political manipulation, its outcome would still be totally unreliable. The information available to the Secretary about the costs of protecting the species in question would be extremely incomplete, because no one could know at the time of the cost/benefit analysis what human activities would ultimately threaten the species and whether those activities could be modified through the ESA consultation process to avoid or reduce economic losses.

Equally important, the Secretary would also have incomplete information about the benefits to people provided by the species. Despite years of research and development, we have only just begun to discover the beneficial uses of species. Of the estimated five to 30 million species living today on Earth, scientists have identified and named only about 1.6 million species, and most of these have never been screened for beneficial uses. As species become extinct, we simply don't know what we are losing. The species that become extinct today might have provided the chemical for a miracle cancer treatment or the gene that saves the U.S. wheat crop from the next potentially devastating disease. A cost/benefit analysis of the penicillin fungus in the years prior to the discovery of its antibiotic qualities would have been a surefire recipe for extinction because no one could foresee its future role in the development of wonder drugs that would save and enhance the lives of millions of people.

There is yet another reason why we should not attempt to decide the fate of species based on a prediction of their future benefits. Species within an ecosystem are interdependent, and thus the extinction of one species potentially disrupts other species and the functioning of the entire ecosystem. As reported by the Missouri Botanical Garden, the loss of one plant species can cause a chain reaction leading to the extinction of up to 30 other species, including insects, higher animals, and other plants. Like pulling a single bolt from an airplane wing, we cannot know beforehand what effect the loss of a single species might have on the entire ecosystem.

A final flaw with the cost/benefit approach is that it is based on a false premise that the ESA lacks opportunities for consideration of economic and social impacts of listings. In fact, numerous ESA provisions require that economic and social consequences be balanced with species protection goals. Once a species is listed, the ESA provides for the consideration of socioeconomic factors in the designation of critical habitat, the development of special regulations for threatened species and experimental populations, the issuance of incidental take permits, the development of reasonable and prudent alternatives during Federal agency consultations, and the existence of the Endangered Species Committee to resolve any conflicts between conservation and economic goals.

Reauthorization of an effective Endangered Species Act is in the best interest of everyone involved. Species provide untold benefits to humans and are essential to our quality of life. By making thoughtful improvements to the ESA, we can enable private landowners and other stakeholders to take a greater conservation role and thereby provide for both species conservation and sustainable development—for the benefit of each of us and generations to come.

NO ↵

Mark L. Plummer

Is the Endangered Species Act Fundamentally Sound? Con

In 1973, when Congress passed the Endangered Species Act, its Members believed the goal of banishing extinction was imperative and within quick reach. The assumption, echoed by many conservationists today, was that endangered species can be saved without significant sacrifice. If development affects a species here, we can just move the development or the species somewhere else —an easy thing to do.

Over the past 21 years, it has become increasingly clear that the opposite is the case. From loggers in the Pacific Northwest to orange growers in Florida, from backyard barbecuers in upper-State New York to real estate developers in southern California, it is ordinary men and women doing ordinary things that threaten species, not trivialities. Still, the good reasons for endangering biodiversity might still not be enough if losing even one species would indeed "tinker with our future."

Yet that belief also stands in ruins. Losing a species may be tragic, but the result is rarely, if ever, catastrophic.

The problem of endangered species, then, presents us with few automatic solutions. As uncomfortable as facing the prospect may be, we must make choices that will have profound consequences for the future of our natural heritage. Ignoring this necessity, as the Endangered Species Act does, will not make the difficult choices go away. Instead, we will make them poorly.

The goal of the Endangered Species Act is to bring species to "recovery," which the Act defines as the point "at which the measures provided pursuant to this Act are no longer necessary." If a species attains recovery, Fish and Wildlife is supposed to remove it from the official list. At the end of 1973, the list consisted of 122 species. By the end of 1994, 21 years later, the agency had added another 833 domestic species, an average of almost 40 species a year. In that time, the agency delisted 21 species, an average of one species a year.

In fact, the 40 to 1 ratio of listings to delistings overstates the progress rate, because few of the latter were due to recovery. Seven species left the list when Fish and Wildlife declared them extinct. Of these, only one species with a good chance of survival—the dusky seaside sparrow—disappeared on the

From Mark L. Plummer, "Is the Endangered Species Act Fundamentally Sound? Con," in "Saving Endangered Species: Wildlife Conservation and Property Rights," *Congressional Digest*, vol. 75, no. 3 (March 1996). Copyright © 1996 by The Congressional Digest Corp., Washington, DC, (202) 333-7332. Reprinted by permission.

agency's watch. The others were either on the verge of extinction at the time of listing because of their extreme rarity, or long thought to be extinct, but placed on the endangered list in the hope that the action would spur biologists to discover new populations.

Another eight of the 21 delisted species were removed because they should not have been on it to begin with. The data on which the agency decided to list them turned out to be mistaken.

Finally, even the remaining balance of six domestic species delisted by Fish and Wildlife because their status had improved did not always owe that improvement to the Endangered Species Act. Consider the arctic peregrine falcon, which Fish and Wildlife struck from the list in October 1995. Although the Endangered Species Act banned hunting the falcon or harming its habitat, these actions, according to the official notice of delisting, were not "pivotal" to its recovery. Instead, the bird owes its improvement largely to the ban on pesticides like DDT [dichloro-diphenyl-trichloro-ethane]—an action that predated the Endangered Species Act.

By other measures of success, the Act shows similarly poor results. Reclassifying species from endangered to threatened has occurred less often than delisting: Between 1973 and 1994, Fish and Wildlife reclassified 13 species. And according to the 1992 biennial report from Fish and Wildlife on the recovery of listed species, the latest available, only 69 of the 711 species then listed—not quite 10 percent of the total—could be described as "improving," indicating active progress toward full recovery. Twenty-eight percent had "stable" populations, a sign that their declines had been halted. But a full 33 percent were "declining"; another 27 percent were "unknown." (The remaining 2 percent were believed to be extinct.) And species with stable populations were being held in a precarious position: almost three-fifths had achieved fewer than 25 percent of their recovery objectives.

The failure of the law to achieve full recovery means that once a species joins the list, it is almost certain to remain there for an indefinite period of time. Any private or public action that threatens the species no matter how praiseworthy in other circumstances, becomes tainted. Private landowners and Federal agency managers live under the perpetual shadow of the Endangered Species Act. In this way, an endangered species becomes a permanent liability for anyone unlucky enough to be host to one.

Understandably, landowners have responded by trying to free themselves from these restraints, sometimes in ways that work against the goal of protecting biodiversity. In the Austin area, for example, some landowners keep their property clear of the vegetation that could provide homes for the black-capped vireo or the golden-cheeked warbler, two endangered birds. In the Pacific Northwest, some timber owners have adopted forest practices that ensure conditions inimical to the northern spotted owl.

Most perverse of all is the fight over the attempts by the Department of Interior to launch a nationwide biological survey. What possible objection could landowners have to this survey? The answer is simple. The knowledge that a parcel of land houses a listed or potentially listable species puts that land under a cloud. The better course of action is to keep the government in the dark, and quietly scrape the land bare of vegetation.

These responses point to a central defect in the current law. In principle, the Endangered Species Act creates a two-step process. Science is the first step, policy is the second—except that the second step admits only one goal, full recovery. The scientific determination that a species is endangered effectively locks in the duty to save it, almost no matter what.

Because full recovery has turned out to be an impossibly difficult task, the political conflicts that should naturally be resolved in the second step find their expression instead in the first, where they cannot be debated on any but scientific grounds. Biologists, not government or elected officials, are the ones who set policy, assuming the role of economic mandarins with the power to bless or condemn a wide variety of land uses. In this way, science becomes embroiled in what are essentially policy questions, and the actions of scientists, just like those of landowners, are greeted with suspicion, fostered by a belief that their values, not their data, hold sway.

Reforming the Endangered Species Act must begin with restoring the separate domains of science and policy. The endangered list should remain as a scientific tally of this Nation's threatened wildlife. But it should no longer be tied to the single goal of full recovery for each of its entries. Instead, we must acknowledge that the choices of how much and what forms of protection an endangered species receives profoundly affect people's lives, and are therefore inherently political.

Crying "no more extinctions" produces a noble sound, but it does nothing to ensure that extinction will stop. And it has the potential for worsening the status of biodiversity, because aspiring to the perfect may prevent us from obtaining the merely good. The absolute duty of the Federal Government to stop any action that threatens a listed species must be relinquished. Otherwise, attempts to resolve conflicts between species and humans will wither under the eternal shadow of the Endangered Species Act.

POSTSCRIPT

Is the U.S. Endangered Species Act Fundamentally Sound?

A 1995 Supreme Court decision upheld regulations that interpret the Endangered Species Act (ESA) as prohibiting the destruction of habitat necessary to sustain a species, as well as prohibiting actions that directly kill or harm members of the species. This decision has heightened the concern of those who claim that the sanctions against habitat destruction should require compensation to any property owner whose economic interests are negatively impacted by them. Opponents of compensation claim that such a requirement would make wildlife protection prohibitively expensive and that there is no obligation to pay property owners to act in the public interest.

Langhorst's support for the ESA implies that the goal of protecting all species is achievable. Most defenders of the act admit that this is not actually possible and that decisions have to be made about the most effective way to spend funds allocated for wildlife protection. Plummer's negative assessment of the ESA is based on its failure to bring listed species to full recovery. This conclusion, however, is contradicted by a recently proposed delisting of more than two dozen plants and animals from the endangered species list as a result of recoveries attributed to the ESA. The National Research Council recently released a generally favorable evaluation of the act based on a review of its overall impact rather than on the overly ambitious initial legislative goal. Like many proponents of the use of cost/benefit analysis as a means to decide which species to protect, Plummer fails to explain how the benefits of preventing the extinction of any particular plant or animal species could possibly be scientifically evaluated. Bruce Babbitt, Secretary of the Interior, favors preserving biodiversity by protecting critical ecosystems rather than by focusing on near-extinct species. Such a strategy, which was authorized by Congress in 1982 but only recently implemented, is described by Suzanne Winckler in "Stopgap Measures," *The Atlantic Monthly* (January 1992). The reasons why many environmentalists consider the recent popularity of habitat conservation plans an unacceptable compromise that undermines the intent of the ESA are detailed in Jon Luoma's article "Habitat-Conservation Plans: Compromise or Capitulation?" in the January/February 1998 issue of *Audubon*.

For additional arguments favoring reauthorization of the ESA, see "A Tough Law to Solve Tough Problems," by Frances Hunt and William Robert Irvin, *Journal of Forestry* (August 1992); "Making Room in the Ark," by John Volkman, *Environment* (May 1992); articles by Ted Williams and Paul Rauber in the January/February 1996 issue of *Sierra*; and "What's Wrong With the Endangered Species Act?" by T. H. Watkins, in the January/February 1996 issue of

Audubon. Arguing that the act is economically too costly are John Heissenbuttel and William Murray in "A Troubled Law in Need of Revision," *Journal of Forestry* (August 1992). Omar N. White addresses the question of the act's constitutionality in "The Endangered Species Act's Precarious Perch: A Constitutional Analysis Under the Commerce Clause and the Treaty Power," *Ecology Law Quarterly* (February 2000).

ISSUE 4

Is the Precautionary Principle a Sound Basis for Environmental Policy?

YES: Patti Goldman and J. Martin Wagner, from "Trading Away Public Health: WTO Obstacles to Effective Toxics Controls," *Multinational Monitor* (October/November 1999)

NO: Ronald Bailey, from "Precautionary Tale," *Reason* (April 1999)

ISSUE SUMMARY

YES: Environmental attorneys Patti Goldman and J. Martin Wagner argue that the precautionary principle is essential to protecting public health and the environment; it must not be abandoned in the pursuit of international policy.

NO: Science correspondent Ronald Bailey counters that the precautionary principle slows development and is a luxury affordable only by "those who live in societies already replete with technology."

The traditional approach to the regulation of a potential environmental pollutant has been to impose restrictions only when there is reasonable proof that the agent in question poses a considerable threat to human or environmental health. The precautionary principle reverses this regulatory requirement by requiring reasonable proof that a potential pollutant will not harm humans or the environment before its release is permitted. This principle has played an increasingly important part in environmental law ever since it first appeared in Germany in the mid-1960s. On the domestic and international scene, it has been applied to climate change, hazardous waste management, ozone depletion, biodiversity, and fisheries management. In 1992 the Rio Declaration on Environment and Development (Principle 15) codified it as:

> In order to protect the environment, the precautionary approach shall be widely applied by States according to their capabilities. When there are threats of serious or irreversible damage, lack of full scientific certainty shall not be used as a reason for postponing cost-effective measures to prevent environmental degradation.

Other versions of the principle also exist, but all agree that when there is reason to think—but not absolute proof—that some human activity is harming the environment, precautions should be taken. This has come to be broadly accepted as a basic tenet of ecologically or environmentally sustainable development.

It also contributes to thinking in the areas of risk assessment and risk management in general. Human activities—the manufacture of chemicals and other products; the use of pesticides, drugs, and fossil fuels; the construction of airports and shopping malls; even agriculture—can damage health and the environment. Some insist that we do not need to take action against any particular activity unless and until we have solid, scientific proof that it is doing harm. Even then the risks must be weighed against each other. See, for instance, Wendy Cleland-Hamnett, "The Role of Comparative Risk Analysis," *EPA Journal* (January–March 1993). Others insist that mere suspicion should be enough for action and that there is a broad middle ground. Robert Costanza and Laura Cornwell, in "The 4P Approach to Dealing With Scientific Uncertainty," *Environment* (November 1992), argue that when uncertainty about potential harm is high, those who are potentially responsible should be required to post in advance a bond sufficient to cover the costs associated with the worst possible results.

Since solid, scientific proof can be very difficult to obtain, the question of just how much proof is needed to justify action is vital. Not surprisingly, if action threatens an industry, that industry's advocates will argue against taking precautions, generally saying that more proof is needed. A good example can be found in Stuart Pape's "Watch out for the Precautionary Principle," *Prepared Foods* (October 1999): "In recent months, U.S. food manufacturers have experienced a rude introduction to the 'Precautionary Principle'.... European regulators have begun to adopt extreme definitions of the Principle in order to protect domestic industries and place severe restrictions on the use of both old and new materials without justifying their action upon sound science." The Europeans in question would disagree with Pape's use of the term *extreme*.

Advocates of the precautionary principle state how difficult it is to predict what the ultimate effect of a new environmental pollutant may be. They point to many cases such as the use of flourochlorocarbons or the widespread introduction of organochlorine pesticides where failure to foresee the harmful effects of new industrial chemicals has produced devastating consequences.

In the following selections, Patti Goldman and J. Martin Wagner contend that the World Trade Organization's efforts to reduce barriers to global trade threaten use of the precautionary principle, which is broadly recognized as essential to protecting public health and the environment. Ronald Bailey argues that the precautionary principle slows economic and technological development and is a luxury that societies that have limited technology cannot afford.

Patti Goldman and J. Martin Wagner  **YES**

Trading Away Public Health

The WTO [World Trade Organization] has initiated a major expansion of trade rules into the realm of public health protection—a matter traditionally within the purview of national and local governments. The new WTO rules collide with many public health and environmental protections and dictate the extent to which a country or state may ban or restrict the use of toxic chemicals to protect public health.

The WTO erects obstacles to government restrictions on exposure to toxic chemicals by:

- Seeking to move toxics standards to the lowest common denominator throughout the world;
- Rejecting the Precautionary Principle with requirements that governments provide definitive proof of harm before acting;
- Foreclosing the most effective means of protecting public health and the environment; and
- Prohibiting restrictions designed to prevent toxic effects of production.

The WTO currently and under proposed expansions may increasingly stand in the way of:

- The consumer right to know which products are environmentally friendly;
- The government right to restrict its purchases of environmentally harmful products; and
- Government regulation of foreign investors to protect public health and the environment.

Downward Harmonization

Many initiatives to protect people and the environment from toxic pollution have been the result of citizen initiatives at the local level, which have, in turn, prodded higher levels of government to take action. International bodies tend

From Patti Goldman and J. Martin Wagner, "Trading Away Public Health: WTO Obstacles to Effective Toxics Controls," *Multinational Monitor* (October/November 1999). Copyright © 1999 by Essential Information, Inc. Reprinted by permission of *Multinational Monitor*.

to move more slowly and lag behind cutting edge initiatives to protect health and the environment.

The WTO threatens to block these grass roots initiatives. It promotes "downward harmonization" of health and environmental standards. Under the WTO, countries must base their standards on relevant international ones. If a country adopts a food safety or product standard that is more protective of public health than the international norm, its standard must satisfy a battery of cumbersome WTO tests.

Existing international standards, however, are generally established with extensive industry input and without the scientific rigor and public participation that characterize U.S. standard-setting. The preferred food safety standard-setting body—the Codex Alimentarius Commission—has standards that tend to lag behind U.S. standards, for example. In the early 1990s, Codex allowed residues of DDT on numerous foods, in sharp contrast to the U.S. ban on DDT imposed in the early 1970s.

In August 1999, the U.S. Environmental Protection Agency (EPA) announced a ban on numerous uses of the pesticide methyl parathion, soon to be followed by a prohibition on residues of this pesticide on these foods.

Just two months prior to EPA's cancellations, Codex concluded its review of methyl parathion and continued to allow residues on many of the foods subject to the EPA cancellations, including cherries, plums and carrots. The United States participated in the Codex meeting that adopted the methyl parathion standard, but did not object. The existence of the Codex authorization will make it harder for the United States to prevail in any future challenge to its bans.

The WTO's foray into public health and environmental standard-setting shifts decision-making power away from local, state and national governments to international trade bureaucrats resolving WTO disputes in secret in Geneva, Switzerland, and to Codex, an obscure international standard-setting organization in Rome, Italy.

Jettisoning the Precautionary Principle

The Precautionary Principle allows countries to protect their citizens based on scientific evidence of risk, but before the scientific proof of harm is conclusive. Under the Precautionary Principle, for example, studies showing that a chemical causes cancer in animals should be sufficient evidence to allow governments to prevent human exposure to it. As a matter of public policy, the Precautionary Principal holds it is more prudent and generally more cost-effective to prevent toxic contamination and exposure rather than try to clean up the mess or treat the injured people after the fact.

The WTO precludes use of the Precautionary Principle. Instead, the WTO requires conclusive scientific evidence of a risk before trade in food products may be restricted. As a WTO panel stated in a ruling against an Australian ban on salmon that might contaminate local stocks with alien species, a country must have identified a probability, not simply a possibility of harm before it can regulate in a way that restricts trade.

In the most notable case blocking a government from acting on the basis of the Precautionary Principle, the WTO in 1997 and 1998 sided with the United States in its challenge to a European Union [EU] ban on beef treated with growth-inducing hormones that have been scientifically linked to cancer and other serious diseases. Although the EU asserted that the ban was necessary to achieve its chosen degree of protection—zero risk to consumers from exposure to hormone-treated meat—the WTO dispute resolution and appellate panels rejected an absolute right to prohibit all such risk. The fact that the hormones caused cancer in laboratory animals—a scientific as well as common-sense basis for suspecting a risk in humans—was not a sufficient basis for a ban on their use in human food, the panel held. By requiring proof of harm, the WTO removed the ability of governments to take precautionary action to protect against risks strongly suggested, but not proven, by scientific evidence. The WTO has granted the United States permission to impose retaliatory trade sanctions until the EU rescinds the ban. The United States recently imposed more than $120 million in trade sanctions for this year.

The WTO's hostility to the Precautionary Principle puts numerous public health protections at risk, including the following examples in the United States:

Pesticide safeguards for children and uncertainties. In 1996, Congress unanimously passed the Food Quality Protection Act, which requires that extra protection be built into U.S. pesticide standards where scientific evidence is incomplete and to account for risks to children. A National Academy of Sciences study showing that children are more susceptible to adverse impacts from pesticides because of their size, metabolism, age and rate of growth precipitated the added protection for children. This extra protection is not based on definitive scientific evidence of harm from the particular pesticide; but rather on the lack of studies deemed necessary to decide whether the pesticide residues will be harmful to children. The Act also calls for extra protection for other gaps in the scientific evidence of harmful effects from a particular pesticide.

Bans on carcinogens in food. The U.S. Delaney Clause prohibits color and food additives that cause cancer in animals. The ban extends to artificial sweeteners, preservatives, chemical processing aids, animal drug residues and packaging materials that leach into food. The zero-risk Delaney Clause standard is based on a policy decision in the face of uncertainties about cancer risks from the consumption of carcinogens even in small amounts. The Delaney Clause constitutes a political determination made by Congress about whether carcinogens should be introduced into our nation's food supply. The European Union has threatened to challenge the Delaney Clause as an unfair trade barrier.

Proposition 65. California's Proposition 65, an initiative adopted in 1986 by a nearly two-to-one majority of California voters, requires a clear warning before exposing anyone to chemicals that cause cancer or reproductive toxic effects. A limited exception to Proposition 65's warning requirement is provided if the one responsible for the exposure can demonstrate that it results in "no significant risk" of cancer and or reproductive harm. Proposition 65 imposes a more stringent public health standard than U.S. federal law. Because of Proposition 65, many products, including cigars, household pest strips, lead ceramic tableware and paint strippers, now contain health warnings. Other products, such as typewriter correction fluids containing a reproductive toxin and a spot remover containing a carcinogen, have been reformulated to remove the listed substances. Proposition 65 precipitated fetal alcohol syndrome labels on alcoholic beverages, which, in turn, spurred the federal government to establish such labeling requirements.

The European Union has identified Proposition 65 as a trade barrier. Because Proposition 65 places the burden of proof on industry to demonstrate the safety of known carcinogens or reproductive toxins in their products, including foods with pesticide residues, it could be challenged for violating the WTO's rules against instituting precautionary measures.

Banning toxic chemicals.   To address the pernicious effects of toxic contamination and exposure, many public health officials have embraced prevention as the best strategy. Rather than impose lead poisoning upon children, countries around the world have decided to phase out many... uses of lead. Instead of contributing to future toxic waste sites, many governments are instituting bans on persistent toxic chemicals. To avert toxic poisoning of birds, like that precipitated by DDT, governments have severely restricted or banned pesticides that kill birds.

The WTO superimposes a rule on toxics measures that may preclude many bans and phase-outs. Under the WTO, a country must use the least trade-restrictive means of achieving its public health or environmental protection goals. When a country decides to protect its citizens by banning a chemical, that decision may be called into question, since bans are the most trade-restrictive measures available.

For example, Canada is challenging a French ban on asbestos, arguing that requiring protective clothing and other measures that limit exposure would be less burdensome on trade than a ban. Under this same logic, the United States has objected to a proposed European Union ban on heavy metals in electronics products, arguing that other less trade-restrictive alternatives are available. The industry has suggested landfill restrictions and eco-taxes as a viable alternative to bans on toxic chemicals and government subsidies for recycling and purchasing policies as alternatives to manufacturer responsibility for these products' waste.

Other countries might lodge challenges to bans on residues of harmful pesticides on the grounds that labeling, washing or limiting the residues permitted on foods for consumption would be less restrictive ways to protect public health—even if such alternatives would not be as effective as a ban in protecting public health.

Disclosure requirements could also be challenged, particularly where they are mandatory or used by governments to guide their purchases. Indeed, both Japan and the European Union have already made claims that the U.S. mandatory nutritional labeling requirements are an unfair trade barrier. They have argued that voluntary labeling, as provided for in Codex guidelines, would suffice or that not all foods need to be covered by mandatory requirements.

State and local standards that go further than national requirements would be vulnerable under the least trade-restrictive test. In fact, a trade dispute panel concluded that a tax law in place in only five U.S. states was not "necessary" because other states had found "alternative, and possibly less trade restrictive... ways of enforcing their tax laws." This rationale could be devastating if it were applied to the federal pesticide regulatory scheme, which permits, but does not require, states to provide greater health or environmental protection than the federal government.

Blinded to Harm From Toxic Production Processes

Many toxic chemicals are used in the production process or become toxic waste that needs disposal after production or use of a product. To reduce exposure to such chemicals, it is necessary to curb the harmful byproducts of the production process and to limit the creation of toxic waste.

The WTO prohibits discrimination between products based on how they are produced. If the physical attributes of two products are the same, the one produced in a manner that depletes natural resources or pollutes the air and water must be treated the same as the one that does not cause such pollution. By extension, many have argued that the WTO prohibits "cradle-to-grave" eco-labeling because the label is based on how a product is produced.

Some toxics restrictions are put in place to protect the environment or workers during the production process. For example, in the United States, bans have been imposed on pesticide use to protect farm workers and water quality. If the United States restricted imports of food produced using these pesticides, it could run afoul of WTO rules because the restrictions would not be based on some tainted characteristic of the food, but rather would be designed to protect workers or the environment where the food was grown.

Obstacles to Eco-Labeling

Consumers are increasingly choosing to use their purchasing power to promote environmentally sound practices. Eco-labeling distinguishes between products based on their relative impact on the environment in an attempt to influence consumer purchasing decisions in favor of "environmentally friendly" products.

Under the WTO, an eco-label that reflects how a product is produced would be vulnerable to challenge. Similarly, an eco-label could be contested based on its scientific underpinnings, its effect on imports or its stringency.

- The European Union has threatened to challenge U.S. nutritional labeling requirements, which are among the most advanced in the world, as well as California's Proposition 65, which goes further than the national standards and shifts the burden of proof. Because organic labeling reflects how the food is produced, the final U.S. organics standard may be vulnerable to a WTO challenge.

- At the behest of the U.S. paper industry, the United States objected on WTO grounds to a proposed EU eco-label for paper. The industry argued that the EU could not base the label on the environmental effects of the production process. The U.S. government claimed the eco-label would unfairly disadvantage U.S. paper products because it favored recycling, while U.S. producers use virgin timber for pulp and paper production, and it was based on EU pollution standards that were stronger than those in the United States.

- More recently, the United States has objected to initiatives to require labels on foods produced with genetic engineering. "Providing information regarding the method of production on the food label would be highly impractical and inequitable," the United States argued in a 1999 submission to Codex. The United States would require that the food undergo some change in nutritional value or its use before a label could be required.

Stopping Green Procurement

Green procurement is a mechanism for reducing consumption and its harmful environmental effects. In recent years, government procurement has increasingly been used to reduce production of paper, which is the third largest industrial consumer of energy and a large contributor to both air and water pollution. In the United States, federal, state and local governments have extensive recycled paper requirements. Some cities and states direct that some government paper purchases must consist of paper that has not been bleached with chlorine.

These government procurement laws are at risk from current WTO rules and a potential expansion of these rules. By way of example, a trade threat surfaced in connection with recycled paper requirements in the early 1990s. Canada threatened a trade challenge to Minnesota's requirement for recycled paper content in state paper procurement bids. Canada claimed that the requirement had a discriminatory effect on Canadian suppliers because Canada relies on virgin timber and has a smaller supply of recycled paper. To avert a trade challenge, Minnesota allowed nonconforming bids from Canadian suppliers.

A Call for a Moratorium

Environmentalists are demanding that the WTO be reformed, not expanded, so that it cease jeopardizing strong environmental and health protections. Specifically, environmental groups are demanding that the WTO be reformed to protect: (1) the right to have strong environmental standards that use the Precautionary Principle to protect citizen health and the environment; (2) the right to limit the harmful effects of production, such as pesticide poisoning of workers and toxic air and water pollution from factories; (3) the consumer right to know which products are environmentally friendly; and (4) the right to use government purchasing power to protect the environment.

No country should be forced to abandon strong toxics standards because they are ahead of the international *status quo* or, in the view of trade officials, too restrictive of trade. To ensure that no further damage is done before the much-needed reforms are made, a moratorium should be imposed on WTO challenges to food safety, health and environmental protections.

NO ↵

Precautionary Tale

Look before you leap.

Sounds reasonable, doesn't it? But how reasonable would it be to take such proverbial wisdom and turn it into a Federal Leaping Commission? The environmentalist movement is seeking to create the moral equivalent of just that. In effect, before you or anybody else can leap, you will not only have to look beforehand in the prescribed manner, you will have to prove that if you leap, you won't be hurt, nor will any other living thing be hurt, now and for all time. And if you can't prove all of that, the commission will refuse to grant you a leaping license.

At [the 1999] annual meeting of the prestigious American Association for the Advancement of Science [AAAS] in Anaheim, California, in a symposium titled "The Precautionary Principle: A Revolution in Environmental Policymaking?", environmentalist advocates and academics insisted that a principle of ultimate precaution should trump all other considerations in future environmental and technological policy making. They pointed out that the Principle has already been incorporated into several international treaties, including the Framework Convention on Climate Change and the Kyoto Protocol, which require developed nations to cut back dramatically on the burning of fossil fuels to reduce the putative threat of global warming. The U.S. Environmental Protection Agency is already using it to help guide its promulgation of new regulations on synthetic chemicals.

Jeff Howard, a panel member who once worked on Greenpeace's International Toxics Campaign and now has a gig at the Center for Science and Technology Policy and Ethics at Texas A&M University, defined the Principle: It calls for precaution in the face of any actions that may affect people or the environment, no matter what science is able—or unable—to say about that action.

Before examining this concept, it's worth pausing to see where it came from. Howard's version of the Principle was formalized . . . by environmentalist advocates who convened at the Wingspread Conference Center in Wisconsin. Gathering in such a place allowed them to give their ruminations a sonorous title: "The Wingspread Consensus Statement." (After all, you wouldn't want to call such a document "The Bronx Consensus Statement.")

From Ronald Bailey, "Precautionary Tale," *Reason* (April 1999). Copyright © 1999 by The Reason Foundation. Reprinted by permission of *Reason* Magazine, 3415 S. Sepulveda Blvd., Suite 400, Los Angeles, CA 90034. http://www.reason.com.

That the Wingspread delegates achieved "consensus" on precaution might imply to some that their meeting was a strenuous, perhaps even contentious, effort by experts of diverse views to find a balance between the demands of scientific inquiry and the well-being of nature. That's certainly how the AAAS meeting treated this "consensus": as though it had arisen from a symposium presenting peer-reviewed scientific data.

But Wingspread's delegates were not exactly diverse; rather, they were a panel of activists with an agenda. They included representatives from an array of like-minded groups, including Greenpeace, Physicians for Social Responsibility, the Toxics Use Reduction Institute in Massachusetts, Britain's Centre for Social and Economic Research on the Global Environment, the Environmental Research Foundation, the Science and Environmental Health Network, the Environmental Network, the Silicon Valley Toxics Coalition, the Environmental Health Coalition, the Indigenous Environmental Network, and the Center for Health, Environment, and Justice. It's not hard to reach "consensus" when you gather a group of people who all share your values and views. If I hand-pick my delegates, I can achieve a consensus on just about anything. (How about the "Miami Beach Consensus Statement on Abolishing Social Security"?)

･✿･

What did the Wingspread activists finally recommend? The actual text of the Principle that Howard offered at the AAAS meeting reads: "When an activity raises threats of harm to human health or environment, precautionary measures should be taken even if some cause-and-effect relationships are not fully established scientifically."

The Wingspreaders and their followers on the AAAS panel want to apply the Principle solely to environmentalist concerns, but, in fact, their formula is essentially an empty vessel into which anyone can pour whatever values they prefer. It simply codifies a very risk-averse version of standard cost-benefit analysis; the Wingspread participants think that certain activities, such as manufacturing plastics or burning fossil fuels, are unacceptably risky. In other words, very conservative environmentalist values are being privileged over what, to other people, may be equally or more compelling values.

The formula can be adapted to fit many different agendas. Try this, for example: "When an activity (say, employment tests) raises threats of harm to equality and equal access, precautionary measures should be taken even if some cause-and-effect relationships are not fully established sociologically." Or this: "When an activity (say, higher taxes) raises threats of harm to private property or economic growth, precautionary measures should be taken even if some cause-and-effect relationships are not fully established economically." We could do this all day.

The heart of the Principle, of course, is the admonition that "precautionary measures should be taken even if some cause-and-effect relationships are not fully established scientifically." As one biomedical researcher in the audience objected, *all* scientific conclusions are subject to revision, and none is ever "fully established." Since that is the case, the researcher pointed out, the Precautionary Principle could logically apply to every conceivable activity, since their outcomes are always in some sense uncertain. Furthermore, David Murray, the director of the Statistical Assessment Service in Washington D.C., points out another possible—and disquieting—interpretation of the Principle. Anyone who merely raises "threats of harm" with no more evidence than their fearful imagination gets to invoke precautionary measures. Precautionists would not need to establish any empirical basis for their fears; they may simply posit that something might go wrong and thus stymie any proposed action.

Ah, so. Just what these activists had in mind all along, as we shall see.

But let's parse the Principle a bit more. One troublesome issue is that some activities that promote human health might "raise threats of harm to the environment," and some activities that might be thought of as promoting the environment might "raise threats of harm to human health."

Take the use of pesticides. Humanity has used them to better control disease-carrying insects like flies, mosquitoes, and cockroaches, and to protect crops. Clearly, pesticide use has significantly improved the health of scores of millions of people. But some pesticides have had side effects on the environment, such as harming nontargeted species. The Precautionary Principle gives no guidance on how to make this tradeoff between human health and the protection of nonpest species (though I suspect I know how the panel members would choose).

During the discussion period, another audience member asked panelist Steve Breyman, a professor in the Department of Science and Technology Studies at Rensselaer Polytechnic Institute, if he thought the last 200 years had been all bad. Breyman revealingly responded with something like, Oh sure, some things like life expectancy and living standards have improved, but there have been losses too. The quality of drinking water, Breyman asserted, has gone down.

Really? Two hundred years ago, drinking from any stream, well, or spring could expose one to typhoid, typhus, cholera, and other diseases. In fact, chlorination has so improved drinking water quality with regard to health that people in the West no longer even think twice about drinking tap water. Unfortunately, more than a billion people in the developing world can't say the same; millions still die of water-borne diseases each year.

Proponents of the Precautionary Principle are trying to smuggle in a default position: The environment trumps all other values. Yet the panelists all pretended that the Principle is a value-neutral scientific procedure for determining which policies humanity should pursue. The fact is that the Precautionary Principle incorporates the values of the most extreme versions of know-nothing environmentalism. When challenged from the audience on this point, Breyman fumed, "We're talking about the survival of the planet and the human race here."

Breyman sees the Precautionary Principle as an essential part of a radical agenda to reshape human culture. He writes in his AAAS presentation, "Introduced as part of an overall green plan that included conservation and renewable energy, grass roots democracy, green taxes, defense conversion, deep cuts in military spending, bioregionalism, full cost accounting, the cessation of perverse subsidies, the adoption of green materials, designs and codes, green purchasing, pollution prevention, industrial ecology and zero emissions, etc., the PP could be an essential element of the transition to sustainability."

⋅❀⋅

Jeff Howard later offered some corollaries to the Precautionary Principle that reveal just how sweeping a proposal it is.

The first corollary is that "the proponent of an activity, rather than the public, should bear the burden of proof (reverse onus)." This means that "proponents would have to demonstrate through an open process that a technology is safe or necessary and that no better alternatives were available." Unlike the members of the AAAS panel, Boston University law professor George Annas, a prominent bioethicist who favors the Precautionary Principle, clearly understands that it is not a value-neutral concept. He gleefully told me, "The truth of the matter is that whoever has the burden of proof loses."

The result: Anything new is guilty until proven innocent. It's like demanding that a newborn baby prove that it will never grow up to be a serial killer, or even just a schoolyard bully, before the baby is allowed to leave the hospital. Under this corollary, inventors, scientists, and manufacturers would have to prove that their creations wouldn't cause harm—ever—to the environment or human health before they would be allowed to offer them to the public. This is asking them to prove a negative. How can someone prove that a new plastic will never, ever interact with any metabolic pathway in any plant, animal, microbe, or person? There is simply no way to test for all possible effects given the millions of different species living on the earth.

But is this inability to test for everything really dangerous? Howard thinks it's murderous. He warned the audience that humanity has been engaged in a "great global experiment since the dawn of the chemical age" and predicted that "death and disease will increase as a result."

The plain fact is that the introduction of thousands of synthetic chemicals has not resulted in increased levels of death and disease but *has* resulted in substantial health benefits and greater convenience and efficiency. Life expectancy has never been higher and, as just reported by the National Cancer Institute, even cancer incidence rates are going down. In addition, the Food and Drug Administration estimates that less than 2 percent of cancers are the result of exposure to man-made substances. Finally, the few bad actors, like some organochlorine compounds, have been replaced.

Under the "reverse onus" corollary, would-be innovators would have to demonstrate that a technology was "necessary" because no alternatives were available. Necessary? Like air, water, and food? This is potentially a *very* high threshold. Are antibiotics necessary? Computers? Microwave ovens? What makes something "necessary" or not depends on the goals that individuals are trying to achieve. Necessity is the mother of invention only to the degree that it is in the eye of the inventor.

This requirement of demonstrable necessity ignores a vital fact about progress: All technologies serve as bridges to other technologies, to ever-better alternatives. For example, without the production of fossil fuels, humanity would not be in the position to make the costly, knowledge-intensive transition to the solar/hydrogen future that environmentalists wish to subsidize into existence. One technology leads to another. As dirty as burning fossil fuels may be, they aren't a tenth as dirty as burning wood.

Embedded in the Precautionary Principle is the notion that we can anticipate all of the ramifications of a technology in advance and can tell whether on balance it will be a net benefit or cost to humanity and the environment. That's complete nonsense. To cite a single example, when the optical laser was invented in 1960, it was dismissed as "an invention looking for a job." No one could imagine of what possible use this interesting phenomenon might be. Of course, now it is integral to the operation of hundreds of everyday products: It runs our printers, runs our optical telephone networks, performs laser surgery to correct myopia, removes tattoos, plays our CDs, opens clogged arteries, helps level our crop fields, etc. It's ubiquitous. Yet no one anticipated—no one could have anticipated—how incredibly useful lasers would turn out to be, not even the wisest tribunal of environmentalist seers or panel of Federal Leaping Commissioners.

The same thing goes for items which eventually turned up on the environmentalist hit list: organochlorine pesticides. After all, it is not as though evil chemical corporations invented pesticides for the purpose of polluting the environment. When these compounds were introduced they were a genuine miracle; they saved millions of lives that would have been lost to malaria and malnutrition. No one could have anticipated that their persistence in the environment would allow them to accumulate in animal fat, leading to some reproductive problems in eagles and falcons. The data simply weren't there. Indeed, there was not even a theory of bioaccumulation. Only by gaining experience with these substances were we able to learn about their downside and eventually decide that other, less persistent pesticides achieved a better tradeoff between human benefits and harm to the natural environment.

⚜

A second vexed corollary is that "the process of applying the Precautionary Principle must be open, informed and democratic and must include potentially affected parties." At one point, panel member Breyman declared that we had to get environmental decisions out of the hands of EPA regulators. Sounds good, right? But what if the open, democratic process ended with a choice to exploit a natural resource in ways that environmentalists don't like?

The deputy administrator for the National Marine Fisheries Service, Andy Rosenberg, happened to be in the audience and offered an illustration of realpolitik to the panel's starry-eyed egalitarians. Rosenberg pointed out that if you allowed New England fishermen to vote on whether or not to keep the cod fishery open, they would fish it until the last fish was gone. Breyman responded lamely that if the fisherman did that, they didn't have enough information.

Other panelists suggested that the "affected parties" aren't just fishermen, but all of us. If we don't get the result we like at one democratic level, these panelists implied, we'll just keep shifting the definition of "affected parties" until we do get the result we like. But wait a minute. Does this mean that when one of us wants to engage in an activity that someone thinks may result in harm, we *all* get to vote on it?

This problem—deciding who gets to decide what—is just one of many slippery slopes that the Precautionary Principle teeters over. Of course, it quickly became apparent that, for the AAAS panel, the only democratic decisions that are acceptable are those consistent with environmentalist goals. But other obvious problems were never acknowledged.

For example, democratic decision making concerning any and all environment-affecting actions could have the effect of ratifying extraordinarily conservative choices. That is, a community could use its environmental veto to say, No, we don't want a new store, a new housing development, a new factory, a new road. Basically, it means that the vested interests of the present can strangle the future. After all, as one wag noted, an environmentalist is somebody who already owns his second home in the woods.

Of course, neither the regulators at the meeting nor the environmental activists on the AAAS panel considered a real solution: removing the decisions about resources from the political process entirely. Politics is always win/lose, while market decisions are generally win/win. Give fishermen, loggers, and cattlemen secure property rights to the resources, and that shifts their incentives toward trying to protect and enhance *their* resource, rather than merely plundering somebody else's resource.

❧❦❧

Draconian as the Wingspread proposals are, Jeff Howard doesn't think they are strong enough. He fears that wily capitalists and innovators will find ways around them, so he suggests five additional tenets:

- Precaution must become the default mode of all technological decision making.
- Even the most fundamental of past decisions must be subject to re-examination and precautionary reform.
- The primary mode of regulation and regulatory science should be at the macroscale.
- Knowledge of broad patterns trumps ignorance of detail.
- Human society must identify and accommodate itself to broad patterns in natural processes.

Consider for a moment the tenet that "even the most fundamental of past decisions must be subject to re-examination and precautionary reform." Actually, the process of technological innovation constantly "re-examines" past decisions, but that's not what Howard has in mind. He wants to create a political process, which he naturally insists would be open, that would eliminate technologies of which he disapproves: nuclear power plants, organochlorines, most plastics, etc. But what I find intriguing is the idea that "even the most fundamental of past decisions" could be "reformed."

How fundamental is fundamental? Decisions like the invention of the automobile? The use of fossil fuels? The development of agriculture? Fire? Look at Howard's last tenet, that society must accommodate itself "to broad patterns in natural processes." What violates the broad patterns in natural processes? Medicine? City building? Farming?

Before the AAAS session ended, Howard offered a third corollary to the Principle: "Precaution requires consideration of the full range of social and technological alternatives" to what is being proposed. It is very much in line with the Wingspread Consensus Statement, which declares that precaution "must also involve an examination of the full range of alternatives, including no action."

Environmentalists often liken technology and economic growth to a car careening down a foggy road. They suggest that it would be better if we slowed before we crashed into a wall hidden in the fog. The Precautionary Principle, its champions believe, "would serve as a 'speed bump' in the development of technologies and enterprises."

Unfortunately, these principles and tenets may sound sensible to many people, especially those who live in societies already replete with technology. These people already have their centrally heated house in the woods; they already enjoy the freedom from want, disease, and ignorance that technology can provide. They may think they can afford the luxury of ultimate precaution. But there are billions of people who still yearn to have their lives transformed. For them, the Precautionary Principle represents not a speed bump but a wall.

Should we look before we leap? Sure we should. But every utterance of proverbial wisdom has its counterpart, reflecting both the complexity and the variety of life's situations and the foolishness involved in applying a short list of hard rules to them. For some people in some situations, "Look before you leap" is good advice. Others might be wiser to heed the equally proverbial, "He who hesitates is lost."

People have understood this maxim for millennia, and the chances are that its message will eventually reach even Wisconsin's Wingspread Conference Center. And when it does, I want the Wingspreaders to understand that the moral equivalent of a Federal Anti-Hesitation Commission isn't such a good idea, either.

POSTSCRIPT

Is the Precautionary Principle a Sound Basis for Environmental Policy?

Bailey defines the precautionary principle as "precaution in the face of any actions that may affect people or the environment, no matter what science is able—or unable—to say about that action." Some assert that widespread use of the precautionary principle would indeed hamstring the development of the Third World. Yet the1992 Rio Declaration emphasizes that the precautionary principle should be "applied by States according to their capabilities" and that it should be applied in a cost-effective way. These provisions would seem to preclude the severe interpretations of the precautionary principle. Turned around, however, these same provisions are what led to the World Trade Organization's efforts to keep developed nations from applying their own stringent environmental regulations to Third World trading partners. This alarms those who favor the precautionary principle. See Hilary French, "Challenging the WTO," *World Watch* (November/December 1999). French takes a somewhat broader view in "Coping With Ecological Globalization," in *State of the World 2000* (W. W. Norton, 2000), where she notes that the "clash between two different spheres of international law [environmental and trade] present[s] the world with a major legal challenge, as it is not always clear which agreement trumps the other in cases where two treaties are in conflict."

In "Are Decision-Makers Too Cautious With the Precautionary Principle?" *Environmental and Planning Law Journal* (February 2000), Paul L. Stein's focus is Australia, but he notes that the precautionary principle obliges "decision-makers to consider the likely harmful effects of their activities on the environment before they pursue those activities." A very similar sense of obligation informed the passage of the United States' National Environmental Policy Act (NEPA) in 1969, regarding its requirement for Environmental Impact Statements.

The precautionary principle finds application in many areas. It is part of the debate over the storage of nuclear waste. It enters into discussions of risk to human health or the environment. It even comes into discussions of wildlife protection: see Barnabas Dickson, "The Precautionary Principle in CITES: A Critical Assessment," *Natural Resources Journal* (Spring 1999). CITES is the Convention on International Trade in Endangered Species of Wild Flora and Fauna; it aims to protect species such as elephants, rhinoceroses, monkeys, and parrots, which are threatened by intense demand for pets, ivory, and other body parts. Under CITES, international trade in ivory was shut down in 1989, and as a result, the elephant population in many African nations increased to the point where local authorities now wish to reduce it. There was considerable debate over whether or not they should be allowed to market the resulting meat, hides, and ivory prior to the April 2000 Conference of the Parties to CITES in Nairobi, Kenya, but the ban remained in effect. See "Excitement at CITES," *The Economist [US]*, April 15, 2000. See also M. Lynne Corn and Susan R. Fletcher, *African Elephant Issues: CITES and CAMPFIRE*, Congressional Research Service, Report for Congress (August 5, 1997).

ISSUE 5

Does Environmental Regulation Unnecessarily Limit Private Property Rights?

YES: Bruce Yandle, from "Property Rights and Constitutional Order: Paradoxes and Environmental Regulation," *Vital Speeches of the Day* (June 15, 1998)

NO: Doug Harbrecht, from "A Question of Property Rights and Wrongs," *National Wildlife* (October/November 1994)

ISSUE SUMMARY

YES: Bruce Yandle, a professor of economics and legal studies, argues that technological development is transforming the world into a Garden of Eden. He maintains that environmental regulation is an unnecessary, misguided effort that threatens private property rights.

NO: *Business Week*'s Washington correspondent Doug Harbrecht contends that it is absurd to have to pay owners of private property for obeying environmental regulations.

The question of possible conflicts between the public interest and a private landowner's development plans for his or her property is not a new issue. It has, however, taken on new meaning over the past decade as a result of a growing U.S. property rights movement that has been attempting, with some success, to prevent local, federal, and state governments from imposing environmentally motivated restrictions on the use of private property without due compensation for any resulting loss of value to the landholder.

Those who defend the absolute development rights of landowners frequently justify their position by quoting from the writings of seventeenth-century political philosopher John Locke. They further cite as a legal basis for their position the eminent domain clause of the Fifth Amendment, which prohibits the government from "taking" private property for public use without just compensation.

Opponents of property rights activists maintain that Locke based his position on the abundant availability of land that prevailed in his day and that he considered undeveloped land to be of no value, a position that would find few supporters today. They argue further that the environmental problems that have resulted from population pressures coupled with inappropriate development could hardly have been foreseen by a seventeenth-century philosopher. In their view the language of the Fifth Amendment was meant to preclude actual seizure of property by the government, not loss of value resulting from regulatory restrictions.

The courts have historically supported this latter interpretation. However, in a few recent cases, court rulings have interpreted the "taking" of property as applicable, particularly in those cases where the regulatory restriction was such as to render the property totally worthless or where the public benefits resulting from the restriction could not be shown to be at least commensurate with the loss of value suffered by the property owner.

Only the most extreme property rights advocates argue that all regulations require compensation to a landowner for the potential value of activities they preclude. Few would support the rights of a landowner to profit from a project that results in a serious pollution problem that is not confined to the owner's property. It is also generally recognized that investment in property is a speculative activity and that the purchaser should be aware that potential actions by governments or private parties can result in either the enhancement or the diminution of the worth or value of the property.

Despite the legislative and judicial progress that has been made, the property rights movement has not had much success at the ballot box. Well-publicized property rights initiatives were soundly defeated by the voters in Arizona in 1994 and again in the state of Washington in 1995. Recent efforts to pass broad property rights legislation in Congress have also failed thus far.

Bruce Yandle is a strong supporter of free markets and unfettered private development. In the following selection, he bemoans the many environmental regulations that he views as unnecessary restrictions on the rights of property owners. In the second selection, Doug Harbrecht argues that environmental laws do more to protect than to reduce property rights. He fears that legislative proposals that would require compensation to property owners for decreases in value resulting from such laws will make it economically impossible for the government to require an ecologically sound development policy.

Bruce Yandle

 **YES**

Property Rights and Constitutional Order

Delivered to The Philadelphia Society, Chicago, Illinois, April 25, 1998

What a wonderful world! In vital ways, we seem to be living in a Garden of Eden. In spite of horrible ongoing civil wars and terrorism that plague large numbers of people on the planet, we know that more people are experiencing freedom than ever before. Just a short time ago, Eastern Europe was a walled community; ordinary people were unable to move their families from one county to another, let alone to cross country lines or venture forth to the land of the free.

I shall never forget seeing Czech families making their way on the autostrade in Italy in 1991, proudly displaying cardboard signs telling the world that they were free. The story has been repeated for millions. Free markets are burgeoning where command and control once dictated the who, what, and when of production and consumption. The power of contracts, property rights, and gains from trade are lifting and spreading incomes and wealth.

Consider our routine activities. We have almost unlimited access to information, goods, and travel. There are spot and futures markets for things as diverse as natural gas, electricity, stock market averages, and even permits to emit sulfur dioxide.

We go "on line" to buy automobiles, airplane tickets, financial derivatives.

We can swap email messages in the dead of the night, laying plans for meetings.... Scientists and engineers in Europe collaborate in real time with manufacturing operators in the U.S. Component parts for new BMWs are in the air traveling to their home plant in S.C. Ten hours later, they happily power shining Z3 convertibles.

In split seconds, contracts are negotiated, deals are made, and the rights of contracting parties are respected in the rough and tumble of the market place. Goods, capital, and people move to the Four Corners of the globe, with less government interference and friction than ever before. Incomes are rising rapidly; wealth is being created. Life expectancies are rising, and the world is getting cleaner.

With expanded knowledge and information, liberty has expanded and markets are flourishing in one part of our world. This is the case on one side of the ledger. But alas, there is another side.

From Bruce Yandle, "Property Rights and Constitutional Order: Paradoxes and Environmental Regulation," *Vital Speeches of the Day* (June 15, 1998). Copyright © 1998 by Bruce Yandle. Reprinted by permission of City News Publishing Company, Inc.

While we can design, build, and sell new BMWs with relative ease, we must think more than twice before cutting down a juniper tree that may be listed as a potential habitat for an endangered species. The wiser course of action says we should first get permission from a federal agent. Unloading a truck of builders' sand on a new home construction site can land us in the federal penitentiary, unless we first have federal permits. Building duck ponds without permission can yield penalties more severe than manslaughter. Sending a small amount of chemical waste to a local recycler can generate liability for cleaning up an entire landfill.

Taking steps to build a home on land initially zoned residential but later rezoned can lead to an uncompensated loss of the entire investment. Making an effort to plow and plant a dry field that was temporarily covered by a flooding river can lead to lessons learned about wetlands violations. Setting aside land for a retirement home can be frustrated, if there is a possibility that a golden-cheeked warbler just may decide to locate there. And one must exercise restraint before pointing a rifle and pulling the trigger when a menacing grizzly bear turns to attack. Taking such common sense action may put one in the position of calling on the Mountain States Legal Foundation for assistance when charged with killing an endangered species. With the Foundation's help, you may ultimately win after enduring a costly court battle, but the record says it will take nine years.

While we live in a veritable Garden of Eden on the one hand, we live as peasants on a feudalistic manor on the other. And the baneful lord of the manor is not bashful when it comes to command and control.

Superfund, wetlands, endangered species. How did the lord of the manor become so powerful? What has happened to property rights? What happened to liberty?

Property Rights and the Constitutional Order

Precious time and my hope of keeping your goodwill prevent a complete rendering of the story. Doing so requires us to reconsider the important and precious foundation stones of this wonderful republic, to remember the vital constitutional constraints that set in motion the most powerful story of human action the world has ever witnessed.

Our Founders knew much that is now seemingly forgotten. They knew from bitter experience that for human beings to be free, for the free spirit of man to soar, there must be basic protections that limit the powers of the government Leviathan. They understood the difference between rules of law based on just conduct and wisdom of the ages and rules of politics that were based on expediency, concentrated power, and special interest struggles.

They knew that for markets to flourish, for wealth to be created, preserved, and accumulated, property rights had to be protected. More than words on parchment, the Constitution they penned formed moral and legal constraints believed necessary to limit the heavy hand of government. They wrote in the Constitution: "Nor shall private property be taken for public use without just compensation."

Property rights. Why are they so important? Deep in the genetic material that forms human life, instincts lie buried calling for survival, reproduction, and conservation of energy and resources. Over the millennia, human beings that survived and prospered devised rules, customs, and traditions that protected the stuff of life. The rules were simple: You do not reap where you have not sown. You do not eat unless you have contributed. You do not impose unwanted costs on your neighbor or he on you. Property rights formed the foundation for moral behavior and markets. And government was invented for the purpose of protecting those rights, knowing full well that its naturally destructive tendencies were to be tolerated so long as property rights were made secure.

The founders did not put a happy face on government as we do now. They knew that government was not Santa Claus. They understood that self-interest, the natural human desire to improve position and family, could be channeled beneficially by rules of law, property rights, and free market forces. They knew from experience that those same motivations are perverted by the rule of politics. Special interest groups driven by self-interest will tilt the hand of government in their direction. Instead of working the soil, factory, and field to produce new wealth, the special interest groups will work the halls of Congress and transfer wealth to themselves from the unorganized, unwitting common man.

Leviathan One and Two

The constitutional containment delivered by the founders lasted for almost 100 years. Then, at the turn of the century, roughly one hundred years ago, the rules changed. Marked by the 1887 Act to Regulate Commerce and the rise of the Interstate Commerce Commission, then by rules for regulating food and other products, the federal government took on new powers that limited competition, provided monopoly protection, and transferred wealth across special interest groups. The genie was out of the bottle. The first American Leviathan was born. But while government flexed its new regulatory muscles, the industry regulation that emerged did not systematically destroy bedrock private property rights. That was to come later, delivered by the environmental revolution in the late 1960s and early 1970s, the advent of Leviathan Two.

The environmental revolution cut through and across the property rights fabric that formed the market economy. The new social regulation that emerged regulated industries and functions within each and every plant, factory, field, and mine. Command and control, technology-based, regulation emerged in the fine print of the Federal Register. The rule book grew thick with regulations affecting water and air, waste disposal, reconstruction of closed mines and wells, landfills, tree cutting, waste disposal, home building, and chemicals from birth to grave. The creative energies of America's brightest and best entrepreneurs were diverted. Energy previously devoted to finding new and better products and services, exploration, making markets, and reducing costs became focused on regulators, regulation, and ways to avoid costs and gain political favors.

And who supported this massive effort? Dedicated environmentalists with true concerns about the biological envelope that supports all life. Other quasi environmentalists who saw an opportunity to stop economic development in its tracks. Bureaucrats who found enriched careers in regulating. Politicians who found new markets for their services. And some industrialists who found it easier to compete in Washington for political favors than to slug it out in competitive markets. The famous iron triangle emerged full-blown.

And what was the glue that held it together? Bootleggers and Baptists, those who take the high ground populate the environmental revolution and appeal for rules they believe will protect the environment—the Baptists in the story. Then, the bootleggers, those who let the Baptists fight their battles for them as they obtain regulations that raise competitors' costs, limit the entry of new competition, and shield them from common law suits that previously protected the property rights of ordinary people. Yes, some firms saw opportunity in the emerging rules. If government was going to regulate, then let's get the right regulation, the one that favors us.

What do we have? Rules that limit the emissions of sulfur dioxide in the name of acid rain, even though the scientists who studied the issue said there was no acid rain problem. Rules that expand the market for low sulfur coal and natural gas, but reduce the demand for petroleum and ordinary coal. We have toxic release inventory rules that give a black eye to industries that must publicize emissions of more than 600 chemicals, in pounds, even though all industry taken together produces just 17 percent of the total emissions. Rules that set stricter controls on new and expanding plants while grandfathering older ones. Regulations that favor one fuel over others. Endangered species protection that limits timber cutting on government land, raising the price of timber and the profits of firms that cut on their own land.

The list could go on, and if it did, it would contain the fallout from Kyoto [an agreement to limit greenhouse gas emissions]. The agreement itself seeks to raise competitor's cost, with Europe pushing the U.S. to cut back even more, and challenging the idea of seeking lower cost emission reductions from countries that can do it at lower cost. Following in the bootlegger lineup are two major international oil companies that announced support for Kyoto. Why? A combination of reasons to be sure. But they had made major investments in alternative energy products.

The bootlegger list includes corn farmers and producers of corn-based ethanol, assisted by none other than the U.S. Department of Agriculture. And what do they want? Continuation of the rich subsidies on ethanol blends of gasoline. All this in the name of reducing global warming.

Quite understandably, firms and industries that live in this age of regulated capitalism have invested in the rules and stand to lose when the rules are revised. They quite logically support the Leviathan, hoping always to bring down costs once the favors have been won. And just as understandable, there are the politicians who seek to harvest a political commission each time they pass favors to special interest groups.

Yes, an effort to deregulate raises the wrath of the favored special interest groups. Is it any wonder that Congress is unable to take meaningful action to eliminate Superfund, the program that produces lots of litigation and hardly any cleanup? Is it any wonder that Congress is unable to pass a property rights protection bill? Should we be surprised that Congress cannot completely modify the endangered species act in ways that reverse incentives, making exploration, production, and conservation of species attractive endeavors instead of deadly activities to be avoided at all costs?

Houses Built on the Sand of Politics Will Fall

But every house built on sand will eventually fall. Societies that fail to protect fundamental property rights are destined to fail. We know that. So it is with the property rights destruction of the Environmental Revolution. But why?

Rapidly expanding global markets, advances in technology, unavoidable needs to reduce costs, outright failings of old smokestack regulation, rising incomes, and a veritable property rights rebellion are combining to force change. In the absence of meaningful international competition, U.S. firms can gain from costly regulations that cartelize industries. When technology change is slow and obsolescence rates even slower, firms can gain with command-and-control technology-based regulation. Eventually, the true environmentalist begins to wonder what it's all about. If air and water quality is no longer improving, if Superfund sites are caught in a deadly gridlock that accomplishes nothing, if firms are required to spend huge amounts on trivial problems and little on more serious ones, eventually, environmentalists say "enough." Rapidly rising incomes generate demand for real environmental improvements, improvements that cannot be and are not being delivered by central authorities. And when countless ordinary people find that rights to use their land in customary ways are challenged and taken without compensation by regulatory authorities, they rebel. The house built on sand comes crashing down.

The Property Rights Rebellion

Led by Westerners fed up with increased interference with traditional land rights, ordinary people nationwide in hundreds of grassroots organizations took it upon themselves to remind the politicians of the Constitution's Fifth Amendment. "Private property rights will not be taken for public purpose without just compensation." Failing to get satisfaction in Washington, the movement turned to the states. Today, 24 states have passed some form of property rights legislation, and legislation is pending in seven others. The geographic pattern that results is interesting. The Rocky Mountain states, Pacific Northwest, and western states that border Canada have all passed property rights laws. Just four Western states have failed to respond. By contrast, the Northeastern states have done practically nothing. In urban America, it seems, property rights are not an issue. City folks commuting on morning trains to New York, Hartford, and Boston know little about wetlands, Superfund, and endangered species. Indeed, they know little about the importance of basic property rights to land. Opposition to compensated takings comes understandably from the urban areas—the organized interests that support zoning, planning, and government ownership of environmental resources.

Yes, there is friction in the workings of the property rights movement, but its force cannot be denied. Recognizing this, regulators have changed their behavior. They have backed away from the more blatant and arbitrary enforcement of rules. And gridlocked Washington is getting the message.

Following in the wake of the property movement, we see devolution of environmental protection to the states, where politicians recognize that they do not hold monopoly power over investors who might build and expand facilities and who know that they must deliver real environmental benefits to real people who can vote at low cost with their feet.

But if the house of regulation falls, will another, even stouter one, take its place? Will property rights be more secure?

America's Golden Age

I leave these important questions on the table momentarily to tell you about the prospects for America's coming Golden Age. Think with me for a few minutes about this country 100 years ago, and then let us compare that time with the one we now face.

Three major forces played through the economy 100 years ago. First, massive waves of immigration—the largest seen in our history. People came to these shores looking for a better life, seeking opportunity to work and build. They came. Wages fell. Prices fell. Incomes rose. Second. There was a technological revolution. Electricity. Petroleum. Moving assembly lines. Interchangeable parts. Transportation. Communication. And third, massive restructuring of firms and industries—efficiency enhancing mergers, consolidations, alliances, spin-offs. The decades that followed delivered America's Golden Age. Rapid increases in income and wealth. Improvements in basic living conditions. The emergence of the Great American Bread Machine.

What do we observe today, 100 years later? Massive waves of immigration, a technology revolution, and cost-minimizing restructuring, mergers, consolidations, alliances, and the computer-driven virtual corporation.

Is there another Golden Age in the offing? I think so, but that is because I am extremely optimistic. Along with current features common with those one hundred years ago, we have higher taxes, a mountain of regulation, and a threatened system of property rights. To put it simply, we Americans face a daunting challenge. There is a Golden Age in the offing, but its realization is threatened.

Where Do We Stand?

Now, in closing, let me return to the question: Will a stouter regulatory regime replace the one that is falling? This, perhaps, is the most troublesome question, one addressed by the Kyoto Agreement. Out of that agreement comes an international authority, an international cartel manager, if you will, that seeks to coordinate nations worldwide in production and consumption activities that relate to carbon emissions. This parallels the earlier formation of the U.N. Commission on Global Governance, important elements written into the North American Free Trade Agreement, and the new powers of the World Trade Organization. These fledgling bureaucracies will naturally seek to build a global regulatory regime, a potential threat to property rights, free markets, and the coming Golden Age.

Final Thoughts

Paradoxes, property rights, and America's Golden Age. We live in a marvelous time. A time of expanding freedom, increasing wealth, and rapidly improving technical capabilities. It is also a time conditioned by regulation, bureaucratic controls, and property rights destruction. The Golden Age beckons. Will the Golden Age emerge? Will enforcement of constitutional constraints be strengthened? Will this time of environmental feudalism come to an end?

These are the questions that plague us. And the answers? The answers depend on human action, what you and I do to strengthen the case for liberty, individual responsibility, and meaningful protection of the biological envelope which contains and supports all life.

NO

Doug Harbrecht

A Question of Property Rights and Wrongs

Ralph Seidel's livelihood depends on the clean waters of Natrona Creek in rural Pratt County, Kansas, where he owns a golf course, a private fishing resort and a trailer park. But for more than two decades, starting in the late 1960s, neighboring cattle outfit Pratt Feeders dumped livestock wastes into the creek, causing repeated fish kills, according to state environmental officials. When the company's pollution-control permits came up for renewal a few years ago, Seidel organized a public uprising that led to state changes in Pratt Feeders' permits requiring more stringent treatment of its waste water.

Seidel's property rights, not Pratt's, were the issue. "That's what 'property rights' has always meant for conservationists: protections for average Americans and their property," says National Wildlife Federation [NWF] attorney Glenn Sugameli. But a new property-rights movement is afoot, one that could lead polluters like Pratt Feeders—to claim loss of *their* property rights through regulations. All over the country, ranchers, developers, mining companies and others are charging that property owners should be "compensated" if obeying the law lowers the value of private property or results in less-than-anticipated corporate profit.

The notion may seem absurd. "The whole idea that the government needs to pay people not to do bad things is ridiculous," says John Humbach, a property-rights expert at Pace University. "The reason the government exists in the first place is to define what is for the common good and what's not."

Absurd or not, the movement has become a political force to be reckoned with, linked as it is to the powerful notion that landowners should be allowed to do what they want with their property. "People better start taking this movement seriously," says Robert Meltz, a property-law expert at the Congressional Research Service. "This isn't just some fringe element anymore." The proof can be found in Congress, where proposed property-rights amendments are delaying nearly all major environmental legislation.

The new movement has the potential to disrupt a delicate balance between private greed and public need forged over two centuries of U.S. property law, legal experts say. The outcome will affect the survival of endangered

From Doug Harbrecht, "A Question of Property Rights and Wrongs," *National Wildlife* (October/November 1994). Copyright © 1994 by The National Wildlife Federation. Reprinted by permission.

wildlife, and it threatens not only environmental protections like pollution laws, but also zoning regulations and even obscenity laws. "Extremists are trying to take away the ability of Americans to act through their government to protect neighboring private-property owners and the public welfare," says NWF's Sugameli.

In Congress, property-rights debates have held up renewal of the Endangered Species Act, originally slated for 1993, and reauthorization of the Clean Water Act. Property-rights issues have also helped hold up bills to reform the Mining Law of 1872, elevate the Environmental Protection Agency to Cabinet status and reauthorize the Safe Drinking Water Act. The delays are due in large part to property-rights lobbying for amendments such as a ban on volunteers collecting data on private land for the National Biological Survey or a requirement that the government do "loss-of-value" assessments when regulations "could" cause a change in the worth of private property ranging from land to stocks and bonds.

At the state level, "takings" bills similar to those in Congress have been introduced in 37 state legislatures in the past two years; nearly all have been defeated. Many of the bills would require taxpayers to "compensate" landowners, including corporations, for property values diminished because of regulation. Such payments could be extremely costly, and the measures could erode state authority to protect public health and safety—as well as wreak havoc on long-established planning tools such as zoning.

For the most part, the new movement is not faring well in the courts either but it has scored some wins. In one case directly affecting wildlife, last March the U.S. Court of Appeals for the District of Columbia struck down a U.S. Fish and Wildlife Service regulation that prevented private landowners from destroying habitat of federally listed species.

The court declared the provision as "neither clearly authorized by Congress nor a 'reasonable interpretation'" of the Endangered Species Act. The Clinton Administration has asked the court to reconsider the decision. If it stands, the ruling would allow landowners to take actions such as chopping down a tree containing the nest of an endangered red-cockaded woodpecker or bulldozing a beach where threatened sea turtles lay their eggs—*as long as the animals are not around.* No matter that the animals later would return to their habitat; no protections would exist in their absence.

While that case raises the issue of how much conservation laws apply to private land, it is technically a question of the intent of Congress in passing the Endangered Species Act. In contrast, the heart of most of the property-rights debate lies in a Fifth Amendment clause in the Constitution's Bill of Rights: " . . . nor shall private property be taken for public use, without just compensation." Legal historians interpret the original intent as requiring that landowners be paid when the government seizes property for public conveniences like roads and dams.

No one disagrees that if the government takes all of a person's property for public use, then just compensation is required. But the new movement pushes the argument a big step further, contending that regulation of landowners' ability to do as they wish with their property is a "taking" as well. The movement was sparked by the 1987 book *Takings,* by University of Chicago professor Richard Epstein. Epstein argued that the broad definition of a taking "invalidates much of the 20th-century legislation."

Such arguments mask the myriad ways governments increase the value of public property. Partly for this reason, editorial boards at newspapers across the country have condemned property-rights legislation. In one April 1994 editorial, *The Atlanta Constitution* called the demands of property rights forces "pure hypocrisy." As an example, it cited Arizona, "one of the fastest-growing states in the country and a hotbed of property-rights legislation. But its cities and suburbs would still be worthless desert if not for water brought from hundreds of miles away, at huge expense to the federal government."

Other examples: Developers in coastal areas that depend on taxpayer-subsidized insurance and agri-businesses that thrive with federal price support and crop insurance. Property values often exist only because of sewers, roads and other government-paid amenities.

The takings argument has quickly reached the level of the absurd. "Compensation" has been asked for costs incurred in widening restroom doors to allow wheelchair access required by law, losses due to limits on the sale and import of assault rifles—and even losses due to restrictions on "dial-a-porn" services.

In Mississippi and Georgia, religious groups have joined environmentalists in opposing proposed property-rights legislation. The bills would require taxpayers to compensate pornography dealers prevented from locating next to schools and churches. "Where this leads is to the end of government's role as protector of the little guy and provider of amenities the market alone cannot provide," says Jessica Mathews, a senior fellow at the Council on Foreign Relations. "Things like public health, worker safety, civil rights, environment, planning, historic preservation and anti-discrimination measures."

The rule of law in the United States has long been that landowners must not use their land in any way that creates a public or private nuisance (in other words, harms the public or neighbors). "(A)ll property in this country is held under the implied obligation that the owner's use of it shall not be injurious to the community," the Supreme Court ruled 100 years ago. In a string of cases since then, the high court has consistently reaffirmed that bedrock principle.

In 1992, the Supreme Court did rule conditionally in favor of South Carolina developer David Lucas, who had been denied permission to build on two ocean-front lots after the state adopted a coastal-zone management plan. Lucas owned two lots appraised at $1 million. The Court ruled he was entitled to compensation in part *if* the action deprived him of "all economically viable use" of his land. The case then went back to the state courts, where it was eventually settled. Justice Antonin Scalia, who wrote the Supreme Court's majority opinion, warned that anyone who purchases property always takes a risk that government regulation will diminish its value. He wrote that a lakebed owner "would not be entitled to compensation when he is denied the requisite permit to engage in a landfilling operation that would have the effect of flooding others' land." In other words, says NWF attorney Sugameli, "Property ownership does not include the right to flood your neighbors."

Even permits for livestock grazing on public land, claim "takings" advocates, are property. In Nevada, rancher Wayne Hage is suing the U.S. government for $28 million in damages for diminishing the value of his property (his permit) in a number of ways. But the range that Hage's cattle roamed is not his. It's yours: 700,000 acres of the Toiyabe National Forest in Nevada he leased from the federal government. In 1990, the Forest Service warned Hage he was letting his cattle overgraze. The animals were devouring vegetation along clear-running streams among the mountain meadows and piñon pine, birch and aspen trees— and in the process destroying key habitat for fish, birds, elk and other wildlife.

After Hage didn't respond to repeated warnings, in July 1991 contract cowboys protected by armed Forest Service rangers rounded up 73 of Hage's scofflaw bovines. Later, 31 more were taken in. Hage sold off the remainder of his herd of 2,000 and has taken his grievances to court. Among them are his claims that the government ruined his business by introducing elk, which competed with his cattle for grass; allowing backpackers and elk to drink from springs used by his livestock; and restricting how heavily his cattle could graze streamside vegetation. Federal officials say that by ignoring grazing regulations, Hage has only himself to blame for his troubles. On behalf of several environmental groups and the state of Nevada, the National Wildlife Federation is actively participating in the case as a friend of the court.

So-called property-rights advocates like to portray themselves as average Janes and Joes fighting the daunting power of federal bureaucrats and tree-hugging elites. Says J. T. "Jake" Commins, executive vice president of the antiregulation Montana Farm Bureau, "Walt Whitman was speaking to the universal aspiration of humanity throughout history when he said, 'A man is not whole and complete unless he owns a house and the ground it stands on.'"

But loss of regulations often benefits big landowners most. Charles Geisler, a sociologist at Cornell University, has found that the nation's land is concentrated in the hands of the wealthy few. According to the Department of Agriculture, almost three quarters of all the privately owned land in the country is owned by less than 5 percent of the landowning population. "This is important to keep in mind when the property-rights people talk about fighting for the little guy," says Geisler.

Who are the nation's largest private landowners? Timber and mining companies, agri-businesses, developers and energy conglomerates, says Geisler. These owners appear to be prime movers behind the property-rights movement.

One example: When M & J Coal Company of Marion County, West Virginia, was ordered by federal officials not to mine portions of coal deposits because a gas line ruptured and huge cracks opened on the land of homeowners living over the underground mine, the company sued. Though it earned a 34.5 percent annual profit on the mine, the company claimed the restrictions were a "regulatory taking" for which it was entitled to $580,000 in lost profits. The court rejected the claim earlier this year; the company has appealed.

In Wyoming, the Clajon Corporation recently challenged in court state limits on the number of hunting licenses issued to large landowners. As owner of a large ranch, Clajon contended that it owned the right to hunt wildlife on its land and that the state's hunting-license limit was a taking of the company's property rights. The Wyoming Wildlife Federation and NWF's Rocky Mountain Natural Resource Center were leaders in fighting the claims, which were thrown out by Wyoming's federal district court in June.

Despite such cases, legal experts do not dismiss the takings movement in general. Even Humbach of Pace University maintains environmentalists have two decades of their own success partly to blame for the current backlash. "People concerned about the wise use of land should be equally concerned about misapplied and heavy-handed government rules that turn average Americans into poster children for the property-rights movement," he says.

Environmentalists might even agree with that notion—at least up to the mention of poster children. "I am convinced there is no actual case of an 'American poster child' for the property-rights movement," says NWF's Sugameli. NWF has examined hundreds of such purported cases. "And every single case either falls apart or doesn't exist at all," he says.

One example: the well-publicized 1988 jailing of Hungarian immigrant John Pozsgai for filling in a small wetland next to his Pennsylvania diesel mechanic shop. Property-rights advocates portray Pozsgai as a hapless victim who only meant to make his own land useful. But according to the Environmental Protection Agency, engineers told Pozsgai before he bought the property that there were wetlands on it; he even used that information to negotiate a $20,000 reduction in the land's purchase price. He then refused to obtain the required federal permit to fill his wetlands, filled them without the permit and ignored repeated official notification to stop doing so. Said the judge during Pozsgai's sentencing, "It is hard to visualize a more stubborn violator of the laws that were designed to protect the environment."

Such cases aside, environmentalists say they do recognize a need for collaboration. "The whole environmental community should be advancing the view that environmental laws are vital to protecting property rights, not taking them away, says Michael Bean, chairman of the Wildlife Program of the nonprofit Environmental Defense Fund. To that end, the group is exploring a plan to help save the endangered red-cockaded woodpecker in the pine forests of North Carolina, where private landowners own most of the prime habitat. The idea involves "land-use credits" for leaving large stands of trees untouched. The credits could take the form of lower taxes, regulatory relief from parts of the Endangered Species Act or some other tangible asset.

And in Kern County, California, conservationists have long supported a plan aimed at allowing developers to build near habitat of the endangered kit fox in return for undisturbed parcel set-asides and a developer-funded conservation program, both of which would aid the fox.

That sort of thinking may be the best hope for habitat and wildlife in the future. "In recent years, proponents of various private property-rights amendments have come to view the protection of private-property rights and government regulation as mutually exclusive goals," says Senator John Chafee (R-Rhode Island). "That view is wrong." In the end, property rights are as much an issue for Ralph Seidel and the elk in Toiyabe National Forest as for Wayne Hage. The framers of the Constitution wouldn't have had it any other way.

POSTSCRIPT

Does Environmental Regulation Unnecessarily Limit Private Property Rights?

Harbrecht takes comfort in the fact that the legislative program of the property rights movement has not been very successful thus far. Recent congressional attempts to undermine the Endangered Species Act and the Clean Water Act by imposing the requirement for costly compensation to affected landowners have been unsuccessful. But there are few signs that the debate is over. The Private Property Rights Implementation Act of 1999 (H.R. 2372; passed by the House in March 2000) aims to give developers the right to bypass the local and state appeals process for unfavorable zoning and regulatory decisions by going directly to federal court. See Kristin Loiacono, "Developers Want Their Day in (Federal) Court," *Trial* (May 2000). Like most property rights and free-market advocates, Yandle speaks about the need for "real environmental improvements," but he seems to go further than others in suggesting that this can be achieved through a system of marketplace incentives.

University of Chicago law professor Richard Epstein's book *Takings: Private Property and the Power of Eminent Domain* (Harvard University Press, 1985) is credited with providing the property rights movement with legal arguments, which have been used in courtroom challenges, to regulations that have invoked the Fifth Amendment. For a concise debate between Epstein and John Echeverria about the legal aspects of the issue, see the May/June 1992 issue of the *Cato Policy Report,* published by the Cato Institute in Washington, D.C. Another dialogue, specifically on the issue of the environment versus property rights, appears on the op-ed page of the March 15, 1995, issue of *The New York Times.*

For another slant by an environmentalist who opposes the property rights position, see Carl Pope's article in the March 1994 issue of *Sierra*. In a cover story in the November 1995 issue of *The Progressive,* Erik Ness argues that the private property movement is carving up America. David Helvarg, in "Legal Assault on the Environment," *The Nation* (January 30, 1995), presents a critical history of the campaign by property rights activists. For the perspective of those who see environmental restrictions on property rights as part of a leftist-inspired government conspiracy to control all aspects of our lives, see William Norman Rigg's article in the August 9, 1993, issue of *New American.*

Chapter 2 of *Foundations of Environmental Ethics* by Eugene C. Hargrove (Prentice Hall, 1989) and Section VI B of *The Environmental Ethics and Policy Book* by Donald VanDeVeer and Christine Pierce (Wadsworth, 1994) are sources of information on ideological and philosophical attitudes toward land use and property rights. Finally, the Winter 1999 issue of *Environmental Law* has several papers on the takings issue.

ISSUE 6

Should Environmental Policy Attempt to Cure Environmental Racism?

YES: Jan Marie Fritz, from "Searching for Environmental Justice: National Stories, Global Possibilities," *Social Justice* (Fall 1999)

NO: David Friedman, from "The 'Environmental Racism' Hoax," *The American Enterprise* (November/December 1998)

ISSUE SUMMARY

YES: Health planning sociologist Jan Marie Fritz discusses the national and international manifestations of environmental racism and the global imperative in the search for environmental justice.

NO: Writer and social analyst David Friedman denies the evidence of environmental racism. He argues that the environmental justice movement is a government-sanctioned political ploy that will hurt urban minorities by driving away industrial jobs.

$\mathbf{T}$he environmental movement has often been described as reflecting the idealist aspirations of white middle- and upper-income people. Indeed, poor people and minority groups were not well represented among those who gathered for the teach-ins and other events organized to celebrate the first Earth Day in April 1970. Only later did some of the growing environmental organizations even begin to discuss the need to reach out in their organizing efforts to low-income communities and ethnic minority neighborhoods.

The dearth of African Americans, Native Americans, Hispanics, and poor white people among the early environmental activists was passed off as a reflection of the pressing need such people felt to pay attention to more "basic" social concerns such as hunger, homelessness, and safety.

Until very recently, little media attention has been given to publicizing the fact that, after a slow start, the involvement of poor and minority people in grassroots environmental organizing has been growing dramatically for more than a decade. The movement for environmental justice was triggered in 1982 by demonstrations to protest the decision to locate a poorly planned PCB disposal site adjacent to impoverished African American and Native American communities in Warren County, North Carolina. Since then, the movement has grown to encompass local, regional, and national groups organized to protest the systematic discrimination in the setting of environmental goals and the siting of polluting industries and waste disposal facilities in their backyards.

Recognition of the demands of this movement by mainstream environmental organizations and government officials has been slow in coming. It was not until 1990 that the Environmental Protection Agency (EPA) issued the report *Environmental Equity: Reducing Risks for All Communities,* which acknowledged the need to pay attention to many of the concerns being raised by environmental justice activists. In that same year leaders of the Southwest Organizing Project sent a letter demanding a dialogue with U.S. environmental organizations, which, they charged, "emphasize the cleanup and preservation of the environment on the backs of working people in general and people of color in particular." At the 1992 United Nations Earth Summit in Rio de Janeiro, a set of "Principles of Environmental Justice" was widely circulated and discussed. In 1993 the EPA opened an Office of Environmental Equity with plans for cleaning up sites in several poor communities, and in 1994 President Bill Clinton made the cause of environmental equity a national priority by issuing a sweeping Executive Order on Environmental Justice. Since then, many complaints of environmental discrimination have been filed with the EPA under Title VI of the federal Civil Rights Act of 1964. In March 1998 the EPA issued guidelines for investigating those complaints.

The environmental justice movement has given rise to several controversies. Critics of the charges of environmental racism assert that inequities in the siting of sources of pollution are simply the result of market forces that make the poor neighborhoods where minorities live the economically logical choice for the location of such facilities, or that apparent inequities result because once these facilities are built, they depress real estate values, turning the neighborhoods into poor and minority communities. A more fundamental concern is whether or not simplistic efforts to combat environmental racism will simply shift pollution to poor white neighborhoods. To avoid this consequence, some environmentalists have suggested that the principal demand of the environmental justice movement should be general pollution reduction.

In the following selections, Jan Marie Fritz examines the meaning of environmental justice and the consequences of its converse, which has been called environmental racism. She relates the evidence that has fueled the national and international concern about environmental inequities and explores the global possibilities for establishing enforceable principles of environmental justice. David Friedman describes the environmental justice movement as a politically inspired movement that he asserts is unsupported by scientific facts. He labels environmental racism a hoax and argues that its consequences will harm the urban poor by denying them the industrial jobs they need.

Jan Marie Fritz  **YES**

Searching for Environmental Justice

Environmental justice is increasingly a topic of great concern and/or academic interest for community activists and scholars in a number of countries. Within the last four years, for instance, international conferences on environmental justice have been held in Australia, the United States, New Zealand, and South Africa, and a research group based in England studied access to environmental justice in seven cities in Africa and Asia (Harding, Anderson, and Jenkins, 1997).

In this article, definitions of environmental justice are discussed and the term is compared to environmental injustice, environmental equity, and environmental racism. The article also reviews, for the first time, selected activities that have been explicitly labeled as environmental justice initiatives by nongovernmental organizations, community-based organizations, and/or governments in four countries—Canada, Israel, the United States, and South Africa.

Environmental Justice: Definition and Principles

It is probably true that "examples of . . . social justice environmentalism can be found in every country on the planet, organized from the bottom up, linking community-based concerns to national and international political and economic contexts" (Johnston, 1994: 225). This article, however, discusses only organizations or initiatives that specifically define themselves as having an environmental justice emphasis.

Environmental justice, as the term is used most frequently in the United States, focuses on the environmental problems (in terms of programs, policies, and/or activities) disproportionately faced by those with the least power. In the United States, these groups would be minority and low-income populations. Although the term "environmental justice" is becoming a popular concept, the focus really is environmental injustice. The decision to use the term justice rather than injustice is interesting both legally and socially. "Environmental justice" pushes us to think about a negative situation by using a positive term. Equivalent concepts with the same positive ring are "environmental equity" or "human rights and the environment."

From Jan Marie Fritz, "Searching for Environmental Justice: National Stories, Global Possibilities," *Social Justice*, vol. 26, no. 3 (Fall 1999). Copyright © 1999 by *Social Justice*. Reprinted by permission. Notes and references omitted.

Environmental injustice also has been called environmental racism. Environmental racism (Mohai and Bryant, 1992; Bullard, 1994a, 1994b; Westra and Wenz, 1995; Kraft and Scheberle, 1995) is a strong term chosen by some because they believe that racism frequently or always is the root of a problem. Their concern is perhaps that using "environmental justice" dilutes the central analysis of racism or could even prevent an analyst from mentioning the possibility of racism. Others prefer the term environmental justice because what may have been a racist decision at one time, for instance, may have been perpetuated for reasons other than racism. Because environmental problems often develop over a long period of time and are complicated, these analysts prefer to use the general term environmental justice and, when a specific situation warrants, the term environmental racism.

Robert Bullard (1995: 4–8), Ware Professor of Sociology and Director of the Environmental Justice Resource Center at Clark Atlanta University, has written extensively about environmental justice. He has noted that:

> unequal environmental protection undermines three basic types of equity: procedural (rules, regulations, evaluation criteria, and enforcement are applied in a nondiscriminatory way); geographic (proximity to environmental hazards such as landfills, incinerators...), and social (role of sociological factors such as race, ethnicity, class, culture... in environmental decision-making).

Bullard (1995:9–20) has advised governments that they will have to adopt five basic principles of environmental justice if they wish to end unequal environmental protection. The principles are: guarantee the right to environmental protection; eliminate environmental threats before harm occurs; shift the burden of proof from citizens having to show they have been harmed to potential polluters having to prove that their operations will not harm; assure that laws allow "disparate impact and statistical weight" rather than requiring proof of intent to discriminate; and put resources where "environmental and health problems are the greatest."

National Stories

Environmental justice activities, labeled as such, have been identified in at least four countries:

Canada. The Canadian Environmental Defence Fund (1997: 4), or CEDF, defines itself as "a national, charitable, nonprofit organization founded in 1984 in order to help citizens gain access to environmental justice." The organization, supported by 5,000 to 7,000 individual donors and some foundations (Mausberg, 1998a), provides "funding, fundraising assistance, legal, scientific, planning and engineering referrals, and organizational support to grass-roots groups pursuing significant environmental law cases" (Canadian Environmental Defence Fund, 1997: 4). Since 1984, CEDF has provided the equivalent of "$4 million in assistance to more than 150... partner groups" (Canadian Environmental Defence Fund, 1998; Mausberg, 1998b).

Environmental justice, according to the CEDF, refers to any local community or group of citizens being able to bring an environmental law case before a Canadian court or tribunal. CEDF (1997: 1) believes that it can be overwhelming for groups to find "the amount of time, energy, commitment, and funds required" and so CEDF works with selected organizations so they "will not be denied access to justice on the basis of cost alone."

CEDF chooses cases to support based on "the importance of the legal matter, ecological impact, and (the applicant group's) needs" (Mausberg, 1998a). CEDF also considers its organizational capacity (particularly the small size of the CEDF staff) in making decisions. When CEDF decides to fully support a case, it does the legal work as well as all the organizational support for a case, including accepting donations on behalf of the client group. As Burkhard Mausberg (1998a), the executive director, has noted: "Once we take on a case, we are in it to the end."

Among the cases that have been given CEDF's full support (Canadian Environmental Defence Fund, 1997; 1998) are: (1) Manitoba's Future Forest Alliance (MFFA). MFFA is attempting to force a federal environmental assessment of forest clearcutting. (2) Sierra Club of Canada. The Sierra Club argued that an environmental assessment needed to be undertaken before the federal government could complete the sale of CANDU nuclear reactors to China. CEDF undertook fundraising and media campaigns to support the legal challenge. (3) Wastebusters Environmental Watchdogs (Saskatchewan). Wastebusters organized the community against the construction of a hazardous waste treatment facility on farmland. (4) Citizens' Mining Council (Newfoundland and Labrador). The Citizens' Mining Council filed an application in Toronto asking the federal court to require one unified environmental assessment for all of the proposed Voisey's Bay development. (5) Innu Nation. CEDF supported the Innu Nation in its legal battle against low-level military flying over their Labrador environment.

Israel.   The New Israel Fund (1997: 53), a 20-year-old organization with significant resources and accomplishments, describes itself as:

> a joint effort of Israelis, North Americans, and Europeans ... to safeguard civil and human rights, promote Jewish-Arab equality and coexistence, advance the status of women, foster tolerance and religious freedom, bridge social and economic gaps, pursue environmental justice, and increase government accountability.

According to New Israel Fund (NIF) staff member Lauren Erdreich (1998a, 1998b), the NIF definition of environmental justice focuses not only on the environment, but "also is connected to citizen rights." NIF started using the term environmental justice "because we wanted to ... focus on disadvantaged groups. It has turned out that a lot of groups we work with are more middle class, but they still deal with citizens' rights."

Shatil (which means "seedling" in Hebrew) is the New Israel Fund's Capacity building center for social change organizations. Shatil (1997: 5; 7) provides assistance "by increasing the professionalism of Israeli voluntary organizations through strategic training and consultation (and) building coalitions to influence public policy on critical social issues...." Shatil's Environmental Justice Project began in 1995 and is beginning to make "a decisive impact on the budding grassroots environmental community in Israel."

Shatil (1998: 1; 1997: 22–23) states that Israel's government-planning decisions "have too often been dictated by free market forces" and that "marginalized ethnic minorities and residents of disadvantaged areas suffer most from environmental hazards such as toxic-waste dumping, open sewage, and industrial park development." Shatil's Environmental Justice Project is working with 30 different grass-roots environmental groups in disadvantaged and low-income areas to help them undertake effective actions against environmental hazards in their communities and build coalitions with national environmental organizations.

Among the projects undertaken by the Environmental Justice Project (Shatil, 1998: 2; Vardi, 1998) are: (1) Bedouin organizations are being assisted in their attempt to fight "the open sewage running into unrecognized Bedouin villages from neighboring Jewish towns and cities"; (2) citizen-action groups are being helped in their fight against over-development of the Mediterranean coastline; (3) the Committee to Save HaNeviim Street is being assisted in its efforts to keep the historic street from being turned into a six-lane highway; and (4) a coalition has been put in place to combat the pollution generated by the Ramat Hovav Industrial Area, and, if the area continues to be polluted, to block expansion.

Shatil (1997: 16) describes the environment as "one of the most important fronts for building Jewish/Arab coexistence." For example:

> While Shatil was in the process of assisting a grassroots environment group in the Palestinian-Israeli village of Majd el-Krum, a Jewish-owned recycling plant approached the village with an offer to recycle plastic bottles there. A unique cooperative partnership for environmental preservation has since developed between the factory and the town, between Jewish- and Palestinian-Israelis.

The United States. In 1982, the state of North Carolina proposed building a polychlorinated biphenyl (PCB) disposal site in Warren County, where the residents were predominately African American. The United States Environmental Protection Agency approved the proposal even though PCB-contaminated soil would be dumped on land with a water table, which did not meet EPA standards. All the local residents received their water from wells. The citizens argued that the decisions on the Site and the relaxation of standards were racially motivated. Protest rallies and marches were held and more than 500 people were arrested. The PCB facility was built, but the dispute brought national public attention to the possible connection between racism and siting decisions, regulations, and enforcement.

Soon after the well-publicized protest in North Carolina, the U.S. General Accounting Office (GAO, 1983) studied the location of hazardous waste landfills in eight Southern states. The GAO, an investigative arm of Congress, found that three of every four sites were in low-income, minority communities. In the next few years, two more important reports were released—"Toxics and Minority Communities Report," by the Alternative Policy Institute of the Center for Third World Organizing (1986) and "Toxic Waste and Race in the United States: A National Report on the Racial and Socioeconomic Characteristics of Communities with Hazardous Waste Sites," by the United Church of Christ's Commission for Racial Justice (1987). Both reports pointed to substantial, unaddressed environmental problems facing minority communities.

In 1991, the First National People of Color Environmental Leadership Summit was held in Washington, D.C. The five-day meeting attracted over 600 African Americans, Asian Americans, Latino Americans, and Native Americans, as well as representatives from other countries. Those attending the event agreed (Grossman, 1994: 272–273):

> to begin to build a national and international movement of all peoples of color to fight the destruction and taking of our lands and communities...; to respect and celebrate each of our cultures, languages, and beliefs about the natural world and our roles in healing ourselves; to insure environmental justice; to promote economic alternatives which would contribute to the development of environmentally safe livelihoods; and to secure our political, economic, and cultural liberation that has been denied for over 500 years of colonization and oppression, resulting in the poisoning of our communities and land and the genocide of our peoples....

"After some prodding from academics of color and environmental and social justice activists" (Ibid.: 287), the EPA director, William Reilly, established in 1990 an internal Environmental Equity Workgroup. The final report of the group was issued in 1992, the same year the EPA established the Office of Environmental Justice, "in response to public concerns" (U.S. EPA, Office of Environmental Justice, 1998: 1). Although the document was criticized for being very weak by those outside the EPA who were familiar with environmental justice issues, the report did note that "racial minority and low-income populations experience higher than average exposures to selected air pollutants, hazardous waste facilities, contaminated fish, and agricultural pesticides in the workplace" (U.S. Environmental Protection Agency, 1992).

The National Environmental Justice Advisory Council (NEJAC), a federal advisory committee, was established in 1993 to "ensure that the U.S. Environmental Protection Agency receives the viewpoints of affected stakeholders on issues related to environmental justice" (U.S. EPA, Office of Environmental Justice, 1997) by "providing independent advice, consultation, and recommendations to the (EPA) Administrator" (U.S. EPA, Office of Environmental Justice and Office of Enforcement/Compliance Assurance, 1998: 1). NEJAC has a Designated Federal Official and 25 members who represent environmental advocacy groups, business and industry, state, local, and tribal governments, grass-roots community organizations, and the academic community. There are seven subcommittees, which focus on issues related to enforcement, health and research, air and water, indigenous peoples, international issues, public participation and accountability, and waste and facility sitings. In addition to the 25 NEJAC members, each of whom is assigned to a subcommittee, more than 36 additional individuals serve as members of the subcommittees. NEJAC meetings are open to the public and, at the most recent meeting, over 60 individuals testified about environmental problems in their communities during the three public comment periods.

In 1994, President Bill Clinton issued an executive order that "each Federal agency shall make achieving environmental justice part of its mission by identifying and addressing, as appropriate, disproportionately high and adverse human health or environmental effects of its programs, policies, and activities on minority populations and low-income populations." Environmental justice, as defined here, focuses on the environmental problems facing those with the least power in the United States—minority groups and low-income groups.

Numerous organizations or groups in the United States (governmental, nongovernmental, and community-based organizations) have environmental justice missions or concerns. For instance, the Maryland Advisory Council on Environmental Justice was established in 1997 to make recommendations to governmental bodies; the Texas Air Control Board Chair and the Texas Water Commission Chair created the Environmental Equity and Justice Task Force in 1993 and, soon thereafter, an Environmental Equity Office was established within the Texas Natural Resource Conservation Commission; the Southwest Network for Environmental and Economic Justice and the Texas Network for Environmental and Economic Justice assist grass-roots organizations of people of color in addressing environmental justice issues at the state, regional, and national levels. Further examples include the Eco-Justice Working Group of the National Council of Churches, a group that provides program ideas and resources to help congregations involved in environmental justice projects, and the Asian Pacific Environmental Network, an organization that empowers Asian American and Pacific Islander communities throughout the United States to achieve multicultural environmental justice.

There are also some initiatives in the United States that work with or offer support to environmental justice efforts in other countries. The EPA, through its Office of International Activities, Office of Environmental Justice, regional offices, and NEJAC's International Subcommittee, has environmental justice interests in South Africa and Mexico. The EPA also has suggested to the Organization for Economic Cooperation and Development (OECD) that the OECD's second round of environmental performance reviews on member countries should include environmental justice as a standard topic (Morant, 1998).

There are a number of initiatives on the U.S.-Mexico border. For example, the Border Environmental Justice Campaign/Campana para la Justicia Ambiental Fronteriza (1994: 1), under the direction of Cesar Luna with the Environmental Health Coalition/Coalicion de Salud Ambiental in San Diego, California, works with nongovernmental organizations in Tijuana, Mexico, "to reduce toxic pollution in Tijuana caused by maquiladoras and Mexican industries." Another example of work on the U.S.-Mexico border (Bravo, 1996) was the combined action of several organizations—the Environmental Health Coalition, the Southwest Network for Economic and Environmental Justice, and the Comite Ciudadano por Restauracion del Canon del Padre. They filed an administrative petition to push the EPA to "formally issue subpoenas to (94)

U.S. companies" with subsidiaries in Mexico "requiring information on the companies' hazardous materials and waste that could potentially escape into the New River." The New River, which is so polluted that Americans residing near it have been warned to stay away from it (Cass, 1996: 99–100), "flows through communities of color and of low income on both sides of the border" (Bravo, 1996: 8).

Many community-based and nongovernmental organizations have provided testimony at the EPA's NEJAC meetings. Even though there are federal and some state government initiatives in place to direct attention to issues of environmental justice, the effort simply does not yet effectively deal with the long-term, complex environmental problems facing minority and low-income communities. This frustration is evident in the words of some of those who spoke during the public comment periods at a 1998 NEJAC meeting in Oakland, California (May 31 through June 3, 1998): "I want more than promises"; "Why doesn't the EPA in my community see this mess is corrected"; "They tell people don't eat the fish, but they don't tell polluters not to hurt the fish"; "I like to use the word environmental racism—I don't see any justice out there"; "We have to adopt an environmental justice paradigm and place public health above all else"; and "If the EPA (the Environmental Protection Agency) doesn't provide protection—maybe it should change its name."

Then there was the woman from Mossville, Louisiana, who provided emotional testimony about the terrible environmental and health problems in her toxic neighborhood. She tipped her red baseball cap so those attending the NEJAC meeting could see her hair had fallen out from cancer treatment. She said she wanted to remind us that "cancer is all over the world and it isn't just caused by smoking." She said "people are dropping dead like flies" in her community and that she was tired, very tired, but she wasn't giving up.

South Africa.   According to sociologist Jacklyn Cock (1991: 1), a faculty member at the University of Witwatersrand in Johannesburg:

> South Africa, with its mix of First World environmental problems such as acid rain, and Third World environmental problems such as soil erosion, is a microcosm of the environmental challenges facing the planet. These challenges are slowly coming onto the agenda of the liberation struggle in South Africa.

One of the organizations putting environmental issues at the center of the agenda is the Environmental Justice Networking Forum (EJNF). The EJNF, formed in 1992, is an alliance of more than 550 nongovernmental, nonprofit South African organizations. The organizations are often quite different—e.g., trade unions, women's associations, youth groups, rural organizations, religious groups—but they all subscribe to the principles of environmental justice. The EJNF (1998a) says the following about its aims:

> EJNF seeks to mobilise people's environmental concerns and facilitate their articulation within provincial, national, regional, and international processes. It seeks to provide an efficient communications, coordinating, and

networking system whereby civil society can contribute to and enable participatory decision-making and democratic environmental governance. At the same time, it supports workers and marginalised groupings in their efforts to reverse local environmental injustices. EJNF thus situates the environment in relation to social justice. In doing so, it effectively challenges the definition of environment and development in the interests of the elite.

The EJNF is organized in eight of South Africa's nine provinces. It has a National Steering Committee, composed of elected provincial representatives, which undertakes national responsibilities. Each province has a Provincial Steering Committee and the highest decision-making body is the National Congress of provincial delegations. EJNF's national office is in Pietermaritzburg and there are 15 full-time staff members.

Chris Albertyn, EJNF's national coordinator, has stated that he is a "flexible green." His personal philosophy, he has said, is "dark green but that the radical language of dark green is inclined to frighten off newcomers to the environmental movement—and that would be counterproductive" (Cock and Koch, 1991: 18).

According to an article in Johannesburg's Mail and Guardian (Koch, 1996), Albertyn:

> is best known—and feared—for his fiery confrontations with government ministers over the import of toxic waste into (South Africa). (In 1995) his organization announced, on the very day that former environment minister Dawie de Villiers was due to open a major national conference, that a ship laden with Finish toxic waste was on the high seas and headed for (South Africa). The ship was forced to turn around, De Villiers retired into relative obscurity after his party left the government....

Koch (1996) indicates that this event and at least one other high-profile incident involving toxic waste being "sneaked" into the country "have established a reputation for the EJNF as being one of the most vibrant 'watchdog' organizations in the country." Albertyn, however, has pointed out that the real success of EJNF is "in creating a movement that allows ordinary men and women to realize that their struggle for a better life is inextricably linked to the environmental abuse they experience."

The EJNF, according to Albertyn (Koch, 1996), "walks a tightrope between being independent and opposing government and cooperating." The goal is to see that "civil society has coherent access to government." For example, a small home for the disabled in Mpumalanga experienced difficulties when a nearby mine diverted the water. EJNF used its network to get through the government bureaucracy and within six days engineers arrived to solve the problem. As Albertyn noted, "they did it themselves, but through the solidarity that the EJNF offers."

In November 1996, EJNF hosted a conference on Regional Cooperation in Environmental Governance. The conference was attended by representatives of the countries of the Southern African Development Community. The delegates agreed to have a strategic planning workshop early in 1997 to guide the process of building a regional environmental justice network. The draft charter of the group (Hallowes, 1997: 6–8) indicates that "people in Southern Africa have been excluded from decision-making about development, first by colonialism and apartheid and now by the processes of economic globalization." The delegates expressed deep concern about "the growing scale of exclusion that is characteristic of unsustainable development" and a commitment to "participatory environmental governance." In setting out the principles they believe are necessary to achieve participatory environmental governance, the following was said about environmental justice:

> Poor people suffer most from environmental degradation and benefit least from the wealth produced.... Environmental justice therefore requires a commitment to equity in the distribution of resources and in decision-making regarding the use and control of resources. Equity in decision-making implies that decisions should be made at the most local level possible... (and that there should be) equity between nations and between people irrespective of their race, gender, ethnicity, class, physical ability, sexuality, religion, custom, age, or geographical location. Decision-making processes should be inclusive and designed to facilitate the participation of social groups which are marginalized in society.

The multi-country Southern Africa Network for Environmental Justice has now been established, but "communication difficulties and minimal resources mean this (organization) is still somewhat fragile" (Albertyn, 1998a). There has been a new development in at least one of the represented countries. A delegate from Swaziland has told Albertyn that a network of organizations has been put in place there under the name Swaziland Environmental Justice Agenda. EJNF has been able to move the multi-country environmental justice network forward and the coordinator of the organization also feels EJNF has had effective dealings with its own national government. According to Albertyn (1998b):

> We have been very successful in introducing the term (environmental justice). The new National Policy on Environmental Management uses the term regularly and includes Environmental Justice as a guiding principle of environmental management.

Global Possibilities

Several points maybe made based on this review of international environmental justice activities:

(1) The term "environmental justice" appears to be fairly new, but attempts to achieve environmental justice began much earlier.

(2) Environmental justice may be defined in different ways. One possible approach is that of Justice for Nature, a Costa Rican organization. This organization focuses on protecting the "legal rights and improvement of laws preserving the environment throughout Central America (and) a good administration of justice in environmental matters" (Justice for Nature, 1998). Another definition of environmental justice could follow the Canadian Environmental Defence Fund approach when it refers to all groups having fair access to a hearing about environmental matters by a court or tribunal. Some may prefer the "procedural" definition used by Harding and his associates in England (Harding, 1998;

Harding, Anderson, and Jenkins, 1997), who talk about "whether any actors in the system regard an (environmental situation) as unjust, or as a/the basis for a potential claim against the state or a private party." The most frequent definition used by the numerous U.S. community-based groups, the U.S. government, and the large coalition group in South Africa is a one that emphasizes fair treatment for racial minorities and low-income groups (marginalized groupings) in regard to environmental siting, planning, policies, and implementation.

If the term "environmental justice" continues to be an important concept, the "neutral" definition, focusing on fair access to all, may become the one of choice for many government and corporate bodies. There is a tendency now for some government agencies and corporate organizations to show an awareness of the term, but they choose not to use it. There appear to be several reasons for this: (a) the agencies or corporate bodies do not think they must use it; (b) they may think it is too direct in that it calls attention to problems, and (c) a government agency might find using this term less effective when it wishes to delicately raise an issue and solve an environmental problem in cooperation with another government or a corporation. The "neutral" definition would not call direct attention to the unfair treatment of those with less power in a society and, in fact, may not address that situation at all.

(3) The environmental justice movements in the United States and South Africa developed from community-based organizations and nongovernmental organizations. It is important to recognize the power of such groups in bringing this topic to the attention of the public and policymakers and the need to open up our participation mechanisms in our countries to let these groups be a direct part of "the capital investment decisions through which environmental burdens are produced and by which communities are affected" (Lake, 1996: 171).

(4) The national environmental justice movements are reaching outside their borders. For instance, the U.S. Environmental Protection Agency has environmental justice interests in South Africa and Mexico; community-based organizations in the United States are working with Mexican organizations, while the South African Environmental Justice Networking Forum has facilitated the establishment of a multi-country environmental justice coalition in Southern Africa.

We cannot be certain if the concept of environmental justice will remain central to our environmental analysis or become only a term of the 1990s. Similarly, it is uncertain whether its definition will focus primarily on the environmental problems facing minority and low-income groups or whether it will refer to the rights of all. What is clear is that there is an important relationship between environmental justice and sustainable development. Development in all countries is being shaped by global forces (Environmental Justice Networking Forum, 1998b). This development needs to be reviewed in terms of the aims, values, relationships, and practices of the involved organizations and countries and the potential short-term and long-term effects of the development initiatives on the community, nation, and planet.

An environmental justice framework would dictate that development be sustainable and based on fair treatment of all affected parties. Sustainable development in any country would need to be reviewed in terms of four independent but related processes: (1) inclusive economic development (how all adults will be able to make a living); (2) social and political development based on widespread public participation (by individuals as well as community and non-governmental groups) in local, regional, and national decision-making; (3) environmental development that respects the spirit of the environment and at a minimum does not undermine the future of our global community, and (4) ethical development (the extent to which individuals, groups, and the national government believe in fair treatment for all—a country's own citizens as well as citizens of other nations—and act accordingly).

If an external party (e.g., another country, international corporation, or international funding source) is involved in a development initiative in a country, it is particularly important to underscore the basic premise of fair treatment to all affected constituencies. An environmental justice approach includes all those who will be affected (e.g., those living in a neighboring country, those who are not yet born) whether or not a group is a signatory to a development agreement.

There is a tragic reluctance to solve and prevent the environmental problems that are facing minority and low-income communities as well as low-income countries. It is very important that governments initiate and publicly support environmental justice frameworks and activities to solve environmental problems and prevent future ones. The frameworks help us to not lose sight of environmental justice issues.

However, many groups are frustrated and angry that they have not been able to effectively solve the long-term environmental problems facing their communities. Even countries that have environmental justice frameworks need to get beyond just recognizing and discussing environmental problems. Governments must put remedies and preventive strategies in place now that are just, effective, timely, and really protect our fragile global future.

NO ⬅

David Friedman

The "Environmental Racism" Hoax

When the U.S. Environmental Protection Agency (EPA) unveiled its heavily criticized environmental justice "guidance" earlier this year, it crowned years of maneuvering to redress an "outrage" that doesn't exist. The agency claims that state and local policies deliberately cluster hazardous economic activities in politically powerless "communities of color." The reality is that the EPA, by exploiting every possible legal ambiguity, skillfully limiting debate, and ignoring even its own science, has enshrined some of the worst excesses of racialist rhetoric and environmental advocacy into federal law.

"Environmental justice" entered the activist playbook after a failed 1982 effort to block a hazardous-waste landfill in a predominantly black North Carolina county. One of the protesters was the District of Columbia's congressional representative, who returned to Washington and prodded the General Accounting Office (GAO) to investigate whether noxious environmental risks were disproportionately sited in minority communities.

A year later, the GAO said that they were. Superfund and similar toxic dumps, it appeared, were disproportionately located in non-white neighborhoods. The well-heeled, overwhelmingly white environmentalist lobby christened this alleged phenomenon "environmental racism," and ethnic advocates like Ben Chavis and Robert Bullard built a grievance over the next decade.

Few of the relevant studies were peer-reviewed; all made critical errors. Properly analyzed, the data revealed that waste sites are just as likely to be located in white neighborhoods, or in areas where minorities moved only after permits were granted. Despite sensational charges of racial "genocide" in industrial districts and ghastly "cancer alleys," health data don't show minorities being poisoned by toxic sites. "Though activists have a hard time accepting it," notes Brookings fellow Christopher H. Foreman, Jr., a self-described black liberal Democrat, "racism simply doesn't appear to be a significant factor in our national environmental decision-making."

⚜

This reality, and the fact that the most ethnically diverse urban regions were desperately trying to *attract* employers, not sue them, constrained the environmental racism movement for a while. In 1992, a Democrat-controlled Congress

From David Friedman, "The 'Environmental Racism' Hoax," *The American Enterprise*, vol. 9, no. 6 (November/December 1998). Copyright © 1998 by *The American Enterprise*. Reprinted by permission of *The American Enterprise*, a national magazine of politics, business, and culture.

ignored environmental justice legislation introduced by then-Senator Al Gore. Toxic racism made headlines, but not policy.

All of that changed with the Clinton-Gore victory. Vice President Gore got his former staffer Carol Browner appointed head of the EPA and brought Chavis, Bullard, and other activists into the transition government. The administration touted environmental justice as one of the symbols of its new approach.

Even so, it faced enormous political and legal hurdles. Legislative options, never promising in the first place, evaporated with the 1994 Republican takeover in Congress. Supreme Court decisions did not favor the movement.

So the Clinton administration decided to bypass the legislative and judicial branches entirely. In 1994, it issued an executive order—ironically cast as part of Gore's "reinventing government" initiative to streamline bureaucracy—which directed that every federal agency "make achieving environmental justice part of its mission."

At the same time, executive branch lawyers generated a spate of legal memoranda that ingeniously used a poorly defined section of the Civil Rights Act of 1964 as authority for environmental justice programs. Badly split, confusing Supreme Court decisions seemed to construe the 1964 Act's "nondiscrimination" clause (prohibiting federal funds for states that discriminate racially) in such a way as to allow federal intervention wherever a state policy ended up having "disparate effects" on different ethnic groups.

Even better for the activists, the Civil Rights Act was said to authorize private civil rights lawsuits against state and local officials on the basis of disparate impacts. This was a valuable tool for environmental and race activists, who are experienced at using litigation to achieve their ends.

Its legal game plan in place, the EPA then convened an advocate-laden National Environmental Justice Advisory Council (NEJAC), and seeded activist groups (to the tune of $3 million in 1995 alone) to promote its policies. Its efforts paid off. From 1993, the agency backlogged over 50 complaints, and environmental justice rhetoric seeped into state and federal land-use decisions.

⋅✿⋅

Congress, industry, and state and local officials were largely unaware of these developments because, as subsequent news reports and congressional hearings established, they were deliberately excluded from much of the agency's planning process. Contrary perspectives, including EPA-commissioned studies highly critical of the research cited by the agency to justify its environmental justice initiative in the first place; were ignored or suppressed.

The EPA began to address a wider audience in September 1997. It issued an "interim final guidance" (bureaucratese for regulation-like rules that agencies can claim are not "final" so as to avoid legal challenge) which mandated that environmental justice be incorporated into all projects that file federal environmental impact statements. The guidance directed that applicants pay particular attention to potential "disparate impacts" in areas where minorities live in "meaningfully greater" numbers than surrounding regions.

The new rules provoked surprisingly little comment. Many just "saw the guidance as creating yet another section to add to an impact statement," explains Jennifer Hernandez, a San Francisco environmental attorney. In response, companies wanting to build new plants had to start "negotiating with community advocates and federal agencies, offering new computers, job training, school or library improvements, and the like" to grease their projects through.

In December 1997, the Third Circuit Court of Appeals handed the EPA a breathtaking legal victory. It overturned a lower court decision against a group of activists who sued the state of Pennsylvania for granting industrial permits in a town called Chester, and in doing so the appeals court affirmed the EPA's extension of Civil Rights Act enforcement mechanisms to environmental issues.

(When Pennsylvania later appealed, and the Supreme Court agreed to hear the case, the activists suddenly argued the matter was moot, in order to avoid the Supreme Court's handing down an adverse precedent. This August, the Court agreed, but sent the case back to the Third Circuit with orders to dismiss the ruling. While activists may have dodged a decisive legal bullet, they also wiped from the books the only legal precedent squarely in their favor.)

Two months after the Third Circuit's decision, the EPA issued a second "interim guidance" detailing, for the first time, the formal procedures to be used in environmental justice complaints. To the horror of urban development, business, labor, state, local, and even academic observers, the guidance allows the federal agency to intervene at any time up to six months (subject to extension) after any land-use or environmental permit is issued, modified, or renewed anywhere in the United States. All that's required is a simple allegation that the permit in question was "an act of intentional discrimination or has the effect of discriminating on the basis of race, creed, or national origin."

The EPA will investigate such claims by considering "multiple, cumulative, and synergistic risks." In other words, an individual or company might not itself be in violation, but if, combined with previous (also legal) land-use decisions, the "cumulative impact" on a minority community is "disparate," this could suddenly constitute a federal civil rights offense. The guidance leaves important concepts like "community" and "disparate impact" undefined, leaving them to "case by case" determination. "Mitigations" to appease critics will likewise be negotiated with the EPA case by case.

This "guidance" subjects virtually any state or local land-use decision—made by duly elected or appointed officials scrupulously following validly enacted laws and regulations—to limitless ad hoc federal review, any time there is the barest allegation of racial grievance. Marrying the most capricious elements of wetlands, endangered species, and similar environmental regulations with the interest-group extortion that so profoundly mars urban ethnic politics, the guidance transforms the EPA into the nation's supreme land-use regulator.

❧

Reaction to the Clinton administration's gambit was swift. A coalition of groups usually receptive to federal interventions, including the U.S. Conference of Mayors, the National Association of Counties, and the National Association of Black County Officials, demanded that the EPA withdraw the guidance. The House amended an appropriations bill to cut off environmental justice enforcement until the guidance was revised. This August, EPA officials were grilled in congressional hearings led by Democratic stalwarts like Michigan's John Dingell.

Of greatest concern is the likelihood the guidance will dramatically increase already-crippling regulatory uncertainties in urban areas where ethnic populations predominate. Rather than risk endless delay and EPA-brokered activist shakedowns, businesses will tacitly "redline" minority communities and shift operations to white, politically conservative, less-developed locations.

Stunningly, this possibility doesn't bother the EPA and its environmentalist allies. "I've heard senior agency officials just dismiss the possibility that their policies might adversely affect urban development," says lawyer Hernandez. Dingell, a champion of Michigan's industrial revival, was stunned when Ann Goode, the EPA's civil rights director, said her agency never considered the guidance's adverse economic and social effects. "As director of the Office of Civil Rights," she lectured House lawmakers, "local economic development is not something I can help with."

Perhaps it should be. Since 1980, the economies of America's major urban regions, including Cleveland, Chicago, Milwaukee, Detroit, Pittsburgh, New Orleans, San Francisco, Newark, Los Angeles, New York City, Baltimore, and Philadelphia, grew at only one-third the rate of the overall American economy. As the economies of the nation's older cities slumped, 11 million new jobs were created in whiter areas.

Pushing away good industrial jobs hurts the pocketbook of urban minorities, and, ironically, harms their health in the process. In a 1991 *Health Physics* article, University of Pittsburgh physicist Bernard L. Cohen extensively analyzed mortality data and found that while hazardous waste and air pollution exposure takes from three to 40 days off a lifespan, poverty reduces a person's life expectancy by an average of 10 *years*. Separating minorities from industrial plants is thus not only bad economics, but bad health and welfare policy as well.

⋅◉⋅

Such realities matter little to environmental justice advocates, who are really more interested in radical politics than improving lives. "Most Americans would be horrified if they saw NEJAC [the EPA's environmental justice advisory council] in action," says Brookings's Foreman, who recalls a council meeting derailed by two Native Americans seeking freedom for an Indian activist incarcerated for killing two FBI officers. "Because the movement's main thrust is toward... 'empowerment'..., scientific findings that blunt or conflict with that goal are ignored or ridiculed."

Yet it's far from clear that the Clinton administration's environmental justice genie can be put back in the bottle. Though the Supreme Court's dismissal of the Chester case eliminated much of the EPA's legal argument for the new rules, it's likely that more lawsuits and bureaucratic rulemaking will keep the program alive. The success of the environmental justice movement over the last six years shows just how much a handful of ideological, motivated bureaucrats and their activist allies can achieve in contemporary America unfettered by fact, consequence, or accountability, if they've got a President on their side.

POSTSCRIPT

Should Environmental Policy Attempt to Cure Environmental Racism?

$\mathbf{F}$ritz and Friedman present strikingly contrary assessments of the environmental justice movement and the justifications for and implications of charges of environmental racism. It is not uncommon for complex, emotion-laden social issues to lead to analyses by investigators that present extremely different perspectives and diametrically opposite conclusions. As in this case, the cause of this potentially confusing situation reflects sharp differences in the social values and political commitments of the analysts. In such cases it is particularly important for the reader to pay careful attention to the data presented in support of the conclusions that are drawn. What unsupported assumptions do the authors make? Are simplistic explanations offered for phenomena that are likely to have much more complicated causes?

A student of the environmental justice movement would do well to read some of the many articles and books written or edited by Robert D. Bullard. A good starting point is his book *Unequal Protection: Environmental Justice and Communities of Color* (Sierra Club Books, 1994). In "Waste Management and Risk Assessment: Environmental Discrimination Through Regulation," *Urban Geography* (vol. 17, no. 5, 1996), Michael Heiman describes aspects of U.S. regulatory policy that may contribute to environmental injustice and proposals by local activists for reform. In denying that environmental racism exists, other critics of the environmental justice movement often refer to recent studies that have produced results contrary to those relied on by the EPA and the leaders of the environmental justice movement. However, other recent studies have confirmed the earlier evidence of environmental discrimination. A detailed analysis by Philip H. Pollock and M. Elliot Vittas in *Social Science Quarterly* (vol. 76, no. 2, 1995), pp. 294–309, confirms that racial and ethnic subpopulations reside closer to toxic sources than other poor people. The results of a study by James T. Hamilton, published in the *Journal of Policy Analysis and Management* (vol. 14, no. 1, 1995), pp. 107–132, indicate that the key factor in siting hazardous facilities in minority communities is a low perceived probability that collective citizen opposition will raise a firm's expected location costs.

ISSUE 7

Is Limiting Population Growth a Key Factor in Protecting the Global Environment?

YES: Paul Harrison, from "Sex and the Single Planet: Need, Greed, and Earthly Limits," *The Amicus Journal* (Winter 1994)

NO: Betsy Hartmann, from "Population Fictions: The Malthusians Are Back in Town," *Dollars and Sense* (September/October 1994)

ISSUE SUMMARY

YES: Author and Population Institute medal winner Paul Harrison argues for family planning programs that take into account women's rights and socioeconomic concerns in order to prevent world population from exceeding carrying capacity.

NO: Betsy Hartmann, director of the Hampshire College Population and Development Program, counters that the "real problem is not human *numbers* but undemocratic human *systems* of labor and resource exploitation, often backed by military repression."

$\mathbf{T}$ he debate about whether human population growth is a fundamental cause of ecological problems and whether population control should be a central strategy in protecting the environment has long historical roots.

Those who are seriously concerned about uncontrolled human population growth are often referred to as "Malthusians" after the English parson Thomas Malthus, whose "Essay on the Principle of Population" was first published in 1798. Malthus warned that the human race was doomed because geometric population increases would inexorably outstrip productive capacity, leading to famine and poverty. His predictions were undermined by technological improvements in agriculture and the widespread use of birth control (rejected by Malthus on moral grounds), which brought the rate of population growth in industrialized countries under control during the twentieth century.

The theory of the demographic transition was developed to explain why Malthus's dire predictions had not come true. This theory proposes that the first effect of economic development is to lower death rates. This causes a population boom, but stability is again achieved as economic and social changes lead to lower birth rates. This pattern has indeed been followed in Europe, the United States, Canada, and Japan. The less-developed countries have more recently experienced rapidly falling death rates. Thus far, the economic and social changes needed to bring down birth rates have not occurred, and many countries in Asia and Latin America suffer from exponential population growth. This fact has given rise to a group of neo-Malthusian theorists who contend that it is unlikely that less-developed countries will undergo the transition to lower birth rates required to avoid catastrophe due to overpopulation.

Biologist Paul Ehrlich's best-seller *The Population Bomb* (Ballantine Books, 1968) popularized his view that population growth in both the developed and developing world must be halted to avert worldwide ecological disaster. Ecologist Garrett Hardin extended the neo-Malthusian argument by proposing that some less-developed nations have gone so far down the road of population-induced resource scarcity that they are beyond salvation and should be allowed to perish rather than possibly sink the remaining world economies.

Barry Commoner, a prominent early critic of the neo-Malthusian perspective, argues in *The Closing Circle* (Alfred A. Knopf, 1971) and his subsequent popular books and articles that inappropriate technology is the principal cause of local and global environmental degradation. While not denying that population growth is a contributing factor, he favors promoting ecologically sound development rather than population-control strategies that ignore socioeconomic realities.

Enthusiasts for population control as a sociopolitical and environmental strategy have always been opposed by religious leaders whose creeds reject any overt means of birth control. Recently, the traditional population control policy planners have also been confronted with charges of sexism and paternalism by women's groups, minority groups, and representatives of developing nations who argue that the needs and interests of their constituencies have been ignored by the primarily white, male policy planners of the developed world. At the September 1994 World Population Conference in Cairo, organizers and spokespeople for these interests succeeded in promoting policy statements that reflected sensitivity to many of their concerns.

In the following selections, Paul Harrison argues that "population growth combined with... consumption and technology damages the environment." He proposes "quality family planning and reproductive health services, mother and child health care, women's rights and women's education" as a four-point program to rapidly decrease population growth. Betsy Hartmann asserts that "the threat to livelihoods, democracy and the environment posed by the fertility of poor women hardly compares to that posed by the consumption patterns of the rich or the ravages of militaries." She proposes greater democratic control over resources rather than narrow population control as an environmental strategy.

Paul Harrison  **YES**

Sex and the Single Planet:
Need, Greed, and Earthly Limits

Population touches on sex, gender, parenthood, religion, politics—all the deepest aspects of our humanity. Start a debate on the topic, and the temperature quickly warms up. In the preparations for next year's World Population Conference in Cairo, the link between population growth and environmental damage is one of the hottest topics.

The sheer numbers involved today make it hard to ignore the link. The last forty years saw the fastest rise in human numbers in all previous history, from only 2.5 billion people in 1950 to 5.6 billion today. This same period saw natural habitats shrinking and species dying at an accelerating rate. The ozone hole appeared, and the threat of global warming emerged.

Worse is in store. Each year in the 1980s saw an extra 85 million people on earth. The second half of the 1990s will add an additional 94 million people per year. That is equivalent to a new United States every thirty-three months, another Britain every seven months, a Washington every six days. A whole earth of 1800 was added in just one decade, according to United Nations Population Division statistics. After 2000, annual additions will slow, but by 2050 the United Nations expects the human race to total just over 10 billion—an extra earth of 1980 on top of today's, according to U.N. projections.

If population growth does not cause or aggravate environmental problems, as many feminists, socialists, and economists claim, then we do not need to worry about these numbers. If it does, then the problems of the last decade may be only a foretaste of what is to come.

At the local level, links between growing population densities and land degradation are becoming clearer in some cases. Take the case of Madagascar. Madagascar's forests have been reduced to a narrowing strip along the eastern escarpment. Of the original forest cover of 27.6 million acres, only 18.8 million acres remained in 1950. Today this has been halved to 9.4 million acres—which means that habitat for the island's unique wildlife has been halved in just forty years. Every year some 3 percent of the remaining forest is cleared, almost all of that to provide land for populations expanding at 3.2 percent a year.

From Paul Harrison, "Sex and the Single Planet: Need, Greed, and Earthly Limits," *The Amicus Journal* (Winter 1994). Copyright © 1994 by *The Amicus Journal*, a quarterly publication of The Natural Resources Defense Council, 40 West 20th Street, New York, NY 10011. Reprinted by permission. NRDC membership dues or nonmember subscriptions: $10 annually.

The story of one village, Ambodiaviavy, near Ranomafana, shows the process at work. Fifty years ago, the whole area was dense forest. Eight families, thirty-two people in all, came here in 1947, after French colonials burned down their old village. At first they farmed only the valley bottoms, which they easily irrigated from the stream running down from the hilltops. There was no shortage of land. Each family took as much as they were capable of working. During the course of the next forty-three years, the village population swelled ten times over, to 320, and the number of families grew to thirty-six. Natural growth was supplemented by immigration from the overcrowded plateaus, where all cultivable land is occupied. By the 1950s, the valley bottom lands had filled up completely. New couples started to clear forest on the sloping valley sides. They moved gradually uphill; today, they are two-thirds of the way to the hilltops.

Villager Zafindraibe's small paddy field feeds his family of five for only four months of the year. In 1990 he felled and burned five acres of steep forest land to plant hill rice. The next year cassava would take over. After that the plot should be left fallow for at least six or seven years.

Now population growth is forcing farmers to cut back the fallow cycle. As land shortage increases, a growing number of families can no longer afford to leave the hillsides fallow long enough to restore their fertility. They return more and more often. Each year it is cultivated, the hillside plot loses more topsoil, organic matter, nutrients.

⋯⊙⋯

The debate over this link between population growth and the environment has raged back and forth since 1798. In that year Thomas Malthus, in his notorious *Essay on Population,* suggested that population tended to grow faster than the food supply. Human numbers would always be checked by famine and mortality.

Socialists from William Cobbett to Karl Marx attacked Malthus's arguments. U.S. land reformer Henry George, in *Progress and Poverty* (1879), argued that the huge U.S. population growth had surged side by side with huge increases in wealth. Poverty, said George, was caused not by overpopulation, but by warfare and unjust laws. Poverty caused population growth, not the other way around.

In modern times, U.S. ecologist Paul Ehrlich has played the Malthus role. "No geological event in a billion years has posed a threat to terrestrial life comparable to that of human overpopulation," he argued back in 1970, urging compulsion if voluntary methods failed. His early extremism (such as suggesting cutting off aid to certain Third World countries) has mellowed into a more balanced analysis (for example, he acknowledges the need for more than just contraceptives to attack the problem). But doomsday rhetoric remains in his 1990 book, *The Population Explosion,* which predicts "many hundreds of millions" of famine deaths if we do not halt human population growth.

Today's anti-Malthusians come in all shades, from far left to far right. For radical writers Susan George and Frances Moore Lappé, poverty and inequality are the root causes of environmental degradation, not population. For Barry

Commoner the chief threat is misguided technology. Economist Julian Simon sees moderate population growth as no problem at all, but as a tonic for economic growth. More people mean more brains to think up more solutions. "There is no meaningful limit to our capacity to keep growing forever," he wrote in 1981 in *The Ultimate Resource.*

Other voices in the debate focus on ethics and human rights. Orthodox Catholics and fundamentalist Muslims oppose artificial contraception or abortion. A wide range of feminists stress women's rights to choose or refuse and downplay the impact of population growth. "Blaming global environmental degradation on population growth," argued the Global Committee on Women, Population and the Environment before Rio, "stimulates an atmosphere of crisis. It helps lay the groundwork for an intensification of top-down population control programs that are deeply disrespectful of women."

There is no debate quite like this one for sound and fury. As the forgoing examples show, positions are emotional and polarized. Factions pick on one or two elements as the basic problem, and ignore all the others. Thinking proceeds in black-and-white slogans.

Often debaters seem to be locked into the single question: Is population growth a crucial factor in environmental degradation—or not? However, if we frame our inquiry in this simplistic way, only two answers are possible—yes or no—and only two conclusions—obsession with family planning, or opposition to family planning. Both of these positions lead to abuse or neglect of women's rights.

There has to be a way out of this blind alley. Perhaps we can make a start by accepting that *all* the factors mentioned by the rival schools are important. All interact to create the damage. Sometimes one factor is dominant, sometimes another. Population is always there. In some fields it plays the lead role, in others no more than a bit part.

Most observers agree that it is not just population growth that damages the environment. The amount each person consumes matters too, and so does the technology used in production and waste disposal. These three factors work inseparably in every type of damage. Each of them is affected by many other factors, from the status of women to the ownership of land, from the level of democracy to the efficiency of the market. If we adopt this complex, nuanced view, much of the crazy controversy evaporates, and the hard work of measuring impact and designing policy begins.

A number of success stories have emerged. One hallmark of these successes is the recognition that population should be an integral part of long-range resource management.

⊷⊙⊷

Take a snapshot at one particular moment, and there is no way of saying which of the three factors carries the main blame for damage. It would be like asking whether brain, bone, or muscle plays the main role in walking. But if we compare changes over time, we can get an idea of their relative strengths. Results vary a lot, depending on which country or which type of damage we are looking at.

In Madagascar, population growth bears the main blame for deforestation and loss of biodiversity. As described before, the island's rain forests have shrunk to a narrow strip. Increased consumption—a rise in living standards—and technology tend to play less and less of a role in this devastation. Incomes and food intake today are lower than thirty years ago. Farming methods have not changed in centuries.

Population growth is running at 3 percent a year. When technology is stagnant, every extra human means less forest and wildlife.

By contrast, population growth played only a minor role in creating the ozone hole. The main blame lay with rising consumption and technology change. Between 1940 and 1980, world chlorofluorocarbon (CFC) emissions grew at more than 15 percent a year. Almost all of this was in developed countries, where population grew at less than 1 percent a year. So population growth accounted for less than 7 percent (one-fifteenth) of increased CFC emissions.

A central issue in the controversy is whether we are on course to pass the earth's carrying capacity—the maximum population that the environment can support indefinitely. Malthusians like Dennis and Donnella Meadows suggest in their book *Beyond the Limits* that we have already passed the limits in some areas such as alteration of the atmosphere. Anti-Malthusians like Julian Simon insist that we can go on raising the limits through technology.

Here, too, a compromise comes closer to reality. Humans *have* raised the ceiling on growth many times in the past. When hunter-gatherers ran short of wild foods, they turned to farming. When western Europeans started to run out of wood in the seventeenth century, they turned to coal. The process continues today. When one resource runs down, its price changes, and we increase productivity or exploration, bring in substitutes, or reduce use. In other words, we do not just stand by and watch helplessly while the world collapses. We respond and adapt. We change our technology, our consumption patterns, even the number of children we have. It is because we can adapt so fast that we are the dominant species on earth.

So far adaptation has kept us well stocked with minerals despite rising use. It has proved Malthus wrong by raising food production roughly in line with the five-and-a-half-fold growth in population since his time. But it has not worked at all well in maintaining stocks of natural resources like forests, water, sea fish, or biodiversity, nor with preserving the health of sinks for liquid and gaseous wastes such as lakes, oceans, and atmosphere. These are common property resources—no one owns them—so what Garrett Hardin called the "tragedy of the commons" applies. Everyone overuses or abuses the source or sink, fearing that if they hold back others will reap the gains.

Problems like erosion, acid rain, or global warming are not easy to diagnose or cure. Sometimes we do not even know they are happening until they are far advanced, as in the case of the ozone hole. Like cancer, they build up slowly and often pass unseen till things come to a head. Farmers in Burkina Faso did not believe their land was eroding away until someone left a ruler stuck into the soil; then they saw that the level had gone down an inch in a year.

Environmental quality follows a U-shaped curve. Things get worse before they get better, on everything from biodiversity and soil erosion to air and water quality. But everything hinges on how long the downswing lasts—and how serious or irreversible are the problems it gives rise to. Given time we will develop institutions to control overfishing or ocean pollution, stop acid rain or halt global warming. But time is the crux of the matter. Adaptable though we are, we rarely act in time to prevent severe damage. In one area after another, from whales to ozone holes, we have let crises happen before taking action.

Over the next few decades we face the risk of irreversible damage on several fronts. If we lose 10 or 20 percent of species, we may never restore that diversity. If the global climate flips, then all our ability to adapt will not stave off disaster. Rather than wait for global crisis, prudence dictates that we should take action now.

However, the way we look at causes deeply affects the search for solutions. That is why the debate on population and environment matters. If we say that damage results only from technology, only from overconsumption, only from injustice, or only from population, we will act on only one element of the equation. But damage results from population, consumption, and technology multiplied together, so we must act on all three. And we cannot neglect the many factors from inequality to women's rights and free markets that influence all others.

⋅⊙⋅

Consumption will be the hardest nut to crack. Reducing overconsumption may be good for the soul, but the world's poorest billion must *increase* their consumption to escape poverty. The middle 3 billion will not willingly rein in their ambitions. The middle classes in India and China are already launched on the consumer road that Europe took in the 1950s. They are moving faster down that road, and their consumer class probably outnumbers North America's already. Even in the rich countries, consumption goes on growing at roughly 2 percent a year, with hiccups during recession. Consumption can be cut if consumers and producers have to pay for the damage they do through higher prices or taxes—but, politically, it is not easy. Politicians who threaten to raise taxes risk electoral defeat.

So technological change must reduce the *impact* of consumption. But it will be a Herculean task for technology to do the job alone. The massive oil price rises of 1973 and 1979–80 stimulated big advances in energy efficiency. Between 1973 and 1988 gasoline consumption per mile in western countries fell by 29 percent. But this technology gain was wiped out by a rise in car numbers of 58 percent, due to the combined growth of population and consumption. The result was a rise in gasoline consumption of 17 percent.

Population and consumption will go on raising the hurdles that technology must leap. By 2050, world population will have grown by 80 percent, on the U.N. medium projection. Even at the low 1980s growth rate of 1.2 percent a year, consumption per person will have doubled. Technology would have to cut the damage done by each unit of consumption by 72 percent, just to keep total damage rising at today's destructive rate.

Yet the International Panel on Climate Change says we ought to *cut* carbon dioxide output by 60 percent from today's levels. If incomes and population grow as above, technology would have to cut the emissions for each unit of consumption by a massive 89 percent by 2050. This would require a 3.8 percent reduction every year for fifty-seven years.

Such a cut is not utterly impossible, but it would demand massive commitment on all sides. Introducing the 85 miles-per-gallon car could deliver a cut of almost exactly this size in the transport sector, if it took ten years to go into mass production, and another fifteen years to saturate the market. But the combined growth of population and car ownership could easily halve the gain.

Technology change will have a far easier job if it is backed by action on the population front. Population efforts are slow-acting at first: for the first fifteen years the difference is slight. The U.N.'s low population projection points to what might be achieved if all countries did their best in bringing birth rates down. Yet for 2010, the low projection for world population is only 1.2 percent less than the medium projection. Over the longer term, though, there are big benefits. By 2025 the low projection is 7.3 percent less than the medium—621 million fewer people, or a whole Europe plus Japan. By 2050 the low figure is 22 percent or 2.206 billion people less—equal to the whole world's population around 1930.

With a concerted effort in all countries (including the United States), world population could peak at 8.5 billion or less in 2050 and, after that, come down. And it is clear that this would reduce environmental impact and lower the hurdles that improved technology will have to leap.

What do we need to do to bring it about? Here, too, the debate rages. Diehard Malthusians talk of the need for crash programs of "population control." Horrified feminists answer that a woman's fertility is her own business, not a target for male policy measures. The objective should be reproductive health and choice, not simply bringing numbers down, they argue.

Yet this conflict, too, is an artificial one. The best way to bring numbers down fast is to pump resources not into crash or compulsory programs narrowly focused on family planning, but into broad women's development programs that most feminists would welcome. How do we get enough resources out of male governments to do this properly? Only by using the arguments about environment and economy that feminists do not allow.

Coercion and crash programs defeat their own aims. "Population control" is impossible without killing people: the term implies coercion and should be dropped forthwith. Coercion rouses protests that sooner or later bring it to an end. India's brief and brutal experiment with forced sterilization in 1975–76 led within a year to the fall of Indira Gandhi's government. The progress of family planning in India was set back a decade.

Mass saturation with just one or two family planning methods is equally doomed to failure. With female contraceptives, side-effects are common: women need good advice and medical backup to deal with them or avoid them. Left to handle them alone, they will stop using contraceptives and go on having five children each. Once mistrust has been aroused, it will make the job harder even when better programs are finally brought in.

If we want to bring population growth rates down rapidly, we must learn from the real success stories like Thailand. In the early 1960s, the average Thai woman was having 6.4 children. Today she is having only 2.2. This represents a drop of 3.5 percent per year—as speedy as the fastest change in technology.

Such success was achieved, without a whiff of coercion, by universal access to a wide and free choice of family planning methods, with good-quality advice and medical backup. Mother and child health was improved, women's rights were advanced, and female education leveled up with male.

All these measures are worthwhile in their own right. They improve the quality of life for women and men alike. And there are economic spin-offs. Thai incomes grew at 6 percent a year in the 1980s. A healthy and educated work force attracts foreign investment and can compete in the modern high-tech world.

Quality family planning and reproductive health services, mother and child health, women's rights, and women's education—this four-point program is the best way to achieve a rapid slowdown in population growth. It can improve the quality of life directly, through health and education benefits, and it improves the status of women. It creates a healthy and educated work force. It gives people the knowledge with which they can fight for their own rights. It might also help to raise incomes, and it will certainly help to slow environmental damage.

With its human, economic, and environmental benefits, there are few programs that will offer better value for money over the coming decades.

NO ↵

Betsy Hartmann

Population Fictions: The Malthusians Are Back in Town

In the corridors of power, the tailors are back at work, stitching yet another invisible robe to fool the emperor and the people. After 12 years in which the Reagan and Bush administrations downplayed population control as a major aim of U.S. foreign policy, the Clinton administration is playing catch-up. World attention will focus on the issue this month in Cairo, when leaders from the United States and abroad gather at the United Nations' third International Conference on Population and Development. Cloaked in the rhetoric of environmentalism and—ironically—women's rights, population control is back in vogue.

At the UN's second International Conference on Population in Mexico City in 1984, the Reagan administration asserted that rapid population growth is a "neutral phenomenon" that becomes a problem only when the free market is subverted by "too much governmental control of economies." Under the Republicans, the U.S. withdrew funding from any international family planning agencies that perform abortions or even counsel women about them. Aid was cut off to the International Planned Parenthood Foundation as well as the UN Fund for Population Activities (UNFPA).

The Clinton administration, by contrast, is requesting $585 million for population programs in fiscal year 1995, up from $502 million the year before. This aid is channelled through the U.S. Agency for International Development (USAID), which has made population control a central element of its new "Sustainable Development" mission for the post Cold War era. The USAID's draft strategy paper of October 1993 identifies rapid population growth as a key "strategic threat" which "consumes all other economic gains, drives environmental damage, exacerbates poverty, and impedes "democratic governance."

Clinton's more liberal stand on abortion is certainly welcome, but even that has yet to translate into effective Congressional action or foreign policy. Announced in April, USAID's new policy on abortion funding overseas is still very restrictive: It will finance abortion only in cases of rape, incest, and life endangerment, the same conditions the Hyde amendment puts on federal Medicaid funds. Along with the mainstream environmental movement, the administration pays lip service to women's rights but continues to back practices

From Betsy Hartmann, "Population Fictions: The Malthusians Are Back in Town," *Dollars and Sense* (September/October 1994). Copyright © 1994 by Economic Affairs Bureau, Inc. Reprinted by permission. *Dollars and Sense* is a progressive economics magazine published six times a year. First-year subscriptions cost $18.95 and may be ordered by writing to *Dollars and Sense,* 1 Summer Street, Somerville, MA 02143.

—such as promoting long-acting contraceptive methods like Norplant without follow-up medical care—that are actually harmful to women's health.

Population Myths

It is true that population growth (which is actually slowing in most areas of the world) can put additional pressure on resources in specific regions. But the threat to livelihoods, democracy and the global environment posed by the fertility of poor women is hardly comparable to that posed by the consumption patterns of the rich or the ravages of militaries.

The industrialized nations, home to 22% of the world's population, consume 60% of the world's food, 70% of its energy, 75% of its metals, and 85% of its wood. They generate almost three-quarters of all carbon dioxide emissions, which in turn comprise nearly half of the manmade greenhouse gases in the atmosphere, and are responsible for most of the ozone depletion. Militaries are the other big offenders. The German Research Institute for Peace Policy estimates that one-fifth of all global environmental degradation is due to military activities. The U.S. military is the largest domestic oil consumer and generates more toxic waste than the five largest multinational chemical companies combined.

What about the environmental degradation that occurs within developing countries? The UNFPA's *State of World Population 1992* boldly claims that population growth "is responsible for around 79% of deforestation, 72% of arable land expansion, and 69% of growth in livestock numbers." Elsewhere it maintains that the "bottom billion," the very poorest people in developing countries, "often impose greater environmental injury than the other 3 billion of their citizens put together."

Blaming such a large proportion of environmental degradation on the world's poorest people is untenable, scientifically and ethically. It is no secret that in Latin America the extension of cattle ranching—mainly for export, not domestic consumption—has been the primary impetus behind deforestation. And it is rich people who own the ranches, not the poor, as most countries in Latin America have a highly inequitable distribution of land. In Southeast Asia the main culprit is commercial logging, again mainly for export.

In developing countries, according to USAID, rapid population growth also "renders inadequate or obsolete any investment in schools, housing, food production capacity and infrastructure." But are increasing numbers of poor people really the main drain on national budgets? The UN's 1993 *Human Development Report* estimates that developing countries spend only one-tenth of their national budgets on human development priorities. Their military expenditures meanwhile soared from 91% of combined health and education expenditures in 1977 to 169% in 1990. And in any case, the social spending that there is often flows to the rich. A disproportionate share of health budgets frequently goes to expensive hospital services in urban areas rather than to primary care for the poor, and educational resources are often devoted to schools for the sons and daughters of the wealthy.

The "structural adjustment" programs imposed by the World Bank have not helped matters, forcing Third World countries to slash social spending in order to service external debts. The burden of growing inequality has fallen disproportionately on women, children, and minorities who have borne the brunt of structural adjustment policies in reduced access to food, health care and education. But in USAID's view, population growth is at the root of their misery: "As expanding populations demand an even greater number of jobs, a climate is created where workers, especially women and minorities, are oppressed."

A Costly Consensus

In the collective psyche of the national security establishment, population growth is now becoming a great scapegoat and enemy, a substitute for the Evil Empire. A 1992 study by the Carnegie Endowment for International Peace warned that population growth threatens "international stability" and called for "a multilateral effort to drastically expand family planning services." A widely cited February 1993 *Scientific American* article by Thomas Homer-Dixon, Jeffrey Boutwell and George Rathjens identifies rapidly expanding populations as a major factor in growing resource scarcities that are "contributing to violent conflicts in many parts of the developing world."

In the pages of respectable journals, racist metaphors are acceptable again, as the concept of noble savage gives way to post-modern barbarian. In an *Atlantic Monthly* article on the "coming anarchy" caused by population growth and resource depletion, Robert Kaplan likens poor West African children to ants. Their older brothers and fathers (and poor, nonwhite males in general) are "re-primitivized" men who find liberation in violence, since their natural aggression has not been "tranquilized" by the civilizing influences of the Western Enlightenment and middle-class existence.

The scaremongering of security analysts is complemented by the population propaganda of mainstream environmental organizations. U.S. environmentalism has long had a strong neo-Malthusian wing which views Man as the inevitable enemy of Nature. The Sierra Club backed Stanford biologist Paul Ehrlich's 1968 tract *The Population Bomb,* which featured lurid predictions of impending famine and supported compulsory sterilization in India as "coercion in a good cause."

By the late 1980s, population growth had transformed from just one of several preoccupations of the mainstream environmental movement into an intense passion. Groups such as the National Wildlife Federation and the National Audubon Society beefed up their population programs, hoping to attract new membership. Meanwhile, population lobbyists such as the influential Population Crisis Committee (renamed Population Action International) seized on environmental concerns as a new rationale for their existence.

The marriage of convenience between the population and environment establishments led to many joint efforts in advance of the 1992 UN Conference on Environment and Development (UNCED) in Rio de Janeiro. In 1990, Audubon, National Wildlife, Sierra Club, Planned Parenthood Federation of America, and the Population Crisis Committee began a joint Campaign on Population and the Environment. Its major objective was "to expand public awareness of the link between population growth, environmental degradation and the resulting human suffering."

Despite their efforts, the U.S. population/environment lobby had a rude awakening at Rio. In the formal intergovernmental negotiations, many developing nations refused to put population on the UNCED agenda, claiming it would divert attention from the North's responsibility for the environmental crisis. At the same time the nongovernmental Women's Action Agenda 21, endorsed by 1500 women activists from around the world, condemned suggestions that women's fertility rates were to blame for environmental degradation.

In the aftermath of Rio, "the woman question" has forced the population/environment lobby to amend its strategy. Many organizations are emphasizing women's rights in their preparations for the Cairo conference. Women's empowerment—through literacy programs, job opportunities, and access to health care and family planning—is now seen as a prerequisite for the reduction of population growth.

While this is a step forward, the population/environment lobby largely treats the protection of women's rights as a means to population reduction, rather than as a worthy pursuit in itself. Its inclusion—and co-optation—of feminist concerns is part of a larger strategy to create a broad population control "consensus" among the American public. Behind this effort is a small group of powerful actors: the Pew Charitable Trusts Global Stewardship Initiative; the U.S. State Department through the office of Timothy Wirth, Undersecretary for Global Affairs; the UNFPA; and Ted Turner of the powerful Turner Broadcasting System, producer of CNN.

Although the Pew Initiative's "White Paper" lists "population growth and unsustainable patterns of consumption" as its two targets, population growth is by far its main concern. Among Pew's explicit goals are to "forge consensus and to increase public understanding of, and commitment to act on, population and consumption challenges." Its targeted constituencies in the United States are environmental organizations, religious communities, and international affairs and foreign policy specialists.

Pew and the Turner Foundation have sponsored "high visibility" town meetings on population around the country, featuring Ted Turner's wife Jane Fonda, who is also UNFPA's "Goodwill Ambassador." At the Atlanta meeting, covered on Turner's CNN, Fonda attributed the collapse of two ancient Native American communities to overpopulation.

To prepare for the Cairo conference, the Pew Initiative hired three opinion research firms to gauge public understanding of the connections between population, environment and consumption so as to "mobilize Americans" on these issues. The researchers found that the public generally did not feel strongly about population growth or see it as a "personal threat." Their conclusion: An "emotional component" is needed to kindle population fears. Those interviewed complained that they had already been overexposed to "images of stark misery, such as starving children." Although the study notes that these images may in fact "work," it recommends finding "more current, targeted visual devices." One strategy is to build on people's pessimism about the future:

"For women, particularly, relating the problems of excess population growth to children's future offers possibilities."

Sacrificed Rights

Whatever nods the new "consensus" makes towards women's broader rights and needs, family planning is its highest priority. USAID views family planning as "the single most effective means" of reducing population growth; it intends to provide "birth control to every woman in the developing world who wants it by the end of the decade."

The promotion of female contraception as the technical fix for the "population problem" ignores male responsibility for birth-control and undermines the quality of health and family planning services. The overriding objective is to drive down the birth rate as quickly and cheaply as possible, rather than to address people's broader health needs.

In Bangladesh, for example, at least one-third of the health budget is devoted to population control. The principal means is poor-quality female sterilization with incentives for those who undergo the procedure, including cash payments for "wages lost" and transportation costs, as well as a piece of clothing (justified as "surgical apparel"). The World Bank and population specialists are now heralding Bangladesh as a great family planning success story. But at what human cost? Because of the health system's nearly exclusive emphasis on population control, most Bangladeshis have little or no access to primary health care, and infant and maternal death rates remain at tragically high levels.

Lowering the birth rate by itself has hardly solved the country's problems. Poverty in Bangladesh has much more to do with inequitable land ownership and the urban elite's stranglehold over external resources, including foreign aid, than it does with numbers of people. The great irony is that many people in Bangladesh wanted birth control well before the aggressive and often coercive sterilization campaign launched by the government with the help of the World Bank and AID. A truly voluntary family planning program, as part of more comprehensive health services, would have yielded similar demographic results, without deepening human suffering.

The prejudice against basic health care is also reflected in the UN's first draft of the "Program of Action" for the Cairo conference. It asks the international community to spend $10.2 billion on population and family planning by the year 2000, and only $1.2 billion on broader reproductive health services such as maternity care. After pressure from women's groups and more progressive governments, the UN raised this figure to $5 billion, but family planning still has a two-to-one advantage. Meanwhile, the Vatican is attacking women's rights by bracketing for further negotiation any language in the Cairo document which refers to abortion, contraception or sexuality. Women are caught between a rock and a hard place, bracketed by the Vatican, and targeted by the population establishment.

The current focus of population programs is on the introduction of long-acting, provider-dependent contraceptive technologies. The hormonal implant Norplant, for example, which is inserted in a woman's arm, is effective for five years and can only be removed by trained medical personnel. But often, these methods are administered in health systems that are ill-equipped to distribute them safely or ethically; In population programs in Indonesia, Bangladesh and Egypt, researchers have documented many instances of women being denied access to Norplant removal, as well as receiving inadequate counselling, screening, and follow-up care.

A number of new contraceptives in the pipeline pose even more serious problems, in terms of both health risks and the potential for abuse at the hands of zealous population control officials. The non-surgical quinacrine sterilization pellet, which drug specialists suspect may be linked to cancer, can be administered surreptitiously (it was given to Vietnamese women during IUD checks without their knowledge in 1993). Also potentially dangerous are vaccines which immunize women against reproductive hormones. Their long-term reversibility has not yet been tested, and the World Health Organization has expressed some concern about the drugs' interaction with the immune system, especially in people infected with the AIDS virus. Simpler barrier methods, such as condoms and diaphragms, which also protect against sexually transmitted diseases, continue to receive considerably less attention and resources in population programs since they are viewed as less effective in preventing births.

Recently, a network of women formed a caucus on gender issues in order to pressure USAID to live up to its rhetoric about meeting women's broader reproductive health needs. The caucus emerged in the wake of a controversial USAID decision to award a $9 million contract for studying the impact of family planning on women's lives to Family Health International, a North-Carolina-based population agency, rather than to women's organizations with more diverse and critical perspectives.

Progressive environmentalists also intend to monitor USAID's planned initiative to involve Third World environmental groups in building "grass roots awareness around the issue of population and family planning." They fear that USAID funds will be used to steer these groups away from addressing the politically sensitive root causes of environmental degradation—such as land concentration, and corporate logging and ranching—toward a narrow population control agenda.

Trouble at Home

Within the United States, the toughest battle will be challenging the multi-million dollar public opinion "consensus" manufactured by Pew, the State Department, and CNN. Not only does this consensus promote heightened U.S. involvement in population control overseas, but by targeting women's fertility, it helps lay the ground, intentionally or not, for similar domestic efforts.

The Clinton administration is considering whether to endorse state policies that deny additional cash benefits to women who have babies while on welfare. (This despite the fact that women on welfare have only two children on average.) A number of population and environment groups are also fomenting dangerous resentment against immigrant women. The Washington-based

Carrying Capacity Network, for example, states that the United States has every right to impose stricter immigration controls "as increasing numbers of women from Mexico, China and other areas of the world come to the United States for the purpose of giving birth on U.S. soil." And in many circles, Norplant is touted as the wonder drug which will cure the epidemic of crime and poverty allegedly caused by illegitimacy.

Such simple solutions to complex social problems not only don't work, they often breed misogyny and racism, and they prevent positive public action on finding real solutions. Curbing industrial and military pollution, for example, will do far more to solve the environmental crisis than controlling the wombs of poor women who, after all, exert the least pressure on global resources.

The real problem is not human *numbers* but undemocratic human systems of labor and resource exploitation, often backed by military repression. We need to rethink the whole notion of "carrying capacity"—are we really pressing up against the earth's limits because there are too many of us? It would make more sense to talk about "political carrying capacity," defined as the limited capacity of the environment and economy to sustain inequality and injustice. Viewed this way, the solution to environmental degradation and economic decline lies in greater democratic control over resources, not in a narrow population control agenda.

POSTSCRIPT

Is Limiting Population Growth a Key Factor in Protecting the Global Environment?

Harrison extols the virtues of Thailand's population-control program, which he claims has achieved success in significantly reducing birthrates without coercion while promoting women's health care and female education. He implies that this policy has contributed to a growth in average income and the ability to "compete in the modern high-tech world." He does not, however, respond to Hartmann's argument that such policies alone do not ensure a reduction in environmental degradation.

Anyone with a serious interest in environmental issues should certainly read Paul Ehrlich's *The Population Bomb* (Ballantine Books, 1968) and Barry Commoner's *The Closing Circle* (Alfred A. Knopf, 1971). Ehrlich was so distressed by the arguments contained in Commoner's popular book that he coauthored a detailed critique with environmental scientist John P. Holden, which Commoner answered with a lengthy response. These two no-holds-barred pieces were published as a "Dialogue" in the May 1972 issue of the *Bulletin of the Atomic Scientists*. They are interesting reading not only for their technical content but also as a rare example of respected scientists airing their professional and personal antagonisms in public.

Another frequently cited, controversial essay in support of the neo-Malthusian analysis is "The Tragedy of the Commons," by Garrett Hardin, which first appeared in the December 13, 1968, issue of *Science*. For a thorough attempt to justify his authoritarian response to the world population problem, see Hardin's book *Exploring New Ethics for Survival* (Viking Press, 1972).

An economic and political analyst who is concerned about the connections among population growth, resource depletion, and pollution—but who rejects Hardin's proposed solutions—is Lester Brown, director of the Worldwatch Institute. His worldview is detailed in *The Twenty-Ninth Day* (W. W. Norton, 1978). Anthropologist J. Kenneth Smail argues that we have already exceeded the carrying capacity of the planet and need to reduce the world's population in "Beyond Population Stabilization: The Case for Dramatically Reducing Global Human Numbers," *Politics and the Life Sciences* (September 1997).

Anyone willing to entertain the propositions that pollution has not been increasing, natural resources are not becoming scarce, the world food situation is improving, and population growth is actually beneficial might find the late economist Julian Simon's *The Ultimate Resource* (Princeton University Press, 1982) amusing, if not convincing.

For an assessment of needs and strategies to control population growth by several international authorities, including Commoner, see "A Forum: How Big Is the Population Factor?" in the July/August 1990 issue of *EPA Journal*. Ehrlich's present views, which have been somewhat modified in response to criticism by feminists and people from less-developed countries, are presented in an article that he coauthored with Anne Ehrlich and Gretchen Daily in the September/October 1995 issue of *Mother Jones*. A series of articles on the connections among population, development, and environmental degradation are included in the February 1992 issue of *Ambio*.

Harrison's essay is part of a special section entitled "Population, Consumption and Environment" in the Winter 1994 issue of *The Amicus Journal*, which includes other articles focusing on the needs and concerns of people of less-developed countries, along with brief statements representing the views of people from all over the world about the issues that were to be debated at the 1994 Cairo population conference. The Spring 1994 issue of that journal includes an essay by Jodi L. Jacobson that addresses some of the same concerns raised by Hartmann. Distinguished environmentalist Michael Brower addresses the population debate in the Fall 1994 issue of *Nucleus*, the magazine of the Union of Concerned Scientists. A provocative response to the Cairo meeting is the article by Norway's prime minister and sustainable development advocate Gro Harlem Brundtland in the December 1994 issue of *Environment*. Gita Sen, in "The World Programme of Action: A New Paradigm for Population Policy," *Environment* (January/February 1995), describes and analyzes the World Programme of Action, which is the main working document emanating from the Cairo conference. Robin Morgan, in "Dispatch from Beijing," *Ms.* (January/February 1996), reports on the follow-up UN Fourth World Conference on Women, which was held in Beijing in 1995.

In her book *Reproductive Rights and Wrongs: The Global Politics of Population Control*, rev. ed. (South End Press, 1995), Hartmann offers a radical critique of the extent to which the women's rights movement has accepted the politics and rhetoric of what she refers to as the "population establishment."

133

ISSUE 8

Will Pollution Rights Trading Effectively Control Environmental Problems?

YES: Byron Swift, from "A Low-Cost Way to Control Climate Change," *Issues in Science and Technology* (Spring 1998)

NO: Brian Tokar, from "Trading Away the Earth: Pollution Credits and the Perils of 'Free Market Environmentalism'," *Dollars and Sense* (March/April 1996)

ISSUE SUMMARY

YES: Environmental attorney Byron Swift advocates the use of emission trading systems, such as those contained in the Kyoto Protocol, to reduce the emission of greenhouse gases and other pollutants.

NO: Author, college teacher, and environmental activist Brian Tokar maintains that pollution credits and other market-oriented environmental protection policies do nothing to reduce pollution while transferring the power to protect the environment from the public to large corporate polluters.

$\mathbf{F}$ollowing World War II the United States and other developed nations experienced an explosive period of industrialization accompanied by an enormous increase in the use of fossil fuel energy sources and a rapid growth in the manufacture and use of new synthetic chemicals. In response to growing public concern about the pollution and other forms of environmental deterioration resulting from this largely unregulated activity, the U.S. Congress passed the National Environmental Policy Act of 1969. This legislation included a commitment on the part of the government to take an active and aggressive role in protecting the environment. The next year the Environmental Protection Agency (EPA) was established to coordinate and oversee this effort. During the next two decades an unprecedented series of legislative acts and administrative rules were promulgated, placing numerous restrictions on industrial and commercial activities that might result in the pollution, degradation, or contamination of land, air, water, food, and the workplace.

Such forms of regulatory control have always been opposed by the affected industrial corporations and developers as well as by advocates of a free-market policy. More moderate critics of the government's regulatory program recognize that adequate environmental protection will not result from completely voluntary policies. They suggest that a new set of strategies is needed. Arguing that "top down, federal, command and control legislation" is not an appropriate or effective means of preventing ecological degradation, they propose a wide range of alternative tactics, many of which are designed to operate through the economic marketplace. The first significant congressional response to these proposals was the incorporation of tradable pollution emission rights into the 1990 Clean Air Act amendments as a means for achieving the set goals for reducing acid rain-causing sulfur dioxide emissions. More recently, the 1997 international negotiations on controlling global warming in Kyoto, Japan, resulted in a protocol that includes emissions trading as one of the key elements in the plan to limit the atmospheric buildup of greenhouse gases.

Despite past difficulties in obtaining compliance with or enforcing strict statutory pollution limits, the idea of using such market-based strategies as the trading of pollution control credits or the imposition of pollution taxes has won limited acceptance from some major mainstream environmental organizations. Many environmentalists, however, continue to oppose the idea of allowing anyone to pay to pollute, either on moral grounds or because they doubt that these tactics will actually achieve the goal of controlling pollution. Don Munton, in "Dispelling the Myths of the Acid Rain Story," *Environment* (July–August 1998), argues that other control measures, such as switching to low-sulfur fuels, deserve much more of the credit for reducing sulfur dioxide emissions than emission rights trading.

In the following selections, Byron Swift asserts that the "cap-and-trade" feature of the U.S. Acid Rain Program has been very successful. He advocates a similar system for implementing the Kyoto Protocol's emissions trading mandate as a cost-effective means of controlling greenhouse gases. Brian Tokar has a much more negative assessment of sulfur dioxide pollution credit trading. He argues that such "free-market environmentalism" tactics fail to reduce pollution while turning environmental protection into a commodity that corporate powers can manipulate for private profit.

Byron Swift

 **YES**

A Low-Cost Way to Control Climate Change

In December 1997 in Kyoto, Japan, representatives of 159 countries agreed to a protocol to limit the emissions of greenhouse gases. Now comes the hard part: how to achieve the reductions. Emissions trading offers a golden opportunity for a company or country to comply with emissions limits at the lowest possible cost.

Trading allows a company or country that reduces emissions below its preset limit to trade its additional reduction to another company or country whose emissions exceed its limit. It gives companies the flexibility to choose which pollution reduction approach and technology to implement, allowing them to lessen emissions at the least cost. And by harnessing market forces, it leads to innovation and investment. The system encourages swift implementation of the most efficient reductions nationally and internationally; provides economic benefit to those that aggressively reduce emissions; and gives emitters an economically viable way to meet their limits, leading to worldwide efficiency in slowing global warming.

Benefits to the United States from emissions trading would most likely be achieved domestically. However, trading between developed nations and between developed and developing nations has much to offer. It can accelerate investment in developing countries. And it gives developed countries the flexible instruments they say they need to garner the political support necessary to agree to large emissions reductions. In a recent speech in Congress, Sen. Robert Byrd (D-W. Va.) stated that, "reducing projected emissions by a national figure of one-third does not seem plausible without a robust emissions-trading and joint-implementation framework."

If effective trading systems are to be designed, tough political and technical issues will need to be addressed at the Conference of the Parties in Buenos Aires in November 1998—the next big meeting of the nations involved in the Kyoto Protocol. This is especially true for international trading, because different nations have significantly different approaches to reducing greenhouse gases and because many developing countries are opposed to the very notion of

From Byron Swift, "A Low-Cost Way to Control Climate Change," *Issues in Science and Technology* (Spring 1998). Copyright © 1998 by The University of Texas at Dallas, Richardson, TX. Reprinted by permission.

trading. However, if trading systems can be worked out, the United States and the world could meet emissions commitments at the lowest possible cost.

The Challenge

The Kyoto Protocol requires developed countries to reduce greenhouse gas (GHG) emissions to an average of 5 percent below 1990 levels in the years from 2008 to 2012. The United States has agreed to cut emissions by 7 percent below its 1990 level. Russia and other emerging economies have somewhat lesser burdens. However, estimates indicate that at current growth rates, the United States would be almost 30 percent above its 1990 baseline for GHG emissions by 2010. Most emissions come from the combustion of fossil fuels. Carbon dioxide is responsible for 86 percent of U.S. emissions, methane for 10 percent, and other gases for 4 percent. Substantial reductions will be needed.

One strategy would be a tax on the carbon content of fuels, which determines the amount of GHGs emitted when a fuel is burned. Although this may be the most efficient way to reduce GHG emissions, it is politically unrealistic in the United States. Our domestic strategy is more likely to be a choice between a trading system linked with a cap on overall emissions and the more traditional approach of setting emission standards for each sector of the economy.

The strategy in other countries may be different. During the Kyoto debates, a sharp difference was evident between the United States, which favored a trading approach to achieving national emissions targets, and European nations, which are contemplating higher taxes as well as command-and-control strategies such as fuel-efficiency requirements for vehicles and mandated pollution controls for utilities and industry. Nonetheless, all countries can still benefit from international trading.

Why Trading Can Work

An emissions trading system allows emitters with differing costs of pollution reduction to trade pollution allowances or credits among themselves. Through trading, a market price emerges that reflects the marginal costs of emissions reduction. If transaction costs are low, trading leads to overall efficiency in meeting pollution goals, because each source can decide whether it is cheaper to reduce its own emissions or acquire allowances from others.

Trading creates benefits by providing flexibility in technology choices both within and between firms. For example, consider an electric utility that burns coal in its boilers. To comply with its emissions limit, it could add costly scrubbers to its smokestacks or it could buy allowances to tide it over until it is ready to invest in much more efficient capital equipment. The latter option often results in lower or no long-term costs when savings from the new technology and avoidance of the costly quick fix are figured in. It also creates the potential for greater long-term pollution reductions. By not spending money on the quick fix, the utility has more capital to invest in more efficient future processes. This point is critical, because reductions beyond those prescribed in the Kyoto Protocol will be needed in the years after 2010 to stabilize global warming for the rest of the 21st century.

Some political and environmental groups oppose trading, equating it to selling rights to pollute. But this view fails to recognize the substantial differences in business processes and technologies, which may allow one source to reduce emissions much more cheaply than another. It also undervalues the importance of timing in investment decisions; the ability to buy a few years of time through trading may allow companies to install improved equipment or make more significant process changes. Trading leads to the firms with the lowest cost of compliance making the most reductions, creating the most cost-efficient system of meeting pollution goals.

Another concern about trading is that it can create emissions hot spots that result in local health problems. But GHGs have no local effects on human health or ecosystems; they are only problematic at their global concentration levels in the upper atmosphere.

Why a Cap Is Needed

There are two prevailing emissions trading approaches: an emissions cap and allowance system and an open-market system. The cap-and-trade system establishes a hard cap on total emissions, say for a country, and allocates allowances to each emitter that represent its share of the total emissions. Sources could either emit precisely the amount of allowances they are issued, emit fewer tons and sell the difference or store (bank) it for future use, or purchase allowances in order to emit more than their initial allotment. Allowances are freely traded under a private system, much as a stock market operates. A great deal of up-front work must be done to establish baselines for the emitters and to put a trading process in place, but once that work is completed, trades can take place freely between emitters. No regulatory approval is needed. Environmental compliance is ensured because each emitter must have enough allowances to equal its emissions limit each year.

The beauty of the cap-and-trade system is an elegant separation of roles. The government exerts control in setting the cap and monitoring compliance, but decisions about compliance technology and investment choices are left to the private sector.

The best example of such a system in found in the U.S. Acid Rain Program. It has been remarkably effective. An analysis by the Government Accounting Office shows that this cap-and-trade system, created in 1990 to halve emissions of sulfur dioxide by utilities, cut costs to half of what was expected under the previous rate-based standard and well below industry and government estimates. What's more, recent research at MIT indicates that a third of all utilities complied in 1995 at a profit. This happened because there were unforeseen cost savings in switching to burning low-sulfur coal, and because trading enabled a utility to transfer allowances between its own units, allowing it to use low-emitting plants to meet base loads and high-emitting plants only at peak demand periods.

The open-market trading system works differently. Generally, there is no cap. Regulators set limits for each GHG coming from each source of emissions —say, for carbon dioxide from the smokestacks of an electric utility. Therefore, whenever two emitters want to trade, they must get regulatory approval. Although the up-front work may be less than that required for a cap-and-trade system, the need for approval of each trade makes transaction costs high. Also, there is always uncertainty about whether a trade will be approved, and approvals can take weeks or months, all of which reduce the incentive to trade and create an inefficient system.

The most recent results from the U.S. Acid Rain Program show that transaction costs are about 1.5 percent of the value traded, which is about the same as those for trades in a stock market. Transaction costs for open-market trading are an order of magnitude or more higher. Not surprisingly, the results of open-market trading in several U.S. states to reduce emissions of carbon monoxide, nitrogen oxides, and volatile organic compounds have been generally disappointing.

An emissions cap-and-trade system would reduce GHGs within the United States at very low cost. Trading between developed countries and between developed and developing countries could help nations meet their Kyoto Protocol targets, too. Let's consider what is needed for each system.

Trading at Home

The protocol allows a country to use whatever means it wants to achieve its own limit, so there is no restriction on creating a good cap-and-trade system within the United States. The first step would be to allocate the U.S. allotment of carbon emissions among emitters. Emissions come from several major sectors: electricity generation contributes 35 percent; transportation, 31 percent; general industry, 21 percent; and residential and commercial sources, 11 percent. However, because large sources are responsible for most GHGs, the United States could capture between 60 and 98 percent of emissions by including only a few thousand companies in the system.

Possibly the biggest cap-and-trade question for the United States is whom to regulate. The most efficient system would be to impose limits on carbon fuel providers—the coal, oil, and gas industries. These fuels account for up to 98 percent of carbon emissions. Industry groups are concerned, however, that regulating fuel providers is tantamount to a quota on fossil fuels, although similar reductions in fossil fuels would be required by any GHG regulation.

The alternative is to impose limits on fuel consumers—utilities, manufacturers, automobiles, and residential and commercial establishments. This method is less efficient, covering 60 to 80 percent of emissions, because it cannot practically handle the thousands of small industrial or commercial firms, not to mention residences, and because it does not provide incentives to reduce vehicle miles traveled. These inefficiencies will lead to higher overall costs and less burden-sharing.

However, political considerations will be as important as technical ones in choosing whom to regulate, and a hybrid system is possible. The most likely hybrid would be direct regulation of electric utilities and industrial boilers, capturing most of the country's combustion of coal and natural gas. A fuel-provider system would then be used to regulate sales of petroleum products and fossil fuels to residential and commercial markets. This may be politically expedient and could be almost as efficient as a pure fuel-provider model.

The design of the cap-and-trade program should follow the basic features of the U.S. Acid Rain Program. That program creates a gold standard with three key elements: a fixed emissions cap, free trading and banking of allowances, and strict monitoring and penalty provisions.

Several added benefits could be incorporated. First, the cost of continuous emissions monitoring could be reduced because emissions of carbon dioxide are very accurately measured by the carbon content of fuel. Second, the system could allow trading between gases. This could spur significant reductions of methane, which contributes 10 percent of the warming potential of U.S. emissions. A methane molecule has 21 times the warming potential of a carbon dioxide molecule, and certain sources of methane—landfills, coal mines, and natural gas extraction and transportation systems—could be included. Methane control can be low-cost or even profitable, because the captured methane can be sold; thus, trading between carbon dioxide and methane sources could be a cheap way to reduce the U.S. contribution to global warming.

A third design option would hinge on whether to allocate allowances to existing emitters for free or to auction them. Allocating allowances, as in the Acid Rain Program, is the most politically expedient option, but burdens later entrants, who must buy allowances from others who have already received them. Auctioning allowances would make them available to all and could have a dual benefit if the monies are used to reduce employment taxes or spur investment.

The U.S. Acid Rain Program's cap-and-trade system has cut the cost of sulfur dioxide compliance to $100 per ton of abated emissions, compared to initial industry estimates of $700 to $1,000 per ton and Environmental Protection Agency (EPA) estimates of $400 per ton. The same kind of cost reductions can be expected in a GHG system. The National Academy of Sciences has estimated that the United States could reduce 25 percent of its carbon emissions at a profit and 25 percent at very low or no cost, because of the hundreds of opportunities to achieve energy efficiency or switch fuels in our economy. Examples given by the Academy include switching from coal to natural gas in electricity generation, improving vehicle fuel economy, and creating energy-efficient buildings. The low net costs of GHG abatement would be further enhanced by the Clinton administration's recent proposal to speed the development of efficient high-end technologies.

As the world's largest emitter of GHGs, the United States should begin to implement a cap-and-trade system now. Market signals need to be sent right away to start our economy moving toward a less carbon-intensive development path. To prompt action, EPA should set an intermediate cap, perhaps for the year 2005, because the Kyoto Protocol requires countries to show some form of "significant progress" by that year.

Trading Between Developed Countries

International emissions trading could contribute substantially to curbing many nations' cost of compliance with the Kyoto Protocol. An assessment by the Clinton administration concluded that compliance costs could fall from $80 per ton of carbon to $10 to $20 per ton if full trading between all countries were allowed. A more realistic analysis done by the World Resources Institute examined 16 leading economic models and concluded that overall costs are much lower, but that international trading could still reduce the cost by around 1 percent of gross national product over a 20-year period.

Rules for trading between nations must begin to be drawn at the Conference of the Parties this November. However, there are key contentious issues, such as how to ensure the high credibility of trades through good compliance and monitoring systems and how to create a privately run system in which transactions can be made in minutes, not the months or even years required for government approval mechanisms. Whether transaction costs are high or low will probably determine the success of international trading....

Trading With Developing Countries

Trading between developed and developing countries has been hotly debated throughout the treaty process. For a developed country, the appeal is that investments made in developing countries, which are generally very energy-inefficient, can result in emissions reductions at very low cost, making allowances available. For a developing country, trading could be attractive because its sale of allowances could generate capital for projects that help it shift to a more prosperous but less carbon-intensive economy.

However, most developing countries, led by China and India, are opposed to trading. First, they simply distrust the motives of developed nations. Second, they rightly point out that the developed world has created the global warming problem and should therefore clean it up. Although legitimate, this second view ignores the many benefits that trading can bring to developing countries.

Many nongovernmental organizations (NGOs) are also wary of trading, claiming that the availability of allowances from developing countries will allow developed countries to avoid having to reduce their own emissions. This is unlikely, however. The United States, for example, will have to reduce its emissions by 37 percent by 2010 to reach its target. Developing countries that are willing to trade will simply not be able to accumulate enough tons to offset this large reduction. Indeed, trading with developing countries is likely to account for at most 10 to 20 percent of the reductions needed by a developed country.

Another major problem in trying to trade with developing countries lies in the weak emissions monitoring and compliance systems currently in place in many of them. Strengthening the basic institutional and judicial framework for environmental law may be necessary in many countries and could take considerable investment and many years....

An effective cap-and-trade system implemented within the United States would allow this country to comply with the GHG reductions it has committed to in the Kyoto Protocol at low or no cost. Because the system is a market instrument, it can rapidly bring about the adaptation, innovation, and investment needed to reduce emissions.

International trading can contribute substantially to achieving cost reductions, particularly if a cap-and-trade model with private trading mechanisms can be built into the protocol. Although such a system is unlikely to be fully mapped out at the Buenos Aires meeting in November, the first critical steps must be taken there.

NO ↵

Trading Away the Earth: Pollution Credits and the Perils of "Free Market Environmentalism"

The Republican takeover of Congress has unleashed an unprecedented assault on all forms of environmental regulation. From the Endangered Species Act to the Clean Water Act and the Superfund for toxic waste cleanup, laws that may need to be strengthened and expanded to meet the environmental challenges of the next century are instead being targeted for complete evisceration.

For some activists, this is a time to renew the grassroots focus of environmental activism, even to adopt a more aggressively anti-corporate approach that exposes the political and ideological agendas underlying the current backlash. But for many, the current impasse suggests that the movement must adapt to the dominant ideological currents of the time. Some environmentalists have thus shifted their focus toward voluntary programs, economic incentives and the mechanisms of the "free market" as means to advance the cause of environmental protection. Among the most controversial, and widespread, of these proposals are tradeable credits for the right to emit pollutants. These became enshrined in national legislation in 1990 with President George Bush's amendments to the 1970 Clean Air Act.

Even in 1990, "free market environmentalism" was not a new phenomenon. In the closing years of the 1980s, an odd alliance had developed among corporate public relations departments, conservative think tanks such as the American Enterprise Institute, Bill Clinton's Democratic Leadership Council (DLC), and mainstream environmental groups such as the Environmental Defense Fund. The market-oriented environmental policies promoted by this eclectic coalition have received little public attention, but have nonetheless significantly influenced debates over national policy.

Glossy catalogs of "environmental products," television commercials featuring environmental themes, and high profile initiatives to give corporate officials a "greener" image are the hallmarks of corporate environmentalism in the 1990s. But the new market environmentalism goes much further than these showcase efforts. It represents a wholesale effort to recast environmental protection based on a model of commercial transactions within the marketplace.

From Brian Tokar, "Trading Away the Earth: Pollution Credits and the Perils of 'Free Market Environmentalism'," *Dollars and Sense* (March/April 1996). Copyright © 1996 by Economic Affairs Bureau, Inc. Reprinted by permission. *Dollars and Sense* is a progressive economics magazine published six times a year. First-year subscriptions cost $18.95 and may be ordered by writing to *Dollars and Sense,* One Summer Street, Somerville, MA 02143.

"A new environmentalism has emerged," writes economist Robert Stavins, who has been associated with both the Environmental Defense Fund and the DLC's Progressive Policy Institute, "that embraces . . . market-oriented environmental protection policies."

Today, aided by the anti-regulatory climate in Congress, market schemes such as trading pollution credits are granting corporations new ways to circumvent environmental concerns, even as the same firms try to pose as champions of the environment. While tradeable credits are sometimes presented as a solution to environmental problems, in reality they do nothing to reduce pollution —at best they help businesses reduce the costs of complying with limits on toxic emissions. Ultimately, such schemes abdicate control over critical environmental decisions to the very same corporations that are responsible for the greatest environmental abuses.

How It Works, and Doesn't

A close look at the scheme for nationwide emissions trading reveals a particular cleverness; for true believers in the invisible hand of the market, it may seem positively ingenious. Here is how it works: The 1990 Clean Air Act amendments were designed to halt the spread of acid rain, which has threatened lakes, rivers and forests across the country. The amendments required a reduction in the total sulfur dioxide emissions from fossil fuel burning power plants, from 19 to just under 9 million tons per year by the year 2000. These facilities were targeted as the largest contributors to acid rain, and participation by other industries remains optional. To achieve this relatively modest goal for pollution reduction, utilities were granted transferable allowances to emit sulfur dioxide in proportion to their current emissions. For the first time, the ability of companies to buy and sell the "right" to pollute was enshrined in U.S. law.

Any facility that continued to pollute more than its allocated amount (roughly half of its 1990 rate) would then have to buy allowances from someone who is polluting less. The 110 most polluting facilities (mostly coal burners) were given five years to comply, while all the others would have until the year 2000. Emissions allowances were expected to begin selling for around $500 per ton of sulfur dioxide, and have a theoretical ceiling of $2000 per ton, which is the legal penalty for violating the new rules. Companies that could reduce emissions for less than their credits are worth would be able to sell them at a profit, while those that lag behind would have to keep buying credits at a steadily rising price. For example, before pollution trading every company had to comply with environmental regulations, even if it cost one firm twice as much as another to do so. Under the new system, a firm could instead choose to exceed the mandated levels, purchasing credits from the second firm instead of implementing costly controls. This exchange would save money, but in principle yield the same overall level of pollution as if both companies had complied equally. Thus, it is argued, market forces will assure that the most cost-effective means of reducing acid rain will be implemented first, saving the economy billions of dollars in "excess" pollution control costs.

Defenders of the Bush plan claimed that the ability to profit from pollution credits would encourage companies to invest more in new environmental technologies than before. Innovation in environmental technology, they argued, was being stifled by regulations mandating specific pollution control methods. With the added flexibility of tradeable credits, companies could postpone costly controls—through the purchase of some other company's credits—until new technologies became available. Proponents argued that, as pollution standards are tightened over time, the credits would become more valuable and their owners could reap large profits while fighting pollution.

Yet the program also included many pages of rules for extensions and substitutions. The plan eliminated requirements for backup systems on smokestack scrubbers, and then eased the rules for estimating how much pollution is emitted when monitoring systems fail. With reduced emissions now a marketable commodity, the range of possible abuses may grow considerably, as utilities will have a direct financial incentive to manipulate reporting of their emissions to improve their position in the pollution credits market.

Once the EPA actually began auctioning pollution credits in 1993, it became clear that virtually nothing was going according to their projections. The first pollution credits sold for between $122 and $310, significantly less than the agency's estimated minimum price, and by 1995, bids at the EPA's annual auction of sulfur dioxide allowances averaged around $130 per ton of emissions. As an artificial mechanism superimposed on existing regulatory structures, emissions allowances have failed to reflect the true cost of pollution controls. So, as the value of the credits has fallen, it has become increasingly attractive to buy credits rather than invest in pollution controls. And, in problem areas air quality can continue to decline, as companies in some parts of the country simply buy their way out of pollution reductions.

At least one company has tried to cash in on the confusion by assembling packages of "multi-year streams of pollution rights" specifically designed to defer or supplant purchases of new pollution control technologies. "What a scrubber really is, is a decision to buy a 30-year stream of allowances," John B. Henry of Clean Air Capital Markets told the *New York Times,* with impeccable financial logic. "If the price of allowances declines in future years," paraphrased the *Times,* "the scrubber would look like a bad buy."

Where pollution credits have been traded between companies, the results have often run counter to the program's stated intentions. One of the first highly publicized deals was a sale of credits by the Long Island Lighting Company to an unidentified company located in the Midwest, where much of the pollution that causes acid rain originates. This raised concerns that places suffering from the effects of acid rain were shifting "pollution rights" to the very region it was coming from. One of the first companies to bid for additional credits, the Illinois Power Company, canceled construction of a $350 million scrubber system in the city of Decatur, Illinois. "Our compliance plan is based almost totally on purchase of credits," an Illinois Power spokesperson told the *Wall Street Journal.* The comparison with more traditional forms of commodity trading came full circle in 1991, when the government announced that the entire system for trading and auctioning emissions allowances would be administered by the Chicago Board of Trade, long famous for its ever-frantic markets in everything from grain futures and pork bellies to foreign currencies.

Some companies have chosen not to engage in trading pollution credits, proceeding with pollution control projects, such as the installation of new scrubbers, that were planned before the credits became available. Others have switched to low-sulfur coal and increased their use of natural gas. If the 1990 Clean Air Act amendments are to be credited for any overall improvement in the air quality, it is clearly the result of these efforts and not the market in tradeable allowances.

Yet while some firms opt not to purchase the credits, others, most notably North Carolina-based Duke Power, are aggressively buying allowances. At the 1995 EPA auction, Duke Power alone bought 35% of the short-term "spot" allowances for sulfur dioxide emissions, and 60% of the long-term allowances redeemable in the years 2001 and 2002. Seven companies, including five utilities and two brokerage firms, bought 97% of the short term allowances that were auctioned in 1995, and 92% of the longer-term allowances, which are redeemable in 2001 and 2002. This gives these companies significant leverage over the future shape of the allowances market.

The remaining credits were purchased by a wide variety of people and organizations, including some who sincerely wished to take pollution allowances out of circulation. Students at several law schools raised hundreds of dollars, and a group at the Glens Falls Middle School on Long Island raised $3,171 to purchase 21 allowances, equivalent to 21 tons of sulfur dioxide emissions over the course of a year. Unfortunately, this represented less than a tenth of one percent of the allowances auctioned off in 1995.

Some of these trends were predicted at the outset. "With a tradeable permit system, technological improvement will normally result in lower control costs and falling permit prices, rather than declining emissions levels," wrote Robert Stavins and Brad Whitehead (a Cleveland-based management consultant with ties to the Rockefeller Foundation) in a 1992 policy paper published by the Progressive Policy Institute. Despite their belief that market-based environmental policies "lead automatically to the cost-effective allocation of the pollution control burden among firms," they are quite willing to concede that a tradeable permit system will not in itself reduce pollution. As the actual pollution levels still need to be set by some form of regulatory mandate, the market in tradeable allowances merely gives some companies greater leverage over how pollution standards are to be implemented.

Without admitting the underlying irrationality of a futures market in pollution, Stavins and Whitehead do acknowledge (albeit in a footnote to an Appendix) that the system can quite easily be compromised by large companies' "strategic behavior." Control of 10% of the market, they suggest, might be enough to allow firms to engage in "price-setting behavior," a goal apparently sought by companies such as Duke Power. To the rest of us, it should be clear that if pollution credits are like any other commodity that can be bought, sold and traded, then the largest "players" will have substantial control over the entire "game." Emissions trading becomes yet another way to assure that large corporate interests will remain free to threaten public health and ecological survival in their unchallenged pursuit of profit.

Trading the Future

Mainstream groups like the Environmental Defense Fund (EDF) continue to throw their full support behind the trading of emissions allowances, including the establishment of a futures market in Chicago. EDF senior economist Daniel Dudek described the trading of acid rain emissions as a "scale model" for a much more ambitious plan to trade emissions of carbon dioxide and other gases responsible for global warming. This plan was unveiled shortly after the passage of the 1990 Clean Air Act amendments, and was endorsed by then-Senator Al Gore as a way to "rationalize investments" in alternatives to carbon dioxide-producing activities.

International emissions trading gained further support via a U.N. Conference on Trade and Development study issued in 1992. The report was co-authored by Kidder and Peabody executive and Chicago Board of Trade director Richard Sandor, who told the *Wall Street Journal,* "Air and water are simply no longer the 'free goods' that economists once assumed. They must be redefined as property rights so that they can be efficiently allocated."

Radical ecologists have long decried the inherent tendency of capitalism to turn everything into a commodity; here we have a rare instance in which the system fully reveals its intentions. There is little doubt that an international market in "pollution rights" would widen existing inequalities among nations. Even within the United States, a single large investor in pollution credits would be able to control the future development of many different industries. Expanded to an international scale, the potential for unaccountable manipulation of industrial policy by a few corporations would easily compound the disruptions already caused by often reckless international traders in stocks, bonds and currencies.

However, as long as public regulation of industry remains under attack, tradeable credits and other such schemes will continue to be promoted as market-savvy alternatives. Along with an acceptance of pollution as "a by-product of modern civilization that can be regulated and reduced, but not eliminated," to quote another Progressive Policy Institute paper, self-proclaimed environmentalists will call for an end to "widespread antagonism toward corporations and a suspicion that anything supported by business was bad for the environment." Market solutions are offered as the only alternative to the "inefficient," "centralized," "command-and-control" regulations of the past, in language closely mirroring the rhetoric of Cold War anti-communism.

While specific technology-based standards can be criticized as inflexible and sometimes even archaic, critics choose to forget that in many cases, they were instituted by Congress as a safeguard against the widespread abuses of the Reagan-era EPA. During the Reagan years, "flexible" regulations opened the door to widely criticized—and often illegal—bending of the rules for the benefit

of politically favored corporations, leading to the resignation of EPA administrator Anne Gorsuch Burford and a brief jail sentence for one of her more vocal
legal assistants.

The anti-regulatory fervor of the present Congress is bringing a variety
of other market-oriented proposals to the fore. Some are genuinely offered to
further environmental protection, while others are far more cynical attempts
to replace public regulations with virtual blank checks for polluters. Some
have proposed a direct charge for pollution, modeled after the comprehensive pollution taxes that have proved popular in Western Europe. Writers as
diverse as Supreme Court Justice Stephen Breyer, American Enterprise Institute
economist Robert Hahn and environmental business guru Paul Hawken have
defended pollution taxes as an ideal market-oriented approach to controlling
pollution. Indeed, unlike tradeable credits, taxes might help reduce pollution
beyond regulatory levels, as they encourage firms to control emissions as much
as possible. With credits, there is no reduction in pollution below the threshold established in legislation. (If many companies were to opt for substantial
new emissions controls, the market would soon be glutted and the allowances
would rapidly become valueless.) And taxes would work best if combined with
vigilant grassroots activism that makes industries accountable to the communities in which they operate. However, given the rapid dismissal of Bill Clinton's
early plan for an energy tax, it is most likely that any pollution tax proposal
would be immediately dismissed by Congressional ideologues as an outrageous
new government intervention into the marketplace.

Air pollution is not the only environmental problem that free marke-
teers are proposing to solve with the invisible hand. Pro-development interests
in Congress have floated various schemes to replace the Endangered Species
Act with a system of voluntary incentives, conservation easements and other
schemes through which landowners would be compensated by the government
to protect critical habitat. While these proposals are being debated in Congress,
the Clinton administration has quietly changed the rules for administering the
Act in a manner that encourages voluntary compliance and offers some of the
very same loopholes that anti-environmental advocates have sought. This, too,
is being offered in the name of cooperation and "market environmentalism."

Debates over the management of publicly-owned lands have inspired
far more outlandish "free market" schemes. "Nearly all environmental problems are rooted in society's failure to adequately define property rights for
some resource," economist Randal O'Toole has written, suggesting a need for
"property rights for owls and salmon" developed to "protect them from pollution." O'Toole initially gained the attention of environmentalists in the Pacific
Northwest for his detailed studies of the inequities of the U.S. Forest Service's
long-term subsidy programs for logging on public lands. Now he has proposed
dividing the National Forest system into individual units, each governed by its
users and operated on a for-profit basis, with a portion of user fees allocated
for such needs as the protection of biological diversity. Environmental values,
from clean water to recreation to scenic views, should simply be allocated
their proper value in the marketplace, it is argued, and allowed to out-compete
unsustainable resource extraction. Other market advocates have suggested far

more sweeping transfers of federal lands to the states, an idea seen by many in the West as a first step toward complete privatization.

Market enthusiasts like O'Toole repeatedly overlook the fact that ecological values are far more subjective than the market value of timber and minerals removed from public lands. Efforts to quantify these values are based on various sociological methods, market analysis and psychological studies. People are asked how much they would pay to protect a resource, or how much money they would accept to live without it, and their answers are compared with the prices of everything from wilderness expeditions to vacation homes. Results vary widely depending on how questions are asked, how knowledgeable respondents are, and what assumptions are made in the analysis. Environmentalists are rightfully appalled by such efforts as a recent Resources for the Future study designed to calculate the value of human lives lost due to future toxic exposures. Outlandish absurdities like property rights for owls arouse similar skepticism.

The proliferation of such proposals—and their increasing credibility in Washington—suggest the need for a renewed debate over the relationship between ecological values and those of the free market. For many environmental economists, the processes of capitalism, with a little fine tuning, can be made to serve the needs of environmental protection. For many activists, however, there is a fundamental contradiction between the interconnected nature of ecological processes and an economic system which not only reduces everything to isolated commodities, but seeks to manipulate those commodities to further the single, immutable goal of maximizing individual gain. An ecological economy may need to more closely mirror natural processes in their stability, diversity, long time frame, and the prevalence of cooperative, symbiotic interactions over the more extreme forms of competition that thoroughly dominate today's economy. Ultimately, communities of people need to reestablish social control over economic markets and relationships, restoring an economy which, rather than being seen as the engine of social progress, is instead, in the words of economic historian Karl Polanyi, entirely "submerged in social relationships."

Whatever economic model one proposes for the long-term future, it is clear that the current phase of corporate consolidation is threatening the integrity of the earth's living ecosystems—and communities of people who depend on those ecosystems—as never before. There is little room for consideration of ecological integrity in a global economy where a few ambitious currency traders can trigger the collapse of a nation's currency, its food supply, or a centuries-old forest ecosystem before anyone can even begin to discuss the consequences. In this kind of world, replacing our society's meager attempts to restrain and regulate corporate excesses with market mechanisms can only further the degradation of the natural world and threaten the health and well-being of all the earth's inhabitants.

POSTSCRIPT

Will Pollution Rights Trading Effectively Control Environmental Problems?

Swift does not address Tokar's concern about the ability of the major corporate polluters to control and manipulate the market for emission credits. This is one of the key issues that continues to inspire developing countries to withhold their endorsement of the greenhouse gas emissions trading provisions of the Kyoto Protocol. The evidence that Tokar cites, which is primarily based on short-term experience with trading in sulfur dioxide pollution credits, does not appear to fully justify the broad generalizations he makes about the inherent perils in market-based regulatory plans. Recent assessments of the acid rain program by the EPA and such organizations as the Environmental Defense Fund are more in line with Swift's evaluation.

Many environmental leaders see the present antienvironmental backlash, funded by entrepreneurial advocates of unrestrained development and their political allies, as a serious threat to the prevention of continued ecosystem degradation. They fear that market-based regulatory strategies may be part of the campaign to undermine imperfect but necessary governmental pollution-control efforts. The antienvironmental backlash has been promoted by several organizations with what some consider deceptive names, the most prominent of which is "Wise Use," which is heavily funded by forest product, mining, agricultural, and real estate interests. In his article "Determined Opposition," *Environment* (October 1995), Phil Brick argues that the populist appeal of such organizations represents a serious threat to environmentalism, and he suggests a strategy for the environmental movement to adopt in response to this threat.

The position of those who are ideologically opposed to pollution rights is concisely stated in professor of government Michael J. Sandel's op-ed piece "It's Immoral to Buy the Right to Pollute," *The New York Times* (December 15, 1997). In "Selling Air Pollution," *Reason* (May 1996), Brian Doherty supports the concept of pollution rights trading but argues that the kind of emission cap imposed in the case of sulfur dioxide and advocated by Swift is an inappropriate constraint on what he believes should be a completely free-market program. A strong denunciation of a specific pollution rights program is the subject of the December 8, 1994, news release *Environmentalists Reject Faulty Pollution Trading Program*, which is available from the Environmental Defense Fund. Richard A. Kerr, in "Acid Rain Control: Success on the Cheap," *Science* (November 6, 1998), reports that emissions trading has greatly reduced acid rain and that the annual cost has been about a tenth of the $10 billion initially forecast. Byron Swift, in "Allowance Trading and Potential Hot Spots—Good News From the Acid Rain Program," *Environment Reporter* (May 12, 2000), argues that the success of the

EPA's emission trading program has not led to the creation of pollution "hot spots" as feared by some critics. On the other hand, EPA researchers recently reported that even though acid emissions were down dramatically, lakes remain affected by past emissions. See Leslie Roberts, "Acid Rain: Forgotten, Not Gone," *U.S. News & World Report* (November 1, 1999). It will be many years before the damage already done can heal.

The Environmental Defense Fund

The Environmental Defense Fund (EDF) is "dedicated to protecting the environmental rights of all people, including future generations." The EDF is "guided by scientific evaluation of environmental problems" and "will work to create solutions that win lasting political, economic, and social support because they are bipartisan, efficient, and fair." The EDF home page contains links to its numerous reports and other informational publications.

```
http://www.edf.org
```

The EcoJustice Network

The EcoJustice Network is devoted to promoting environmental equity and justice.

```
http://www.igc.apc.org/envjustice/
```

Biotechnology Information Center

This U.S. Department of Agriculture site contains many references to current information about various aspects of agricultural biotechnology.

```
http://www.nal.usda.gov/bic/
```

Organic Farming Research Foundation

This site provides information about the potential environmental value of alternative agriculture from the Organic Farming Research Foundation, an organization whose goal is "to foster the improvement and widespread adoption of organic farming practices."

```
http://www.ofrf.org/index.html
```

Environmental Estrogens and Other Hormones

Maintained by scientists at Tulane University, this is an excellent educational service and interactive forum on all aspects of the environmental estrogen issue.

```
http://www.som.tulane.edu/cbr/ecme/eehome/default.html
```

Air Quality: EPA's Proposed New Ozone and Particulate Matter Standards

This is a Congressional Research Service report to Congress on the pros and cons of the new EPA ozone and particulate matter standards.

```
http://www.cnie.org/nle/air-15.html
```

PART 2

The Environment and Technology

*M*ost *of the environmental concerns that are the focus of current regulatory debates are directly related to modern industrial development —the pace of which has been accelerating dramatically since World War II. Thousands of new synthetic chemicals have been introduced into manufacturing processes and agricultural pursuits. New technology, its byproducts, and the exponential increases in the production and use of energy have all contributed to the release of environmental pollutants. How to continue to improve the standard of living for the world's people without increasing ecological stress and exposure to toxins is the key question that underlies the issues debated in this section.*

- Should Pollution Prevention and Reduction Be a Focus of Agricultural Policy?

- Is Biotechnology an Environmentally Sound Way to Increase Food Production?

- Do Environmental Hormone Mimics Pose a Potentially Serious Health Threat?

- Is the Environmental Protection Agency's Decision to Tighten Air Quality Standards for Ozone and Particulates Justified?

ISSUE 9

Should Pollution Prevention and Reduction Be a Focus of Agricultural Policy?

YES: David E. Ervin, from "Shaping a Smarter Environmental Policy for Farming," *Issues in Science and Technology* (Summer 1998)

NO: Dennis T. Avery, from "Saving Nature's Legacy Through Better Farming," *Issues in Science and Technology* (Fall 1997)

ISSUE SUMMARY

YES: David E. Ervin, policy studies director at the Henry A. Wallace Institute for Alternative Agriculture, argues that water pollution caused by agricultural runoff of fertilizers, pesticides, and silt is an increasingly serious problem. He advocates a variety of new policies designed to prevent and reduce pollution, including setting ambient water quality standards and using market incentives to encourage farmers to reduce chemical use.

NO: Dennis T. Avery, director of the Hudson Institute's Center for Global Food Issues, argues that environmental concerns must give way to future food needs. He maintains that without higher-yielding crops that depend on fertilizers and pesticides, remaining forest lands will have to be used for agriculture, thus threatening biodiversity.

The use of naturally occurring chemicals in agriculture goes back many centuries. After World War II, however, the application of synthetic chemical toxins to croplands became sufficiently intensive to cause widespread environmental problems. DDT, used during the war to control malaria and other insect-borne diseases, was promoted by agribusiness as the choice solution for a wide variety of agricultural pest problems. As insects' resistance to DDT increased, other chlorinated organic toxins—heptachlor, aldrin dieldrin, mirex, and chlordane, for example—were introduced by the burgeoning chemical pesticide industry. Environmental scientists became concerned about the effects of these fat-soluble, persistent toxins, whose concentrations became magnified in carnivorous species at the top of the ecological food chain. The first serious problem to be documented was reproductive failure resulting from DDT ingestion in such birds of prey as falcons, pelicans, osprey, and eagles. Chlorinated pesticides were also found to be poisoning marine life. Marine scientist Rachel Carson's best-seller *Silent Spring* (Houghton Mifflin, 1962) raised public and scientific consciousness about the potential devastating effects of continued, uncontrolled use of chemical pesticides.

In 1966 a group of scientists and lawyers organized the Environmental Defense Fund in an effort to seek legal action against the use of DDT. After a prolonged struggle, they finally won a court ruling in 1972 ending nearly all uses of DDT in the United States. In that same year amendments to the Federal Insecticide, Fungicide, and Rodenticide Act gave the Environmental Protection Agency authority to develop a comprehensive program to regulate the use of pesticides. By 1978 many other chlorinated organic pesticides had been banned in the United States because of evidence linking them to health and environmental problems. Pesticide manufacturers switched to more biodegradable organophosphate and carbamate pesticides. However, the acute human toxicity of many of these chemicals has caused a rise in pesticide-related deaths and illnesses among agricultural field workers. Pesticide manufacture has also resulted in the poisoning of many workers and in serious environmental problems, such as the contamination of Virginia's James River. The 1984 Bhopal disaster, which caused 3,500 deaths and 200,000 injuries, was due to the release of a chemical precursor being used by Union Carbide to manufacture a pesticide.

The role of chemicals in world agriculture is hotly contested. Many high-yield varieties of grain have been developed that are dependent on the intensive use of pesticides and fertilizers. Supporters of this "green revolution" credit it with the large worldwide increase in grain harvests over the past three decades. Other agricultural and environmental policy experts maintain that increased crop yields in developed countries have been accompanied by significant environmental degradation. They also hold that the great cost of this type of farming has limited the value of high-technology agriculture in solving local food problems in less-developed countries, where the principal effect has often been to increase the acreage used to grow export crops.

In the following selections, David E. Ervin and Dennis T. Avery differ sharply on the extent to which we should rely on agricultural chemicals in the future. Ervin advocates creating a system of incentives to reduce the use of fertilizers and pesticides, which he contends are causing serious pollution problems, and substituting ecologically sound, sustainable agricultural practices. Avery maintains that continued intensive use of chemicals will be necessary to provide for food needs without converting more undeveloped land for agricultural use, a prospect that he believes would undermine biodiversity.

David E. Ervin

 **YES**

Shaping a Smarter Environmental Policy for Farming

In the summer of 1997, Maryland Governor Parris Glendening suddenly closed two major rivers to fishing and swimming, after reports of people becoming ill from contact with the water. Tests uncovered outbreaks of a toxic microbe, *Pfiesteria piscicida,* perhaps caused by runoff of chicken manure that had been spread as fertilizer on farmers' fields. Glendening's action riveted national attention on a long-overlooked problem: the pollution of fresh water by agricultural operations. When the governor then proposed a ban on spreading chicken manure, the state's poultry producers lashed back, claiming they would go out of business if they had to pay to dispose of the waste.

The controversy, and others springing up in Virginia, Missouri, California, and elsewhere, has galvanized debate among farmers, ranchers, environmentalists, and regulators over how to control agricultural pollution. The days of relying on voluntary controls and payments to farmers for cutbacks are rapidly ending. A final policy is far from settled, but even defenders of agriculture have endorsed more aggressive approaches than were considered feasible before recent pollution outbreaks.

Maryland's proposed ban is part of a state-led shift toward directly controlling agricultural pollution. Thirty states have at least one law with enforceable measures to reduce contamination of fresh water, most of which have been enacted in the 1990s. Federal policy has lagged behind, but President Clinton's Clean Water Action Plan, introduced in early 1998, may signal a turn toward more direct controls as well. After decades of little effort, state and federal lawmakers seem ready to attack the problem. But there is a serious question as to whether they are going about it in the best way.

The quality of U.S. rivers, lakes, and groundwater has improved dramatically since the 1972 Clean Water Act, which set in motion a series of controls on effluents from industry and in urban areas. Today, states report that the condition of two-thirds of surface water and three-fourths of groundwater is good. But where there is still degradation, agriculture is cited as the primary cause. Public health scares have prompted legislators to take action on the runoff of manure, fertilizer, pesticides, and sediment from farmland.

From David E. Ervin, "Shaping a Smarter Environmental Policy for Farming," *Issues in Science and Technology* (Summer 1998). Copyright © 1998 by The University of Texas at Dallas, Richardson, TX. Reprinted by permission.

Although it is high time to deal with agriculture's contribution to water pollution, the damage is very uneven in scope and severity; it tends to occur where farming is extensive and fresh water resources are vulnerable. Thus, blanket regulations would be unwise. There is also enormous inertia to overcome. For decades, the federal approach to controlling agriculture has been to pay farmers not to engage in certain activities, and agricultural interest groups have resisted any reforms that don't also pay.

Perhaps the most vexing complication is that scientists cannot conclusively say whether specific production practices such as how manure and fertilizer is spread and how land is tiered and tilled will help, because the complex relationship between what runs off a given parcel of land and how it affects water quality is not well understood. Prescribing best practices amounts to guesswork in most situations, yet that is what current proposals do. Unless a clear scientific basis can be shown, the political and monetary cost of mandating and enforcing specific practices will be great. Farmers will suffer from flawed policies, and battle lines will be drawn. Meanwhile, the slow scientific progress in unraveling the link between farm practices and water pollution will continue to hamper innovation that could solve problems in cost-effective ways.

Better policies from the U.S. Department of Agriculture (USDA), the Environmental Protection Agency (EPA), and state agricultural and environmental departments are certainly needed. But which policies? Because the science to prove their effectiveness does not exist, mandating the use of certain practices is problematic. Paying farmers for pollution control is a plain subsidy, a tactic used for no other U.S. industry. A smarter, incentive-based approach is needed. Happily, such an approach does exist, and its lessons can be applied to minimizing agriculture's adverse effects on biodiversity and air pollution as well.

Persistent Pollution

Farms and ranches cover about half of the nation's land base. Recent assessments of agriculture's effects on the environment by the National Research Council (NRC), USDA, and other organizations indicate that serious environmental problems exist in many regions, although their scope and severity vary widely. Significant improvements have been made during the past decade in controlling soil erosion and restoring certain wildlife populations, but serious problems, most notably water pollution, persist with no prospect of enduring remedies.

The biggest contribution to surface water and groundwater problems is polluted runoff, which stems from soil erosion, the use of pesticides, and the spreading of animal wastes and fertilizers, particularly nitrogen and phosphorus. Annual damages caused by sediment runoff alone are estimated at between $2 billion and $8 billion. Excessive sediment is a deceptively big problem: As it fills river beds, it promotes floods and burdens plants for processing municipal drinking water. It also clouds rivers, decreasing sunlight, which in turn lowers oxygen levels and chokes off life in the water.

National data on groundwater quality have been scarce because of the difficulty and cost of monitoring. EPA studies in the late 1980s showed that fewer than 1 percent of community water systems and rural wells exceeded EPA's maximum contaminant level of pesticides. Fewer than 3 percent of wells topped EPA's limit for nitrates. However, the percentages still translate into a large number of unsafe drinking water sources, and only a fraction of state groundwater has been tested. The state inventory data on surface water quality is limited too, covering only 17 percent of the country's rivers and 42 percent of its lakes. A nationally consistent and comprehensive assessment of the nation's water quality does not exist and is not feasible with the state inventory system. We therefore cannot say anything definitive about agriculture's overall role in pollution.

Nonetheless, we know a good deal about water conditions in specific localities, enough to improve pollution policy. Important progress is being made by the U.S. Geological Survey (USGS), which began a National Water Quality Assessment (NAWQA) in the 1980s precisely because we could not construct an accurate national picture. USGS scientists estimated in 1994 that 71 percent of U.S. cropland lies in watersheds where at least one agricultural pollutant violates criteria for recreational or ecological health. The Corn Belt is a prime example. Hundreds of thousands of tons of nutrients—nitrogen and phosphorus from fertilizers and animal wastes—are carried by runoff from as far north as Minnesota to Louisiana's Gulf Coast estuaries. The nutrients cause excessive algae growth, which draws down oxygen levels so low that shellfish and other aquatic organisms die. (This process has helped to create a "dead" zone in the Gulf of Mexico—a several-hundred-square-mile area that is virtually devoid of life.) Investigators have traced 70 percent of the fugitive nutrients that flow into the Gulf to areas above the confluence of the Ohio and Mississippi Rivers. In a separate NAWQA analysis, most nutrients in streams—92 percent of nitrogen and 76 percent of phosphorus—were estimated to flow from nonpoint or diffuse sources, primarily agriculture. USGS scientists also estimated that more than half the phosphorus in rivers in eight Midwestern states, more than half the nitrate in seven states, and more than half the concentrations of atrazine, a common agricultural pesticide, in 16 states all come from sources in other states. Hence those states cannot control the quality of their streams and rivers by acting alone.

Groundwater pollution is another problem. Groundwater supplies half the U.S. population with drinking water and is the sole source for most rural communities. Today, the most serious contamination appears to be high levels of nitrates from fertilizers and animal waste. USGS scientists have found that 12 percent of domestic wells in agricultural areas exceed the maximum contaminant level for nitrate, which is more than twice the rate for wells in nonagricultural areas and six times that for public wells. Also, samples from 48 agricultural areas turned up pesticides in 59 percent of shallow wells. Although most concentrations were substantially below EPA water standards, multiple pesticides were commonly detected. This pesticide soup was even more pronounced in streams. No standards exist for such mixtures.

These results are worrisome enough, and outbreaks of illness such as the *Pfiesteria* scourge have heightened awareness. But what has really focused national attention on agriculture's pollution of waterways has been large spills of animal waste from retention ponds. According to a study done by staff for Sen. Tom Harkin (D-Iowa), Iowa, Minnesota, and Missouri had 40 large manure spills in 1996. When a dike around a large lagoon in North Carolina failed, an estimated 25 million gallons of hog manure (about twice the volume of oil spilled by the *Exxon Valdez* accident) was released into nearby fields and waterways. Virtually all aquatic life was killed along a 17-mile stretch of the New River. North Carolina subsequently approved legislation that requires acceptable animal waste management plans. EPA indicates that as many as two-thirds of confined-animal operations across the nation lack permits governing their pollution discharges. Not surprisingly, a major thrust of the new Clean Water Action Plan is to bring about more uniform compliance for large animal operations.

Dubious Tactics

Historically, environmental programs for agriculture have used one of three approaches, all of which have questionable long-term benefits. Since the Great Depression, when poor farming practices and drought led to huge dust storms that blackened midwestern skies, the predominant model for improving agriculture's effects on the environment has been to encourage farmers to voluntarily change practices. Today, employees of state agencies and extension services and federal conservation agencies visit farmers, explain how certain practices are harming the land or waterways, and suggest new techniques and technologies. The farmers are also told that if they change damaging practices or choose new program X or technology Y, they can get payments from the state or federal government.

Long-term studies indicate that these voluntary payment schemes have been effective in spurring significant change; however, as soon as the payments stop, use of the practices dwindles. The Conservation Reserve Program (CRP) now sets aside about 30 million acres of environmentally vulnerable land. Under CRP, farmers agree to retire eligible lands for 10 years in exchange for annual payments, plus cost sharing to establish land cover such as grasses or trees. About 10 percent of the U.S. cropland base has been protected in this way, at a cost of about $2 billion a year.

Although certain parcels of this land should be retired from intensive cultivation because they are too fragile to be farmed, we may be overdoing it with CRP. Some of this land will be needed to produce more food as U.S. and world demand grows. Much of it could be productively cultivated with new techniques, thereby producing profitable crops, reducing water pollution, and costing taxpayers nothing. One of the most prominent new techniques is no-till farming, which is done with new machines that cut thin parallel grooves in soil and simultaneously plant seeds, which not only minimizes runoff but reduces a farmer's cost. Studies show that no-till farming is usually more profitable than full plowing because of savings in labor, fuel, and machinery.

Evidence suggests that CRP's gains have been temporary. As with the similar Soil Bank program of the 1960s, once contracts expire, virtually all lands are returned to production. Unless the contracts are renewed indefinitely, most of the 30 million acres will again be farmed, again threatening the environment if farmers fail to adopt no-till practices.

The second approach involves compliance schemes. To receive payments from certain agricultural programs, a farmer must meet certain conservation standards. The 1985 Food Security Act contained the boldest set of compliance initiatives in history. Biggest among them was the Conservation Compliance Provision, which required farmers to leave a minimum amount of crop residues on nearly 150 million acres of highly erodible cropland. In effect, these provisions established codes of good practice for farmers who received public subsidies, and they were a first step toward more direct controls. However, these programs are probably doomed. The general inclination of government and the public to eliminate subsidies led to passage of federal farm legislation in 1996 that includes plans to phase out payment programs by 2002.

The third approach to reducing agriculture's impact on the environment involves direct regulation of materials such as pesticides that are applied to the land. These programs have been roundly criticized from all quarters. Farm groups complain that pesticide regulation has been too harsh. Environmental groups counter that although the regulations specify the kinds of pesticides that can be sold and the crops they can be used on, they do not restrict the amount of pesticide that can be spread. Even if regulations did specify quantity, enforcement would be virtually impossible. The registration process for pesticide use has also been miserably slow and promises to get slower as a result of the 1996 Food Quality Protection Act, which requires the reregistration of all pesticides against stricter criteria.

In sum, current approaches to limit the environmental effects of agriculture have cost taxpayers large amounts of money with little guarantee of long-term protection. Unless a steady stream of federal funding continues, many of the gains will evaporate. And the idea of paying people not to pollute is becoming increasingly untenable, especially at the state level.

Getting Smarter

Four actions are needed to establish a smarter environmental policy for agriculture.

Set specific, measurable environmental objectives   Without quantifiable targets, an environmental program cannot be properly guided. To date, most programs have called for the use of specific farming practices rather than setting ambient quality conditions for surface water and groundwater. This is largely because of political precedent and because of the complex nonpoint nature of many pollution problems. However, setting a specific water quality standard, such as nitrate or pesticide concentration in drinking water, presumes that the science exists to trace contaminants back to specific lands. Such research is currently sparse, although major assessments by the NRC and others indicate that clearer science is possible. Setting standards would help stimulate the science.

Several states are taking the lead in setting standards. Nebraska has set maximum groundwater nitrate concentration levels; if tests show concentrations above the standard, restrictions on fertilizer use can be imposed. Florida has implemented controls on the nutrient outflows from large dairies into Lake Okeechobee, which drains into the Everglades. In Oregon, environmental regulators set total maximum daily loads of pollutants discharged into rivers and streams, and the state Department of Agriculture works with farmers to reduce the discharges. Voluntary measures accompanied by government payments are tried first, but if they are not sufficient, civil fines can be imposed in cases of excessive damages.

The federal government can support the states' lead by setting minimum standards for particular pollutants that pose environmental health risks, such as nitrates and phosphorus. The Clean Water Action Plan would establish such criteria for farming by 2000 and for confined animal facilities by 2005. Standards for sediment should be set as well.

There is no easy way around the need for a statutory base that defines what gets done, when it gets done, and how it gets done at the farm, county, state, regional, and national levels. Unless those specific responsibilities are assigned, significant progress on environmental problems will not be made.

Create a portfolio of tangible, significant incentives Without sufficient incentives, we have little hope of meeting environmental objectives. The best designs establish minimum good-neighbor performance, below which financial support will not be provided, and set firm deadlines beyond which significant penalties will be imposed. Incentive programs could include one-stop permitting for all environmental requirements, such as Idaho's "One Plan" program, which saves farmers time and money; "green payments" for farms that provide environmental benefits beyond minimum performance; a system for trading pollution rights; and local, state, or national tax credits for exemplary stewardship.

It is important to stress that a silver-bullet approach to the use of incentives does not exist. The most cost-effective strategy for any given farm or region will be a unique suite of flexible incentives that fit state and local environmental, economic, and social conditions. Although the use of flexible incentives can require substantial administrative expense, they can also trigger the ingenuity of farmers and ranchers, much as market signals have done for the development of more productive crops and livestock.

Although incentives are preferable, penalties and fines will still be needed. Pollution from large factory farms is now spurring states and the federal government to apply to farms the strict limits typically set for other industrial factories. Some of these farms keep more than half a million animals in small areas. The animals can generate hundreds of millions of gallons of wastes per year—as much raw sewage as a mid-sized city but without the sewage treatment plants. The wastes, which are stored in open "lagoons" or spread on fields as fertilizer, not only produce strong odors but can end up in streams and rivers and possibly contaminate groundwater. In 1997, North Carolina, which now generates more economic benefits from hog farms than it does from tobacco, imposed sweeping new environmental rules on hog farming. Under the Clean Water Action Plan, EPA is proposing to work with the states to impose strict pollution-discharge permits on all large farms by 2005. EPA also wants to dictate the type of pollution-control technologies that factory farms must adopt.

Because pollution problems are mostly local, states must do more than the federal government to create a mix of positive and negative incentives, although the federal government must take the lead on larger-scale problems that cross state boundaries. Both the states and the federal government should first focus on places a clear agriculture-pollution link can be shown and the potential damages are severe.

Harness the power of markets   Stimulating as much private environmental initiative as possible is prudent, given the public fervor for shrinking government. The 1996 Federal Agriculture Improvement and Reform Act took the first step by dismantling the system of subsidizing particular crops, which had encouraged farmers to overplant those crops and overapply fertilizers and pesticides in many cases. The potential for using market forces is much broader.

One of the latest and most effective mechanisms may be a trading system for pollution rights. A trading system set up under the U.S. Acid Rain Program has been very effective in reducing air pollution, and trading systems are being proposed to meet commitments made in the recently signed Kyoto Protocol to reduce emissions of greenhouse gases.

Trading systems work by setting total pollution targets for a region, then assigning a baseline level of allowable pollution to each source. A company that reduces emissions below its baseline can sell the shortfall to a company that is above its own baseline. The polluter can then apply that allowance to bring itself into compliance. The system rewards companies that reduce emissions in low-cost ways and helps bad polluters buy time to find cost-effective ways to reduce their own emissions.

A few trading systems are already being tried in agriculture. Farms and industrial companies on the North Carolina's Pamlico Sound are authorized to trade water pollution allowances, but few trades have taken place thus far because of high transaction costs. Experiments are also under way in Wisconsin and Colorado, but the complications of using trading systems for nonpoint pollution will slow implementation.

Pollution taxes can also create incentives for change. Economists have proposed levying taxes that penalize increases in emissions. Some also propose using the proceeds to reward farmers who keep decreasing their emissions below the allowable limit. The tax gives farmers the flexibility to restructure their practices, but political opposition and potentially high administrative costs have hindered development.

One other market mechanism that is cost-effective and nonrestrictive is facilitating consumer purchases of food that is produced by farmers who use minimal amounts of pesticides and synthetic fertilizers. Food industry reports indicate that a growing segment of the public will pay for food and fiber cultivated in environmentally friendly ways. The natural foods market has grown between 15 and 20 percent per year during the past decade, compared with 3 to 4 percent for conventional food products. If this trend continues, natural foods will account for nearly one-quarter of food sales in 10 years. Because organic foods command higher prices, farmers can afford to use practices that reduce pollution, such as crop rotation and biologically based pest controls.

Government can play a stronger role in promoting the sale of natural foods. It should make sure that consumers have accurate information by monitoring the claims of growers and retailers and establishing production, processing, and labeling standards. One experiment to watch is in New York, where the Wegman's supermarket chain is promoting the sale of "IPM" foods grown by farmers who have been certified by the state as users of integrated pest management controls.

Stimulate new research and technology   One of the most overlooked steps needed to establish smarter environmental policy for agriculture is better R&D. Most research to date has focused on remediation of water pollution, rather than forward-looking work that could prevent pollution. Over the years, research for environmental purposes should have increased relative to food production research, but it is not clear that it has.

What is most needed is better science that clarifies the links between agricultural runoff and water quality. As stated earlier, this will be forced as regulations are imposed, but dedicated research by USDA, EPA, and state agricultural and environmental departments should begin right away.

R&D to produce better farm technology is also needed. Despite an imperfect R&D signaling process, some complementary technologies that simultaneously enhance environmental conditions and maintain farm profit have emerged. Examples include no-till farming, mulch-till farming, integrated pest management, soil nutrient testing, rotational grazing (moving livestock to different pastures to reduce the buildup of manure, instead of collecting manure), and organic production. Most of these techniques require advanced farming skills but have a big payoff. No-till and mulch-till farming systems, for example, have transformed crop production in many parts of the nation and now account for nearly 40 percent of planted acres. However, these systems were driven by cost savings from reduced fuel, labor, and machinery requirements and could improve pollution control even further if developed with this goal in mind. Integrated pest management methods generally improve profits while lowering pesticide applications, but they could benefit from more aggressive R&D strategies. A farmer's use of simple testing procedures for nutrients in the soil before planting has been shown to reduce nitrogen fertilizer applications by about one-third in some areas, saving farmers $4 to $14 per acre, according to one Pennsylvania study.

Other technologies are emerging that have unknown potential, including "precision farming" and genetic engineering of crops to improve yield and resist disease. Precision farming uses yield monitors, computer software, and special planting equipment to apply seeds, fertilizers, and pesticides at variable rates across fields, depending on careful evaluation and mapping techniques. This suite of complementary technologies has developed mostly in response to the economic incentive to reduce input costs or increase yields. Their full potential for environmental management has been neglected. It is time to make pollution prevention and control an explicit objective of agricultural R&D policy.

Accountability and Smart Reform

The long-standing lack of public and legislative attention to agricultural pollution is changing. Growing scrutiny suggests that blithely continuing down the path of mostly voluntary-payment approaches to pollution management puts agriculture in a vulnerable position. As is happening in Maryland, a single bad incident could trigger sweeping proposals—in that case, possibly an outright ban against the spreading of chicken manure on fields—that would impose serious costs on agriculture. A disaster could cause an even stronger backlash; the strict clean-water regulations of the 1970s came in torrents after the Cuyahoga River in Ohio actually caught fire because it was so thick with industrial waste.

The inertia that pervades agriculture is understandable. For decades farmers have been paid for making changes. But attempts by agricultural interest groups to stall policy reforms, including some important first steps in the Clean Water Action Plan, will hamper farming's long-term competitiveness, or even backfire. Resistance will invite more direct controls, and slow progress on persistent environmental problems will invite further government intervention.

Under the smarter environmental policy outlined above, farmers, their neighbors and communities, environmental interest groups, government agencies, and the scientific community can create clear objectives and compelling incentives to reduce agricultural pollution. Farmers that deliver environmental benefits beyond their community responsibilities should be rewarded for exemplary performance. Those that fall short should face penalties. We ask no less from other sectors of the economy.

NO ⬅

Dennis T. Avery

Saving Nature's Legacy Through Better Farming

The obvious environmental problems and solutions are not necessarily obvious at all. Organic farming and the time-proven techniques of traditional agriculture hold great emotional attraction. Pure foods without chemical fertilizers and pesticides seem clearly preferable to the methods of large agribusiness. Could they be the cure for the unrelenting destruction of earth's forests and its diverse flora and fauna?

Ironically, developed world demands for these "obvious" solutions may push the world into famine and destroy the planet's biodiversity far faster than chemicals and overpopulation. Only the judicious application of the "evils" of high-yield farming may give us the time to prevent such calamities. Contrary to common wisdom, saving the environment and reducing population growth are likely to come about only if governments significantly increase their support for high-yielding crops and advanced farming methods, including the use of fertilizers and pesticides.

The biggest danger facing the world's wildlife is neither pesticides nor population growth but the potential loss of its habitat. Conversion of natural areas into farmland is the major impact of humans on the natural environment and poses a great threat to biodiversity. About 90 percent of the known species extinctions have occurred because of habitat loss.

Whereas many industrialized countries see their farms occupying less and less of their land, worldwide the opposite is true. The World Bank reports that cities take only 1.5 percent of earth's land, but farms occupy 36 percent. As world population climbs toward 8.5 billion in 2040, it will become even more clear how much food needs govern the world's land use. Unless we bolster our efforts to produce high-yielding crops, we face a plow-down of much of the world's remaining forests for low-yield crops and livestock.

Greens Versus Green Revolution

For decades and certainly since the 1968 publication of Paul Ehrlich's *The Population Bomb,* overpopulation has riven the world's conscience. Each regional

From Dennis T. Avery, "Saving Nature's Legacy Through Better Farming," *Issues in Science and Technology* (Fall 1997). Copyright © 1997 by The University of Texas at Dallas, Richardson, TX. Reprinted by permission.

famine catalyzed by crop failures or weather brings it further to the fore. Yet we seem unaware of how crucial the green revolution has been in forestalling famine and simultaneously saving the environment.

By maximizing land use, the green revolution's high-yield crops and farming techniques have been vital in preserving wildlife. By effectively tripling world crop yields since 1960, they have saved an additional 10 to 12 million square miles of wild lands, according to an analysis that I conducted and which was published in early 1997 in *Choices,* the magazine of the American Agricultural Economics Association. Without the green revolution, the world would have lost wild land equal to the combined land area of the United States, Europe, and Brazil. Instead, with hybrid seeds and chemical fertilizers and pesticides, today we crop the same 6 million square miles of land that we did in 1960 and feed 80 percent more people a diet that requires more than twice as many grain-equivalent calories.

The green revolution, however, has had its detractors. Since the publication of Rachel Carson's *Silent Spring* in 1962, developed-world residents have been bombarded with claims that modern farming kills wildlife, endangers children's health, and poisons the topsoil. Understandably, we love the natural ways of life. For many centuries, humans seemed to grow their crops quite well without deadly chemicals that poison soil, plants, insects, and animals. The organic gardening and farming movements look fondly on that ideal. Unfortunately, those techniques are ill suited to the modern world for two strong reasons.

First, they worked in a much less populous world. Such techniques and the plants they favor require large amounts of relatively fertile land supporting small numbers of people. In modern Europe, Asia, and the developing world, such low-yield farming is impractical. Second, many of those techniques are incredibly destructive to soil and forests, degrading biodiversity quickly and irrevocably. Slash-and-burn agriculture, the time-honored primitive farming method, is perhaps the most harmful to the environment.

Ironically, in a world facing the biggest surge in food demand it will ever see, many environmentalists who want to preserve natural areas are recommending organic and traditional farming systems that have sharply lower yields than mainstream farms. A recent organic farming "success" at the Rodale Institute achieved grain-equivalent yields from organic farming that were 21 percent lower and required 42 percent more labor. Such yields may be theoretically kinder to the environment, but in practice they would lead us to destroy millions of square miles of additional natural areas.

Meanwhile, Greenpeace and the World Wildlife Fund have gathered millions of European signatures on petitions to ban biotechnology in food production. They do not protest the use of biotechnology in human medicine, but only where it could help preserve nature by increasing farm productivity.

No Meat, No Thanks

Humans might be able to meet their nutritional needs with less strain on farming resources by eating nuts and tofu instead of meat and milk. So far, however, no society has been willing to do so. For example, a *Vegetarian Times* poll reported that 7 percent of Americans call themselves vegetarians. Two-thirds of these, however, eat meat regularly; 40 percent eat red meat regularly; and virtually all of them eat dairy products and eggs. Fewer than 500,000 Americans are vegan, foregoing all resource-costly livestock and poultry calories. The vegetarian/vegan percentages are similar in other affluent countries.

The reality is that as the world becomes more affluent, the average person will be eating more meat and consuming more agricultural products. If population growth stopped this hour, we would have to double the world's farm output to provide the meat, fruit, and cotton today's 5.9 billion people will demand in 2030 when virtually all will be affluent. There are no plans, nor any funding, for a huge global vegan recruiting campaign. Nor does history offer much hope of one's success.

Meanwhile, in what used to be the poor countries, the demand for meat, milk, and eggs is already soaring. Chinese meat consumption has risen 10 percent annually in the past six years. India has doubled its milk consumption since 1980, and two-thirds of its Hindus indicate that they will eat meat (though not beef) when they can afford it.

According to the United Nation's Food and Agricultural Organization (FAO), Asian countries provide about 17 grams of animal protein per capita per day for 3.3 billion people. Europeans and North Americans eat 65 to 70 grams. The Japanese not long ago ate less than 20 grams, but are now nearing 60 grams. By 2030, the world will need to be able to provide 55 grams of animal protein per person for four billion Asians, or they will destroy their own tropical forests to produce it themselves. It will not be possible to stave off disaster for biologically rich areas unless we continue to raise farm yields.

To make room for low-yield farming, we burn and plow tropical forests and drive wild species from their ecological niches. Indonesia is clearing millions of acres of tropical forest for low-quality cattle pastures and to grow low-yielding corn and soybeans on highly erodable soils to feed chickens. Similarly, a World Bank study reports that forests throughout the tropics are losing up to one-half of their species because bush-fallow periods (when farm lands are allowed to return to natural states) are shortened to feed higher populations.

Pessimists have said since the late 1960s that we won't be able to continue increasing yields. However, world grain yields have risen by nearly 50 percent in the meantime. If we'd taken the pessimists' advice to scrap agricultural research when they first offered it, the world would already have lost millions of square miles of wildlife habitat that we still have.

Nor is there any objective indication that the world is running out of ways of increasing crop yields and improving farming techniques. For example, world corn yields are continuing to rise as they have since 1960, at about 2.8 percent annually, in what's rapidly becoming the world's key crop. The yield trend has become more erratic, mainly because droughts decrease yield more in an eight-ton field than they do in a one-ton field. U.S. corn breeders are now shooting for populations of 50,000 plants per acre, which is three times the current Corn Belt planting density, and for 300-bushel yields.

Also, the International Rice Research Institute in the Philippines is re-designing the rice plant to get 30 percent more yield. Researchers are putting another 10 percent of the plant's energy into the seed head (supported by fewer but larger stalk shoots). They're using biotechnology techniques to increase resistance to pests and diseases. The new rice has been genetically engineered to resist the tungro virus—humanity's first success against a major virus. The U.S. Food and Drug Administration is close to approving pork growth hormone, which will produce hogs with half as much body fat and 20 percent more lean meat, using 25 percent less feed grain per hog. Globally, that would be equal to another 20 to 30 millions tons of corn production per year.

The world has achieved strong productivity gains from virtually all of its investments in agricultural research. The problem is mainly that we haven't been investing enough. One reason for underinvesting is pessimism about how much can be gained through research. But if humanity succeeds only in doubling instead of tripling farm output per acre, the effort will still save millions of square miles of land. Besides, the more pessimistic we feel about agricultural research, the more eager we should be to raise research investments, because there is no doubt that we will need more food.

Saving the Soil

Throughout history, soil erosion has been by far the biggest problem affecting farming sustainability. Modern high-yield farming is changing that situation dramatically. Simple arithmetic tells us that tripling the yields on the best cropland automatically cuts soil erosion per ton of food produced by about two-thirds. It also avoids pushing crops onto steep or fragile acres.

Relatively new methods such as conservation tillage and no-till farming are also making a big difference. Conservation tillage discs crop residues into the top few inches of soil, creating millions of tiny dams against wind and water erosion. In addition to saving topsoil, conservation tillage produces far more earthworms and subsoil bacteria than any plow-based system. No-till farming involves no plowing at all. The soil is never exposed to the elements. The seeds are planted through a cover crop that has been killed by herbicides. The Soil and Water Conservation Society says that use of these systems can cut soil erosion per acre by 65 to 95 percent.

Organic farmers reject both these systems because they depend on chemical weed killers, not plowing and hoeing, to control weeds. However, these powerful conservation farming systems are already being used on hundreds of millions of acres in the United States, Canada, Australia, Brazil, and Argentina. They have been used successfully in Asia and even tested successfully in Africa.

The model farm of the future will use still more powerful seeds, conservation tillage, and integrated pest management along with still better veterinary medications. It will use global positioning satellites, computers, and intensive soil sampling ("precision farming") to apply exactly the seeds and chemicals for optimum yields, with no leaching of chemicals into streams. Even then, high-yield farming will not offer zero risk to either the environment or to humans.

But it will offer near-zero and declining risk, which will be more than offset by huge increases in food security and wild lands saved.

Food Security and Lower Birthrates

Food availability and modern medicine have lowered the world's death rates, producing a one-time population growth surge. But they are also helping in the longer term to restabilize population by giving parents confidence that their first two or three children will live to adulthood.

Increased food security, for which crop yields are the best proxy, has been a vital element in sharply reducing world fertility rates. Indeed, according to World Bank and FAO statistics, the countries that have raised their crop yields the fastest have generally brought their births per woman down the fastest. For example, Indonesia has increased its rice yields since 1960 by 250 percent and its births per woman have dropped from 5.5 to 2.9. Likewise, Zimbabwe more than doubled its corn yields with Africa's best plant-breeding program, while births per woman have dropped from 8 in 1965 to 3.5 today. In contrast, countries without high-yield trends have kept higher fertility rates. In Ethiopia, which has suffered famine instead of rising yields, births per woman have risen from 5.8 in 1965 to more than 7 today.

Unfortunately, the world is not gearing up its science and technology resources to meet the agricultural and conservation challenge. U.S. funding for agricultural research has declined for decades in real terms, though the cost and complexity of the research projects continue to rise with the size of the challenge. The federal and state governments increased their spending on agricultural research from $1.02 billion in 1970 to $1.65 billion in 1990, a one-third decline in constant dollars. Public funding rose to $1.8 billion in 1996. Likewise, private sector agricultural research spending rose from $1.5 billion in 1970 to $3.15 billion in 1990, a 15 percent real decline.

Overseas, the research funding picture is worse. Europe has never spent heavily on agricultural research. Only a few of the developing world countries, including Brazil, China, and Zimbabwe, have even sporadically spent the few millions of dollars needed to adapt research to their own situations. All told, the entire world's agricultural research investment is probably less than $15 billion a year.

A telling example of the world's cavalier attitude toward agricultural research occurred in 1994, when the United States and other donor nations failed to come up with a large part of the budget for the Consultative Group on International Agricultural Research (CGIAR). CGIAR is the key international vehicle for creating high-yielding crops, supporting a network of 16 agricultural research centers in developing countries. Thus, global agricultural research almost literally went bankrupt at the very moment when the world was pledging another $17 billion for condoms and contraceptive pills at the UN meeting on population in Cairo. The World Bank subsequently stepped in on a conditional basis to keep the CGIAR research network running.

Historically, the U.S. Agency for International Development (AID) provided about 25 percent of CGIAR research funding, or about $60 million per year. Currently, this has fallen to about $30 million per year in much cheaper dollars, or about 10 percent of AID's budget. Indeed, despite the centers' success in raising world crop yields, AID has since shifted its priorities sharply from agricultural research to family planning. Given the sharp downward trends in birthrates in developing countries, additional family planning funds are likely to make only a modest difference in the world's population. However, Western intellectuals and journalists highly approve of population management.

In sum, world spending on agricultural research is tiny, especially if you consider that in 1996, the U.S. food industry alone produced $782 billion in goods and services and that the federal government subsidizes farmers to the tune of nearly $100 billion a year. (The European Union spends another $150 billion a year on farm subsidies.) Meanwhile, agricultural research has saved perhaps one billion lives from famine, increased food calories by one-third for four billion people in the developing world, and prevented millions of square miles of often biologically rich land from being plowed down.

We shouldn't be too surprised at the lack of approval and funding for high-yield agricultural research. Industrialized countries, which have funded most modern farming research, have been surrounded for the past 40 years with highly visible surpluses of grain, meat, and milk. Too many citizens associate the surpluses with science, not with ill-conceived farm price supports and trade barriers.

Western Europe watched its farm population decline from about 20 percent in 1960 to about 5 percent today. This followed an earlier but similar decline in the number of U.S. farmers. Both Europe and the United States associate the decline of the small family farm with the rise in crop yields, not with the rising value of off-farm jobs.

Securing the Future

Feeding the world's people while preserving biologically rich land will require two key things: more agricultural research and freer world trade in farm products. Expanded agricultural research should be the top priority.

Congress should double the federal government's $1.4 billion annual investment in agricultural research and adopt substantially higher farm yields as one of the nation's top research priorities. No other nation has the capacity to step into the U.S. research role in time to save the wild lands. Congress should also release much of the cropland still in the Department of Agriculture's (USDA's) Conservation Reserve Program for farming with conservation tillage, and it should direct AID to make the support of high-yield agriculture at least as important as population management.

In addition, in order to use the world's best farmland for maximum output, farm trade must be liberalized. Farm subsidies and farm trade barriers, although they are beginning to be reduced, have not only drained hundreds of billions of dollars in scarce capital away from economic growth and job creation, they now represent one of the biggest dangers to preservation of biologically diverse lands. The key dynamic in the farm trade arena is Asia's present and growing population density. Without an easy flow of farm products and services, densely populated Asian countries will be tempted to try to rely too much on domestic food production. But it will be extremely difficult to do. By 2030, Asia will have about eight times as many people per acre of cropland as will the Western Hemisphere. It already has the world's most intensive land use. In reality, countries reduce their food security with self-sufficiency. Droughts and plagues that cut crop yields are regional, not global.

The United States must convince the world that free trade in farm products would benefit all, particularly those in developing countries. President Clinton should make free farm trade a top international priority, which could give momentum to the World Trade Organization's scheduled 1999 talks on liberalizing trade in agricultural products.

Changes in Attitude

Finally, a renewed emphasis on high-yield farming aimed at preserving biodiversity will require a change in mind-set on the part of key actors: environmentalists, farmers, and government regulators in particular. The environmental movement must postpone its long-cherished goal of an agriculture free from man-made chemicals and give up its lingering hope that constraining food production can somehow limit population growth. Until we understand biological processes well enough to get ultrahigh yields from organic farming, environmentalists must join with farmers in seeking a research agenda keyed primarily to rapid gains in farm yields whether they are organic or not.

Farmers must accept that environmental goals are valid and urgent in a world that produces enough food to prevent famine. They must collaborate constructively and helpfully in efforts such as protecting endangered species and improving water quality. Without such reasonable efforts, farmers will not get public support for high-yield farming systems and liberalized farm trade.

Government regulators at all levels must realize that chemical fertilizers, pesticides, and biotechnology techniques are powerful conservation tools. For example, the Environmental Protection Agency (EPA) must stop regarding a pesticide banned as a victory for the environment. Having dropped the economic rationale protecting some high-yield pesticide uses, EPA should now take into consideration the potential for new pest-control technologies to save wild lands and wild species through higher yields, both nationally and globally.

Education can play a big role in changing the mind-sets of the various actors. For example, the U.S. Department of State, which has already announced an environmental focus for U.S. foreign policy, could work to ensure that the concept of high-yield conservation is appropriately encouraged in international forums. The U.S. Department of Education could collaborate with USDA to help the nation's students understand the environmental benefits of high farm yields.

On all fronts, this is a time for pragmatism. We know that high-yield farming feeds people, saves land, and fosters biodiversity. We know that agricultural research is the surest path to those same goals. The narrower goals should be subsumed into the larger ones for the short- to mid-term future. A combination of agricultural science and policy can combine for the welfare of the planet, its people, its animals, and its plants. Achieving those crucial aims will mean rethinking population, farming methods, fertilizers, and many related controversial aspects of agriculture.

POSTSCRIPT

Should Pollution Prevention and Reduction Be a Focus of Agricultural Policy?

Avery asserts that the use of chemically intensive, high-yield agriculture has saved many millions of acres of wild lands. Other experts disagree with this assertion, arguing that the destruction of rain forests and other sensitive habitat areas has rarely been motivated by the need to increase agricultural lands to grow crops for local populations.

Some argue that even in developed countries, the environmental impacts of high technology farming are very serious. See, for instance, Andrew Nikiforuk, "When Water Kills: The Dangerous Consequences of Factory Farming Are Being Felt All Across the Country," *Maclean's* (June 12, 2000). J. Clark Ward, P. Lowe, and S. Seymour, in "Keeping Matter in Its Place: Pollution Regulation and the Reconfiguring of Farmers and Farming," *Environment and Planning A* (July 1998), contend that in the United Kingdom, new regulations are needed to "encourage farmers to integrate environmental considerations into their farming systems."

For two articles that focus on the need to develop sustainable agricultural practices that will allow developing countries to increase food production while also protecting the environment, see Donald L. Plucknett and Donald L. Winkelmann, "Technology for Sustainable Agriculture," *Scientific American* (September 1995) and Mark W. Rosegrant and Robert Livernash, "Growing More Food, Doing Less Damage," *Environment* (September 1996). In "Toward a 'Greener' Revolution," *Issues in Science and Technology* (Fall 1997), Ross M. Welch, Gerald F. Combs, Jr., and John M. Duxbury contend that the green revolution's exclusive focus on increasing agricultural yields rather than food quality has contributed to nutritional deficiencies that desperately need to be corrected.

ISSUE 10

Is Biotechnology an Environmentally Sound Way to Increase Food Production?

YES: International Food Information Council, from *Food Biotechnology and the Environment*, <http://ificinfo.health.org/brochure/bioenv.htm> (March 31, 1998)

NO: Brian Halweil, from "The Emperor's New Crops," *World Watch* (July/August 1999)

ISSUE SUMMARY

YES: The International Food Information Council, a food industry–supported education organization, asserts that biotechnology can safely modify crops in ways that will help feed the increasing world population while reducing the resulting toll on the environment.

NO: Worldwatch Institute researcher Brian Halweil argues that the genetic modification of crops threatens to produce pesticide-resistant insect pests and herbicide-resistant weeds, will victimize poor farmers, and is unlikely to feed the world.

In little more than two decades rapid advances in the field of biotechnology have transformed the common science fiction theme of new life forms resulting from the transfer of genes from one organism to another (transgenics) into a commercially successful reality. From its beginning, the field of genetic engineering has raised a wide variety of ethical and scientific controversies. Of specific concern to environmentalists are the possible ecological and human health consequences of developing and using transgenic plants for food and fiber production and the creation of microorganisms that are capable of devouring oil spills or consuming toxic waste components.

In the mid-1970s, when progress in DNA research was making the creation and marketing of genetically modified organisms feasible in the United States, concerned scientists and environmental organizations expressed fears about such possibilities as the unintentional production and proliferation of a hazardous new bacterium or virus. The National Institutes of Health responded to these concerns by imposing stringent controls on genetic engineering research and on the testing of the environmental effects of genetically engineered life forms.

Lobbyists for the growing biotechnology industry mounted a strong protest, claiming that the fears about transgenic organisms were unfounded and that the restrictions were imposing unfair economic burdens that would impede the introduction of a wide variety of extremely valuable new products. The industry promised more productive and healthful crops that could produce their own fertilizers and pesticides. New designer drugs were envisioned that would prevent or cure presently intractable diseases. And it was argued that the development of new organisms through genetic manipulation was no different from the long-accepted practice of modifying the properties of plants and animals through interbreeding.

The Reagan and Bush administrations reacted to industry pressure by relaxing restrictions on the development and commercial introduction of genetically altered organisms. In 1987 this policy received a boost from a controversial report released by the National Academy of Sciences that denied the existence of any unique hazards from the use of recombinant DNA techniques to produce biotechnological products. Nevertheless, ecologists continue to report potentially serious ecological problems, and there have been some demonstrations verifying the truth of predictions of such effects as the unanticipated transfer of introduced genetic modifications to related organisms in the wild. Despite these warnings, the U.S. Office of Science and Technology issued guidelines in 1992 that forbid regulations in most cases that are based on the assumption that the risks of bioengineered crops are greater than those resulting from traditional breeding experiments. This policy has led to faster commercial introduction of genetically engineered organisms in the United States. Most European countries, by contrast, have continued to heed the warnings of many ecologists and the fears of farmers and consumers and have either banned or stringently restricted the marketing of biotechnological products.

The following selections illustrate the two sharply different perspectives on biotechnology. The International Food Information Council (IFIC) describes many potential environmental benefits of biotechnology, such as plants that produce their own fertilizers and pesticides—reducing the need for toxic agricultural chemicals—and more productive crops, which will decrease the need to convert sensitive habitats into farmland. The IFIC asserts that extensive testing will prevent any possible environmental problems. Brian Halweil argues that the genetic modification of crops, as presently controlled by international agribusiness, threatens to produce pesticide-resistant insects and herbicide-resistant weeds, further impoverish poor farmers, and fail to reduce world hunger. He advocates a very different kind of agricultural paradigm.

 **YES**

Food Biotechnology and the Environment

Even under the best of conditions, food production for hundreds of millions of Americans—and billions more around the globe—can take a toll on the environment. Erosion can claim precious topsoil; farm chemicals sometimes pollute streams, rivers and ground water supplies; livestock can deplete grazing lands; American wetlands and other sensitive habitats sometimes get plowed under for use as farmland; while, in the world's tropical forests, where an estimated 90 percent of the world's species exist, poor farmers fell trees in order to provide food and a living for their families.

By improving many aspects of modern agriculture, biotechnology can help alleviate many of these pressures on the land, both by preserving natural resources and reducing environmental stresses.

Increasing Crops' Own Ability to Fight Pests and Diseases

Insect resistance   To protect against insect damage and minimize the amount of chemicals sprayed on crop plants, biotech has modified crop plants to protect themselves against insects, rather than rely solely on surface application of pesticides.

Specifically, researchers have transferred into tomato, potato, corn, eggplant and cotton plants different genes that produce natural proteins that kill specific insects after they take a bite of the modified plants.

Disease resistance   Plant viruses of varying kinds often claim up to 80 percent of many crops. In much the same way vaccines immunize humans against various diseases, biotech allows modern breeders to insert small fragments of plant viruses into crop plants, which develop natural protection or immunity against those viral diseases. The immunity is passed on to future generations of plants.

From International Food Information Council, *Food Biotechnology and the Environment,* March 31, 1998, http://ificinfo.health.org/brochure/bioenv.htm. Copyright © 1998 by The International Food Information Council. Reprinted by permission.

What this will mean for the environment world wide is enormous:

- Growers will only need to plant one or perhaps two acres—instead of five—to ensure one acre's worth of harvest. This obviously means far fewer agricultural inputs like fuel, labor, water and fertilizer.
- Chemical sprays required to kill the aphids that transport most viruses would be reduced or eliminated.
- Viral protection has already been achieved in the crook neck squash and will reach growers of cantaloupes, watermelons, cucumbers, potatoes, tomatoes, lettuce and alfalfa by the year 2000. Put all these crops together with viral resistance and imagine the reduction in the overall chemical use in the farming environment and the scaling back of other agricultural inputs because fewer acres will be necessary to plant.

Reducing Overall Chemical Stress on the Environment

Like any products sold in highly competitive markets, today's fungicides, herbicides, insecticides and pesticides are better, safer and more environmentally sensitive than older versions. Even so, they sometimes pollute the air, soil and ground water when they blow or wash off plants.

Farmers recognize more than anyone that healthy growing environments define their future. Thus, they always seek better ways to control weeds with the least toxic herbicides available that do not damage food crops. It makes good, common sense—not to mention environmental and economic sense—for farmers to choose the chemical product that works best and is safest.

Herbicide-tolerant crops   It will take years before biotechnology reveals enough secrets to develop non-chemical alternatives that can perform as well as farm chemicals do now. Therefore, it makes sense to make today's crop plants work in concert with today's best farm chemicals to require less volume or fewer applications.

Biotech has already made possible crop plants that withstand specific farm chemicals while all the weeds around them die. These crops include corn, canola, cotton, soybeans and some vegetables and fruits. This way farmers will get better yields, farm environments will have less stress placed on them and all of us will enjoy foods grown with the judicious use of agricultural chemicals.

Saving Valuable Topsoil

Erosion of topsoil due to wind and water can be cut by more than 70 percent—in some cases up to 98 percent—when farmers use no-till techniques, meaning they do not plow under weeds after harvesting or before planting. Eager to protect their most important asset—the soil—American farmers have flocked to no-till farming. Acres committed to no-till nearly doubled between 1989 and 1992, reaching 28 million. That figure is expected to top 40 million acres in 1996.

Importantly, no-till increases the need for environmentally-sensitive herbicides, as weeds that do not get plowed under present a pernicious problem. By making big crops like corn, soybeans and others tolerant to today's best herbicides, biotech breeders offer three vital benefits:

- They will increase the total number of acres committed to no-till practices, thus further cutting erosion's toll on America's topsoil.
- Farmers will save money because of less equipment, fuel and labor costs —even with the additional cost for herbicides.
- USDA studies have shown that no-till practices reduce the amount of farm chemical run off because the crop residues trap chemicals that might otherwise spread.

Breeders also seek to improve the situation by breeding crop plants that thrive in no-till settings. This includes plants that yield more to maximize production on cultivated land and lessen the need to put additional land into production. (Viral resistance crops best illustrate this point; without them, farmers often plant up to five acres to get the harvest of one.)

Feedstock Efficiency

As much as 60 percent of American corn and soybean crops each year get eaten by farm animals that become the meat on our dinner plates. Anything that makes animals better able to digest their food will diminish the overall need for feed, without diminishing the overall productivity of the animals. A natural protein from dairy cows called bovine somatotropin (BST) can be given in supplemental doses, which allows the treated cows to absorb more nutrients from their feed, thus producing the same amount of milk from less feed. Developments like these promise to:

- Reduce overall amounts of feed necessary to produce similar yields.
- Reduce the amount of grazing lands needed.
- Reduce water used by livestock and for livestock feed.
- Free up acres used to produce livestock feed for other crops or uses.

The Nitrogen Burden

Even though our atmosphere contains about 78 percent nitrogen, most crops have no mechanism to use this natural nitrogen. Therefore, farmers depend on added fertilizers to provide the nitrogen necessary to boost crop yields. But crops only use about 50 percent of the more than 60 million pounds of nitrogen fertilizers added to them each year. The excess nitrogen can cause environmental problems in soil and water.

Growers have long recognized and used the innate abilities of legumes like soybeans to "fix" nitrogen, which means to use the natural nitrogen in the soil and air. These natural nitrogen fixers actually leave the soils from which they have been harvested with replenished nitrogen. Breeders desire to develop other crop plants that can fix their own nitrogen.

In order to mimic something you must first understand it. And biotech has given breeders the tools they need to understand exactly how crops take up and use nitrogen. Efficiency of nitrogen intake and use rank high on many biotech target lists in many countries. Keep in mind, self-fixing plants remain enticing dreams for the next century, and are not likely in the near future. Should breeders succeed in creating them, they would:

- Allow farmers to decrease their use of synthetic fertilizers while maintaining bountiful yields.
- Result in less nitrogen from fertilizers being left to degrade in the soil and ground water.
- Greatly enhance productivity in many regions of the developing world whose farmers cannot afford nitrogen fertilizers.

Forestry

Trees take decades to reproduce and mature, too long a time for breeders to do much with traditional breeding techniques. New techniques have already shortened certain tree breeding cycles to months, which has resulted in four to six-fold increases in the yields of some rubber, cocoa, teak and pulpwood trees. This means less stress on naturally-occurring trees, which is good for the world's forest environments. Biotech is also helping forest preservation in other ways. Some examples include:

- American breeders have already protected the Dutch elm tree from Dutch elm disease and the American chestnut tree from chestnut blight, making many shade lovers much happier and adding important defensive diversity to American trees.
- In Brazil, a eucalyptus plantation has enjoyed a 112 percent yield increase since it began employing modern cloning techniques in the 1980s; a genetically-enhanced Brazilian yellow pine tree can now produce 50 cubic meters of wood per year, while the average remains about 15 cubic meters.
- Seventy-five percent of 125 medicinal compounds used in drugs today came from plants, animals and trees in the world's rain forests. By enabling scientists to identify the actual genes responsible for these compounds, the forests they came from can be protected because of their enhanced value.
- Because most genes can be spliced into bacteria and their products harvested from fermentation vats, instead of pristine forests, the threat to valuable trees in sensitive forests can be diminished. Taxol from the bark of yew trees often found in old growth forests in America's northwest can now be produced in cell labs, opening up the possibility in the next century that no trees need fall to produce this effective fighter of breast and ovarian cancers.

Through genetic modification, tree breeders seek trees that produce higher-quality wood, as well as trees that better tolerate disease, insects and poor soil conditions. By increasing the productivity of commercial tree farms, biotech can help lessen the need for commercial cutting in rain forests or in national forests and parks around the world.

Plant Biodiversity

Of the more than 80,000 species of edible plants known to exist, humans cultivate only about 300 of them. Of those, only about 12 have emerged as major staples. Through genetic modification, crop breeders can:

- Increase the use of plant diversity by first learning through biotech what genes of interest reside in which plants and then by moving these specific genes into crops now in use around the globe.
- Expanding the genetic variation in staple crops by breeding into them desirable traits from heretofore unavailable sources. This will not affect the relatively narrow genetic lineage of many crops in the near term. By next century it will significantly expand the gene pool used in modern agriculture and thus reduce the relatively low, but real risk of potential crop failures.
- Expand via cloning any wild relative of modern crop plants that might be threatened with extinction.
- Finally, enable scientists to learn what important genes are actually contained in the millions of plant specimens housed in gene banks around the world.

Extensive Testing for Environmental Safety

In addition to the environmental benefits biotechnology will bring, there are potential risks that need to be carefully evaluated for each new plant variety. In order to minimize any environmental risks presented by crop plants modified with biotech, ecologists have conducted years of of outdoor testing under strictly-controlled conditions. In the United States alone, researchers have undertaken more than 1,700 outdoor tests at more then 6,500 sites with no resulting environmental problems.

Before field testing a genetically-modified crop that might introduce a plant pest into the environment, an application must be approved by the USDA Animal and Plant Health Inspection Service (APHIS) and, in some cases, by the Environmental Protection Agency. The federal agencies designate an assessment team that includes geneticists, ecologists and state regulatory staff with expertise in regional environmental conditions to evaluate the technical data required with the application. The test receives approval after the agencies determine that the new crop variety will present no different risks to the environment than the traditional variety.

In March 1993, APHIS announced amended rules that allow six crops with a long history of safe genetic modification to undergo field tests with 30-day advance notification to APHIS. Other crops have since been added to this group. Researchers remain committed to testing and monitoring newly-developed plants to ensure their benefits are safely introduced.

Brian Halweil

 **NO**

The Emperor's New Crops

It's June 1998 and Robert Shapiro, CEO of Monsanto Corporation, is delivering a keynote speech at "BIO 98," the annual meeting of the Biotechnology Industry Organization. "Somehow," he says, "we're going to have to figure out how to meet a demand for a doubling of the world's food supply, when it's impossible to conceive of a doubling of the world's acreage under cultivation. And it is impossible, indeed, even to conceive of increases in productivity—using current technologies—that don't produce major issues for the sustainability of agriculture."

Those "major issues" preoccupy a growing number of economists, environmentalists, and other analysts concerned with agriculture. Given the widespread erosion of topsoil, the continued loss of genetic variety in the major crop species, the uncertain effects of long-term agrochemical use, and the chronic hunger that now haunts nearly 1 billion people, it would seem that a major paradigm shift in agriculture is long overdue. Yet Shapiro was anything but gloomy. Noting "the sense of excitement, energy, and confidence" that engulfed the room, he argued that "biotechnology represents a potentially sustainable solution to the issue of feeding people."

To its proponents, biotech is the key to that new agricultural paradigm. They envision crops genetically engineered to tolerate dry, low-nutrient, or salty soils—allowing some of the world's most degraded farmland to flourish once again. Crops that produce their own pesticides would reduce the need for toxic chemicals, and engineering for better nutrition would help the overfed as well as the hungry. In industry gatherings, biotech appears as some rare hybrid between corporate mega-opportunity and international social program.

The roots of this new paradigm were put down nearly 50 years ago, when James Watson and Francis Crick defined the structure of DNA, the giant molecule that makes up a cell's chromosomes. Once the structure of the genetic code was understood, researchers began looking for ways to isolate little snippets of DNA—particular genes—and manipulate them in various ways. In 1973, scientists managed to paste a gene from one microbe into another microbe of a different species; the result was the first artificial transfer of genetic information across the species boundary. In the early 1980s, several research teams—including one at Monsanto, then a multinational pesticide company—

From Brian Halweil, "The Emperor's New Crops," *World Watch* (July/August 1999). Copyright © 1999 by The Worldwatch Institute. Reprinted by permission of *World Watch*.

succeeded in splicing a bacterium gene into a petunia. The first "transgenic" plant was born.

Such plants represented a quantum leap in crop breeding: the fact that a plant could not interbreed with a bacterium was no longer an obstacle to using the microbe's genes in crop design. Theoretically, at least, the world's entire store of genetic wealth became available to plant breeders, and the biotech labs were quick to test the new possibilities. Among the early creations was a tomato armed with a flounder gene to enhance frost resistance and with a rebuilt tomato gene to retard spoilage. . . .

Transgenic crops are no longer just a laboratory phenomenon. Since 1986, 25,000 transgenic field trials have been conducted worldwide—a full 10,000 of these just in the last two years. More than 60 different crops—ranging from corn to strawberries, from apples to potatoes—have been engineered. From 2 million hectares in 1996, the global area planted in transgenics jumped to 27.8 million hectares in 1998. That's nearly a fifteenfold increase in just two years.

In 1992, China planted out a tobacco variety engineered to resist viruses and became the first nation to grow transgenic crops for commercial use. Farmers in the United States sowed their first commercial crop in 1994; their counterparts in Argentina, Australia, Canada, and Mexico followed suit in 1996. By 1998, nine nations were growing transgenics for market and that number is expected to reach 20 to 25 by 2000.

Ag biotech is now a global phenomenon, but it remains powerfully concentrated in several ways:

In terms of where transgenics are planted.   Three-quarters of transgenic cropland is in the United States. More than a third of the U.S. soybean crop last year was transgenic, as was nearly one-quarter of the corn and one-fifth of the cotton. The only other countries with a substantial transgenic harvest are Argentina and Canada: over half of the 1998 Argentine soybean crop was transgenic, as was over half of the Canadian canola crop. These three nations account for 99 percent of global transgenic crop area. (Most countries have been slow to adopt transgenics because of public concern over possible risks to ecological and human health.)

In terms of which crops are in production.   While many crops have been engineered, only a very few are cultivated in appreciable quantities. Soybeans account for 52 percent of global transgenic area, corn for another 30 percent. Cotton—almost entirely on U.S. soil—and canola in Canada cover most of the rest.

In terms of which traits are in commercial use.   Most of the transgenic harvest has been engineered for "input traits" intended to replace or accommodate the standard chemical "inputs" of large-scale agriculture, especially insecticides and herbicides. Worldwide, nearly 30 percent of transgenic cropland is planted in varieties designed to produce an insect-killing toxin, and almost all of the rest is in crops engineered to resist herbicides. . . .

These two types of crops—the insecticidal and the herbicide-resistant varieties—are biotech's first large-scale commercial ventures. They provide the
first real opportunity to test the industry's claims to be engineering a new
agricultural paradigm.

The Bugs

The only insecticidal transgenics currently in commercial use are "Bt crops."
Grown on nearly 8 million hectares worldwide in 1998, these plants have been
equipped with a gene from the soil organism *Bacillus thuringiensis* (Bt), which
produces a substance that is deadly to certain insects.

The idea behind Bt crops is to free conventional agriculture from the
highly toxic synthetic pesticides that have defined pest control since World War
II. Shapiro, for instance, speaks of Monsanto's Bt cotton as a way of substituting
"information encoded in a gene in a cotton plant for airplanes flying over cotton fields and spraying toxic chemicals on them." ... At least in the short term,
Bt varieties have allowed farmers to cut their spraying of insecticide-intensive
crops, like cotton and potato. In 1998, for instance, the typical Bt cotton grower
in Mississippi sprayed only once for tobacco budworm and cotton bollworm—
the insects targeted by Bt—while non-Bt growers averaged five sprayings.

Farmers are buying into this approach in a big way. Bt crops have had
some of the highest adoption rates that the seed industry has ever seen for new
varieties. In the United States, just a few years after commercialization, nearly
25 percent of the corn crop and 20 percent of the cotton crop is Bt. In some
counties in the southeastern states, the adoption rate of Bt cotton has reached
70 percent. The big draw for farmers is a lowering of production costs from
reduced insecticide spraying, although the savings is partly offset by the more
expensive seed. Some farmers also report that Bt crops are doing a better job
of pest control than conventional spraying, although the crops must still be
sprayed for pests that are unaffected by Bt. (Bt is toxic primarily to members of
the Lepidoptera, the butterfly and moth family, and the Coleoptera, the beetle
family.)

Unfortunately, there is a systemic problem in the background that will almost certainly erode these gains: pesticide resistance. Modern pest management
tends to be very narrowly focused; the idea, essentially, is that when faced with
a problematic pest, you should look for a chemical to kill it. The result has been
a continual toughening of the pests, which has rendered successive generations
of chemicals useless. After more than 50 years of this evolutionary rivalry, there
is abundant evidence that pests of all sorts—insects, weeds, or pathogens—will
develop resistance to just about any chemical that humans throw at them.

The Bt transgenics basically just replace an insecticide that is sprayed on
the crop with one that is packaged inside it. The technique may be more sophisticated but the strategy remains the same: aim the chemical at the pest. Some
entomologists are predicting that, without comprehensive strategies to prevent
it, pest resistance to Bt could appear in the field within three to five years of
widespread use, rendering the crops ineffective. Widespread resistance to Bt
would affect more than the transgenic crops, since Bt is also commonly used in

conventional spraying. Farmers could find one of their most environmentally benign pesticides beginning to slip away.

In one respect, Bt crops are a throwback to the early days of synthetic pesticides, when farmers were encouraged to spray even if their crops didn't appear to need it. The Bt crops show a similar lack of discrimination: they are programmed to churn out toxin during the entire growing season, regardless of the level of infestation. This sort of prophylactic control greatly increases the likelihood of resistance because it tends to maximize exposure to the toxin—it's the plant equivalent of treating antibiotics like vitamins.

Agricultural entomologists now generally agree that Bt crops will have to be managed in a way that discourages resistance if the effectiveness of Bt is to be maintained. In the United States, the Environmental Protection Agency, which regulates the use of pesticides, now requires producers of Bt crops to develop "resistance management plans." This is a new step for the EPA, which has never required analogous plans from manufacturers of conventional pesticides.

The usual form of resistance management involves the creation of "refugia"—areas planted in a crop variety that isn't armed with the Bt gene. If the refugia are large enough, then a substantial proportion of the target pest population will never encounter the Bt toxin, and will not be under any selection pressure to develop resistance to it. Interbreeding between the refugia insects and the insects in the Bt fields should stall the development of resistance in the population as a whole, assuming the resistance gene is recessive.

The biotech companies themselves have been recommending that their customers plant refugia, although the recommendations generally fall short of what most resistance experts consider necessary. This is not surprising, of course, since there is an inherent inconsistency between the refugia idea and the inevitable interest on the part of the manufacturer in selling as much product as possible. An even greater obstacle may be the reactions of farmers themselves, since the refugia concept is counter-intuitive: farmers, who spend much of their lives trying to control pests, are being told that the best way to maintain a high yield is to leave substantial portions of their land vulnerable to pests. The impulse to plant smaller refugia—or to count someone else's land as part of one's own refugia—may prove irresistible. And the possibility of enforcing the planting of larger refugia seems remote, especially once Bt crops are deployed to hundreds of millions of small-scale farmers throughout the developing world....

According to Gary Barton, director of ag biotech communications at Monsanto, "products now in the pipeline which rely on different insecticidal toxins or multiple toxins could replace Bt crops in the event of widespread resistance."...

The result, according to Fred Gould, an entomologist at the University of North Carolina, would be "a crop with a series of silver bullet pest solutions." And each of these solutions, in Gould's view, would be highly vulnerable to pest resistance. This scenario does not differ essentially from the current one: in place of a pesticide treadmill, we would substitute a sort of gene treadmill. The arms race between farmers and pests would continue, but would include an additional biochemical dimension. Transgenic plants, designed to secrete increasingly potent combinations of pesticides, would vie with a host of increasingly resistant pests.

Figure 1

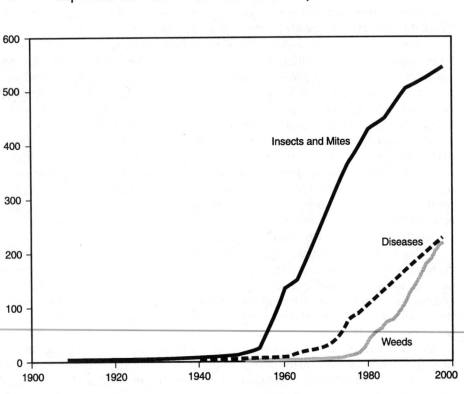

Reported Numbers of Pesticide-Resistant Species, 1908–98

Source: Worldwatch Institute, Vital Signs 1999

The Weeds

The global transgenic harvest is currently dominated, not by Bt crops, but by herbicide-resistant crops (HRCs), which occupy 20 million hectares worldwide. HRCs are sold as part of a "technology package" comprised of HRC seed and the herbicide the crop is designed to resist. The two principal product lines are currently Monsanto's "Roundup Ready" crops—so-named because they tolerate Monsanto's best-selling herbicide, "Roundup" (glyphosate)—and AgrEvo's "Liberty Link" crops, which tolerate that company's "Liberty" herbicide (glufosinate).

It may sound contradictory, but one ostensible objective of HRCs is to reduce herbicide use. By designing crops that tolerate fairly high levels of exposure to a broad-spectrum herbicide (a chemical that is toxic to a wide range of plants), the companies are giving farmers the option of using a heavy, once-in-the-growing-season dousing with that herbicide, instead of the standard practice, which calls for a series of applications of several different compounds. It's not yet clear whether this new herbicide regime actually reduces the amount of material used, but its simplicity is attracting many farmers into the package.

Another potential benefit of HRCs is that they may allow for more "conservation tillage," farming techniques that reduce the need for plowing or even —under "no till" cultivation—eliminate it entirely. A primary reason for plowing is to break up the weeds, but because it exposes bare earth, plowing causes top soil erosion. Top soil is the capital upon which agriculture is built, so conserving soil is one of agriculture's primary responsibilities....

The bigger problem is that HRCs, like Bt crops, are really just an extension of the current pesticide paradigm. HRCs may permit a reduction in herbicide use over the short term, but obviously their widespread adoption would encourage herbicide dependency. In many parts of the developing world, where herbicides are not now common, the herbicide habit could mean substantial additional environmental stresses: herbicides are toxic to many soil organisms, they can pollute groundwater, and they may have long-term effects on both people and wildlife.

And of course, resistance will occur. Bob Hartzler, a weed scientist at Iowa State University, warns that if HRCs encourage reliance on just a few broad-spectrum herbicides, then resistance is likely to develop faster—and agriculture is likely to be more vulnerable to it....

In the U.S. Midwest, heavy use of Roundup on Roundup Ready soybeans is already encouraging weed species, like waterhemp, that are naturally resistant to that herbicide. (As Roundup suppresses the susceptible weeds, the resistant ones have more room to grow.) Thus far, the evolution of resistance in weed species that are susceptible to Roundup has been relatively rare, despite decades of use. The first reported case involved wild ryegrass in Australia, in 1995. But with increasing use, more such cases are all but inevitable—especially since Monsanto is on the verge of releasing Roundup Ready corn. Corn and soybeans are the classic crop rotation in the U.S. Midwest—corn is planted in one year, soy in the next. Roundup Ready varieties of both crops could subject vast areas of the U.S. "breadbasket" to an unremitting rain of that herbicide. As with the Bt crops, the early promise of HRCs is liable to be undercut by the very mentality that inspired them: the single-minded chemical pursuit of the pest.

Transgenes on the Loose

In 1997, just one year after its first commercial planting in Canada, a farmer reported—and DNA testing confirmed—that Roundup Ready canola had cross-pollinated with a related weed species growing in the field's margins, and produced an herbicide-tolerant descendant....

If a transgenic crop is capable of sexual reproduction (and they generally are), the leaking of "transgenes" is to some degree inevitable, if any close relatives are growing in the vicinity. This type of genetic pollution is not likely to be common in the industrialized countries, where most major crops have relatively few close relatives. But in the developing world—especially in regions where a major crop originated—the picture is very different. Such places are the "hot spots" of agricultural diversity: the cultivation of the ancient, traditional varieties—whether it's corn in Mexico or soybeans in China—often involves a subtle genetic interplay between cultivated forms of a species, wild forms, and related species that aren't cultivated at all. The possibilities for genetic pollution in such contexts are substantial.

Ordinary breeding creates some degree of genetic pollution too. But according to Allison Snow, an Ohio State University plant ecologist who studies transgene flow, biotech could amplify the process considerably because of the far more diverse array of genes it can press into service. Any traits that confer a substantial competitive advantage in the wild could be expected to spread widely. The Bt gene would presumably be an excellent candidate for this process, since its toxin affects so many insect species.

There's no way to predict what would happen if the Bt gene were to escape into a wild flora, but there's good reason to be concerned. John Losey, an entomologist at Cornell University, has been experimenting with Monarch butterflies, by raising their caterpillars on milkweed dusted with Bt-corn pollen. Losey found that nearly half of the insects raised on this fare died and the rest were stunted. (Caterpillars raised on milkweed dusted with ordinary corn pollen did fine.) According to Losey, "these levels of mortality are comparable to those you find with especially toxic insecticides." If the gene were to work a change that dramatic in a wild plant's toxicity, then it could trigger a cascade of second- and third-order ecological effects.

The potential for this kind of trouble is likely to grow, since a major interest in biotech product development is "trait-stacking"—combining several engineered genes in a single variety, as with the attempts to develop corn with multiple toxins. Monsanto's "stacked cotton"—Roundup Ready and Bt-producing—is already on the market in the United States. . . .

In the agricultural hot spots, there is an important practical reason to be concerned about any resulting genetic pollution. Plant breeders depend on the genetic wealth of the hot spots to maintain the vigor of the major crops—and there's no realistic possibility of biotech rendering this natural wealth "obsolete." But it certainly is possible that foreign genes could upset the relationships between the local varieties and their wild relatives. . . .

Toward a New Feudalism

The advent of transgenic crops raise serious social questions as well—beginning with ownership. All transgenic seed is patented, as are most nontransgenic commercial varieties. But beginning in the 1980s, the tendency in industrialized countries and in international law has been to permit increasingly broad agricultural patents—and not just on varieties but even on specific genes. Under the earlier, more limited patents, farmers could buy seed and use it in their own breeding; they could grow it out and save some of the resulting seed for the next year; they could even trade it for other seed. About the only thing they couldn't do was sell it outright. But under the broader patents, all of those activities are illegal; the purchaser is essentially just paying for one-time use of the germplasm.

The right to own genes is a relatively new phenomenon in world history and its effects on agriculture—and life in general—are still very uncertain. The biotech companies argue that ownership is essential for driving their industry: without exclusive rights to a product that costs hundreds of millions of dollars to develop, how will it be possible to attract investors?... Val Giddings of the Biotechnology Industry Organization makes this case: "intellectual property rights allow us to harness genetic resources for commercial use, making biodiversity concretely more valuable."...

Patents are clearly an important ingredient in the industry's expansion. Global sales of transgenic crop products grew from $75 million in 1995 to $1.5 billion in 1998—a 20-fold increase. Sales are expected to hit $25 billion by 2010. And as the market has expanded, so has the scramble for patents. Recently, for example, the German agrochemical firm AgrEvo, the maker of "Liberty" herbicide, bought a Dutch biotech company called Plant Genetic Systems (PGS), which owned numerous wheat and corn patents. The patents were so highly valued that AgrEvo was willing to pay $730 million for the acquisition—$700 million more than PGS's annual sales....

This patent frenzy is contributing to an intense wave of consolidation within the industry.... Hoechst recently merged with one of its French counterparts, Rhône-Poulenc, to form Aventis, which is now the world's largest agrochemical firm and a major player in the biotech industry. On the other side of the Atlantic, Monsanto has spent nearly $8 billion since 1996 to purchase various seed companies. DuPont, a major competitor, has bought the world's largest seed company, Pioneer Hi-Bred. DuPont and Monsanto were minor players in the seed industry just a decade ago, but are now respectively the largest and second-largest seed companies in the world.

... Of the 56 transgenic products approved for commercial planting in 1998, 33 belonged to just four corporations: Monsanto, Aventis, Novartis, and DuPont. The first three of these companies control the transgenic seed market in the United States, which amounts to three-fourths of the global market....

But there is far more at stake here than the fortunes of the industry itself: patents and similar legal mechanisms may be giving companies additional control over farmers. As a way of securing their patent rights, biotech companies are requiring farmers to sign "seed contracts" when they purchase transgenic seed —a wholly new phenomenon in agriculture. The contracts may stipulate what brand of pesticides the farmer must use on the crop—a kind of legal cement for those crop-herbicide "technology packages."...

The most troubling aspect of these contracts is the possible effect on seed saving—the ancient practice of reserving a certain amount of harvested seed for the next planting. In the developing world, some 1.4 billion farmers still rely almost exclusively on seed saving for their planting needs. As a widespread, low-tech form of breeding, seed saving is also critical to the husbandry of crop diversity, since farmers generally save seed from plants that have done best under local conditions. The contracts have little immediate relevance to seed saving in the developing world, since the practice there is employed largely by farmers who could not afford transgenic seed in the first place. But even in industrialized countries, seed saving is still common in certain areas and for certain crops, and Monsanto has already taken legal action against over 300 farmers for replanting proprietary seeds.

... The substitution of commercial for farm-saved seed has been a primary reason for the loss of genetic diversity in the agricultural hot spots. Hope Shand, research director for the Rural Advancement Foundation International (RAFI), a farmer advocacy group based in Winnipeg, Canada, regards the extension of patents in general as a means of reducing farmers to "bioserfs," who provide little more than land and labor to agribusiness.

... [B]eyond these control issues, there remains the basic question of biotech's potential for feeding the world's billions. Here too, the current trends are not very encouraging. At present, the industry has funneled its immense pool of investment into a limited range of products for which there are large, secured markets within the capital-intensive production systems of the First World. There is very little connection between that kind of research and the lives of the world's hungry. HRCs, for example, are not helpful to poor farmers who rely on manual labor to pull weeds because they couldn't possibly afford herbicides. (The immediate opportunities for biotech in the developing world are not the subsistence farmers, of course, but the larger operations, which are often producing for export rather than for local consumption.)

Just to get a sense of proportion on this subject, consider this comparison. The entire annual budget of the Consultative Group for International Agricultural Research (CGIAR), a consortium of international research centers that form the world's largest public-sector crop breeding effort, amounts to $400 million. The amount that Monsanto spent to develop Roundup Ready soybeans alone is estimated at $500 million. In such numbers, one can see a kind of financial disconnect. Per Pinstrup-Andersen, director of the International Food Policy Research Institute, the CGIAR's policy arm, puts it flatly: "the private sector will not develop crops to solve poor people's problems, because there is not enough money in it." The very nature of their affliction—poverty—makes hungry people poor customers for expensive technologies.

In addition to the financial obstacle, there is a biological obstacle that may limit the role of biotech as agricultural savior. The crop traits that would be most useful to subsistence farmers tend to be very complex, Miguel Altieri, an entomologist at the University of California at Berkeley, identifies the kind of products that would make sense in a subsistence context: "crop varieties responsive to low levels of soil fertility, crops tolerant of saline or drought conditions and other stresses of marginal lands, improved varieties that are not dependent on agro-chemical inputs for increased yields, varieties that are compatible with small, diverse, capital-poor farm settings." In HRCs and Bt crops, the engineering involves the insertion of a single gene. Most of the traits Altieri is talking about are probably governed by many genes, and for the present at least, that kind of complexity is far beyond the technology's reach.

Beyond the Techno-Fix

... On a 300-acre farm in Boone, Iowa—the heart of the U.S. corn belt—Dick Thompson rotates corn, soybeans, oats, wheat interplanted with clover, and a hay combination that includes an assortment of grasses and legumes. The pests that plague neighboring farmers—including the corn borer targeted by Bt corn—are generally a minor part of the picture on Thompson's farm. High crop diversity tends to reduce insect populations because insect pests are usually "specialists" on one particular crop. In a very diverse setting, no single pest is likely to be able to get the upper hand. Diversity also tends to shut out weeds, because complex cropping uses resources more efficiently than monocultures, so there's less left over for the weeds to consume.... Even without herbicides, Thompson's farm has been on conservation tillage for the last three decades.... [C]attle, a hog operation, and the nitrogen-fixing legumes provide the soil nutrients that most U.S. farmers buy in a bag. The soil organic matter content—the sentinel indicator of soil health—registers at 6 percent on Thompson's land, which is more than twice that of his neighbors.... Thompson's soybean and corn yields are well above the county average and even as the U.S. government continues to bail out indebted farmers, Thompson is making money. He profits both from his healthy soil and crops, and from the fact that his "input" costs—for chemical fertilizer, pesticides, and so forth—are almost nil.

In the activities of people like Herren and Thompson it is possible to see a very different kind of agricultural paradigm, which could move farming beyond the techno-fix approach that currently prevails. Known as agroecology, this paradigm recognizes the farm as an ecosystem—an agroecosystem—and employs ecological principles to improve productivity and build stability. The emphasis is on adapting farm design and practice to the ecological processes actually occurring in the fields and in the landscape that surrounds them. Agroecology aims to substitute detailed (and usually local) ecological knowledge for off-the-shelf and off-the-farm "magic bullet" solutions. The point is to treat the disease, rather than just the symptoms. Instead of engineering a corn variety that is toxic to corn rootworm, for example, an agroecologist would ask why there's a rootworm problem in the first place.

Where would biotech fit within such a paradigm? In the industry's current form, at least, it doesn't appear to fit very well at all. Biotech's first agricultural products are "derivative technologies," to use a term favored by Frederick Buttel, a rural sociologist at the University of Wisconsin. Buttel sees those products as "grafted onto an established trajectory, rather than defining or crystallizing a new one."

There is no question that biotech contains some real potential for agriculture, for instance as a supplement to conventional breeding or as a means of studying crop pathogens. But if the industry continues to follow its current trajectory, then biotech's likely contribution will be marginal at best and at worst, given the additional dimensions of ecological and social unpredictability—who knows? In any case, the biggest hope for agriculture is not something biochemists are going to find in a test tube. The biggest opportunities will be found in what farmers already know, or in what they can readily discover on their farms.

POSTSCRIPT

Is Biotechnology an Environmentally Sound Way to Increase Food Production?

The IFIC makes no mention of the serious ecological and economic risks described by Halweil. Note that Halweil does not deny that biotechnology has the potential to make significant contributions to improving agricultural productivity of the foods that will be needed by the world's burgeoning population. His point is that there are no economic incentives that would induce IFIC member corporations, who now control biotechnological development, to realize that potential.

There is widespread concern among environmentalists and ecologists about the present regulatory situation. In the United States the Environmental Protection Agency is prohibited by federal policy from responding to the possible environmental threats of genetically altered plants. Most developing countries, where transgenic crops are likely to be first tried out, lack the infrastructure for appropriate oversight of these experiments. These and other regulatory issues are discussed in Sheldon Krimsky's essay "Biotechnology Safety," *Environment* (June 1997), which is a critical review of reports by the World Bank and the Institute of Food Technologies. A second critical discussion of this issue by agricultural scientists is "Genetic Engineering in Agriculture and the Environment," by M. G. Paoletti and D. Pimentel, *Bioscience* (vol. 46, no. 9, 1996). The continuing "lack of a rigorous regulatory framework to sort out the risks inherent in agricultural biotech" is discussed by Charles Mann in "Biotech Goes Wild," *Technology Review* (July–August 1999). Margaret Kriz, in "Global Food Fight," *National Journal* (March 4, 2000), describes the January 2000 Montreal meeting where representatives of 130 countries reached "an agreement that requires biotechnology companies to ask permission before importing genetically altered seeds [and] forces food companies to clearly identify all commodity shipments that may contain genetically altered grain."

The Spring 1993 issue of *The Amicus Journal* contains a special section on biotechnology and ecology featuring a series of articles that probe a variety of issues, including those raised by Rifkin. For an entertaining and informative assessment of the Monsanto Corporation's recently marketed, genetically engineered New Leaf Superior potato, read Michael Pollan's cover story "Playing God in the Garden" in the October 25, 1998, issue of *The New York Times Magazine*. For two uncritical articles that extend the promise of the future wonders to be realized from biotechnology far beyond the limits of agriculture, see "The Promise of Genetics," by Joseph F. Coates, John B. Mahaffie, and Andy Hines, *The Futurist* (September/October 1997) and "The Next Biotech Harvest," by David Rotman, *Technology Review* (September/October 1998).

ISSUE 11

Do Environmental Hormone Mimics Pose a Potentially Serious Health Threat?

YES: Theo Colborn, Dianne Dumanoski, and John Peterson Myers, from "Hormone Imposters," *Sierra* (January/February 1997)

NO: Stephen H. Safe, from "Environmental and Dietary Estrogens and Human Health: Is There a Problem?" *Environmental Health Perspectives* (April 1995)

ISSUE SUMMARY

YES: Zoologist Theo Colborn, journalist Dianne Dumanoski, and John Peterson Myers, director of the W. Alton Jones Foundation, present evidence suggesting that low environmental levels of hormone-mimicking chemicals may threaten the health of humans and other animals.

NO: Toxicologist Stephen H. Safe argues that the suggestion that industrial estrogenic compounds contribute to increased cancer incidence and reproductive problems in humans is not plausible.

Following World War II there was an exponential growth in the industrial use and marketing of synthetic chemicals. These chemicals, known as "xenobiotics," were used in numerous products, including solvents, pesticides, refrigerants, coolants, and raw materials for plastics. This resulted in increasing environmental contamination. Many of these chemicals, such as DDT, PCBs, and dioxins, proved to be highly resistant to degradation in the environment; they accumulated in wildlife and were serious contaminants of lakes and estuaries. Carried by winds and ocean currents, these chemicals were soon detected in samples taken from the most remote regions of the planet, far from their points of introduction into the ecosphere.

Until very recently most efforts to assess the potential toxicity of synthetic chemicals to bio-organisms, including human beings, focused almost exclusively on their possible role as carcinogens. This was because of legitimate public concern about rising cancer rates and the belief that cancer causation was the most likely outcome of exposure to low levels of synthetic chemicals.

Some environmental scientists urged public health officials to give serious consideration to other possible health effects of xenobiotics. They were generally ignored because of limited funding and the common belief that toxic effects other than cancer required larger exposures than usually resulted from environmental contamination.

In the late 1980s Theo Colborn, a research scientist for the World Wildlife Fund who was then working on a study of pollution in the Great Lakes, began linking together the results of a growing series of isolated studies. Researchers in the Great Lakes region, as well as in Florida, the U.S. West Coast, and Northern Europe, had observed widespread evidence of serious and frequently lethal physiological problems involving abnormal reproductive development, unusual sexual behavior, and neurological problems exhibited by a diverse group of animal species, including fish, reptiles, amphibians, birds, and marine mammals. Through Colborn's insights, communications among these researchers, and further studies, a hypothesis was developed that all of these wildlife problems were manifestations of abnormal estrogenic activity. The causative agents were identified as more than 50 synthetic chemical compounds that have been shown in laboratory studies to either mimic the action or disrupt the normal function of the powerful estrogenic hormones responsible for female sexual development and many other biological functions.

Concern that human exposure to these ubiquitous environmental contaminants may have serious health repercussions was heightened by a widely publicized European research study, which concluded that male sperm counts had decreased by 50 percent over the past several decades (a result that is disputed by other researchers) and that testicular cancer rates have tripled. Some scientists have also proposed a link between breast cancer and estrogen disrupters.

In response to the mounting scientific evidence that environmental estrogens may be a serious health threat, the U.S. Congress passed legislation requiring that all pesticides be screened for estrogenic activity and that the Environmental Protection Agency (EPA) develop procedures for detecting environmental estrogenic contaminants in drinking water supplies; see http://www.epa.gov/scipoly/oscpendo/index.htm. Government-sponsored studies of synthetic endocrine disrupters and other hormone mimics are also underway in the United Kingdom and in Germany.

In the following article, which was adapted from their highly acclaimed book *Our Stolen Future* (Dutton, 1996), Colborn, Dianne Dumanoski, and John Peterson Myers summarize research connecting numerous instances of wildlife abnormalities to exposure to environmental hormone mimics. Stephen H. Safe's opposing article was chosen even though it includes technical arguments that presume a background in biochemistry that few readers of this book are likely to possess. It is, however, the most often-cited denunciation of the likelihood of a causative relationship between environmental estrogens and human health problems. Safe's principal arguments and the evidence he cites can be understood independent of the more technical information and arguments that he presents.

Theo Colborn, Dianne Dumanoski,
and John Peterson Myers

 **YES**

Hormone Imposters

From the corner of her eye, Theo Colborn caught sight of another scientific report shooting across the floor and settling on the carpet. She didn't even bother to turn her head. In hopes of slowing the tide of paper that was swamping her office, she had taken to closing her door, but it had done little good. With the wrist action of a Frisbee champion, the project director simply flicked them under the door. He had become so adept that he could wing documents to the center of the office. She just let them sit there.

Flick. Another flick. Flick. So much had accumulated since she had begun her 1987 review of scientific papers concerning the health of wildlife and humans in the Great Lakes region that her office looked like a landfill. Some documents, like the text of a speech by a governor from one of the states bordering on the Great Lakes, touted improvements. Public officials on both sides of the border had all but declared total victory in the battle against the severe pollution that had brought the Great Lakes nationwide notoriety in the 1960s and 1970s.

But Colborn, a 58-year-old zoologist working for the Conservation Foundation, had come to doubt that the lakes were truly cleaned up. Biologists working in the field were still reporting vanished mink populations and, in fish-eating birds, unhatched eggs, deformities such as crossed bills, missing eyes, and clubfeet, and a puzzling indifference about incubating eggs. Colborn's eyes wandered across the file boxes lining her office—43 of them, one for each species that had been studied in the Great Lakes. Everywhere in this wildlife data she was finding signs that something was still seriously wrong. And if something was still wrong with wildlife, what were the implications for humans living in the region?

⚜

Colborn dove into the health data, focusing especially on places where wildlife researchers had found cancer in fish. If the wildlife in an area were providing warnings, she reasoned, then that was where one would first expect to find humans with higher cancer rates. Yet no matter which way she cut the data,

From Theo Colborn, Dianne Dumanoski, and John Peterson Myers, "Hormone Imposters," *Sierra* (January/February 1997). Copyright © 1997 by *Sierra* Magazine, 85 Second Street, 2d Floor, San Francisco, CA 94105. Reprinted by permission of *Sierra* Magazine; permission conveyed through Copyright Clearance Center, Inc.

they failed to show that people in the Great Lakes basin were dying of cancer more than people elsewhere in the United States and Canada.

Colborn turned her mind again to the wildlife literature. Sitting there surrounded by the cartons of animal studies, she was suddenly struck by the obvious. For the past three decades, the words "toxic chemical" have become almost synonymous with cancer, not only to the public but to scientists and regulators as well. Colborn now recognized that this preoccupation with cancer had been blinding her to the diversity of data she had collected. Most of the problems reported in wildlife were *not* cancer. Except in fish in highly contaminated areas, cancer was extremely rare.

Moving beyond cancer proved to be the most important step in her investigation. As she looked at the same material from a new perspective, she gradually began to recognize important clues and follow where they led. Birds, mammals, and fish seemed to be experiencing similar problems. Although the adult animals living in and around the Great Lakes were reproducing, their offspring often did not survive. Colborn began to focus on studies that compared lake populations to inland ones. In every case, the lake dwellers, which appeared otherwise healthy, were far less successful in producing surviving offspring. The contamination in the parents was somehow affecting their young.

Because investigations of the effects of exposure to synthetic chemicals on humans had focused largely on cancer in exposed adults, only a handful of scientists had looked for possible effects on the children of exposed individuals. But Colborn recalled a study of the children of women who had regularly eaten Great Lakes fish. She dug it out of her files and read it again. The Wayne State University study had found evidence that the mothers' levels of chemical contamination affected their babies' development. The children of mothers who had eaten two to three meals of fish a month were born sooner, weighed less, and had smaller heads than those of mothers who did not. Moreover, the greater the amount of the family of industrial chemicals called PCBs (polychlorinated biphenyls) in the umbilical-cord blood, the more poorly the child scored on tests assessing neurological development, lagging behind in various measures, such as short-term memory, that tend to predict later performance on intelligence tests.

As Colborn dug deeper, more parallels became apparent. In the tissue analyses done on wildlife, the same chemicals kept showing up in troubled species, among them the pesticides DDT, dieldrin, chlordane, and lindane, as well as PCBs. Studies had found the very same chemicals in human blood and body fat. Colborn was particularly shocked by the high concentrations reported in human breast milk.

She also re-explored the literature on an eerie wasting syndrome seen in young terns. The chicks could look normal and healthy for days, but then suddenly and unpredictably begin to languish, waste away, and finally die. The wasting problem, scientists were learning, was a symptom of disordered metabolism. The young birds could not produce sufficient energy to survive.

What did this all mean? Having plowed through more than two thousand scientific papers and five hundred government documents, Colborn began entering her findings on an electronic spreadsheet. As she made entries under columns headed "population decline," "reproductive effects," "tumors," "wasting," "immune suppressions," and "behavioral changes," her attention came to focus increasingly on the Great Lakes species that seemed to be having the greatest array of problems.

Colborn sat back and looked at the list: the bald eagle, lake trout, herring gull, mink, otter, double-crested cormorant, snapping turtle, common tern, and coho salmon. What did they have in common? Each was a top predator that fed on fish. Although the concentrations of contaminants such as PCBs are so low in the water of the Great Lakes that they cannot be measured using standard water-testing procedures, such persistent chemicals concentrate in the tissue and accumulate exponentially as they move from animal to animal up the food chain. Through this process of magnification, the concentrations of a persistent chemical that resists breakdown and accumulates in body fat can be 25 million times greater in a top predator such as a herring gull than in the surrounding water.

One other vital fact emerged from the spreadsheet. According to the scientific literature, *adult* animals appeared to be doing fine. Health problems were found primarily in their offspring. Although she had been thinking about offspring effects, Colborn had not recognized this stark, across-the-board contrast between adults and young.

Now the pieces were beginning to fit together. If the chemicals found in the parents' bodies were to blame, they were acting as hand-me-down poisons, passed from one generation to the next, victimizing the unborn and the very young. The conclusion was chilling.

But the symptoms did not seem to add up. Some animals, like the gulls, exhibited strange behavior such as same-sex nests, while other species, including the double-crested cormorant, had visible birth defects. Again a pattern emerged. These were all cases of derailed development, a process guided to a significant extent by chemical messengers in the body called hormones. Hormones move about constantly within the body's communication network, regulating metabolism, reproduction, mental processes, and other aspects of development before birth.

The hormone link pointed Colborn's investigation in another direction. She began reading everything she could find about the chemicals that showed up again and again in tissue analyses of animals having trouble producing viable young. She quickly learned that the testing and review done by manufacturers and government regulatory agencies had focused largely on whether a chemical might cause cancer, but she found enough in the peer-reviewed scientific literature to prove that her hunch had been correct: the pass-along poisons found in the fat of the wildlife all disrupted the hormone system.

⋅✦⋅

By 1990 Colborn had expanded her horizons from the Great Lakes to the world. Now a researcher for the World Wildlife Fund, she collected every sort of relevant evidence and filed even the smallest tidbit in an ever-expanding database on hormone disruption. Her synthesis of the far-flung studies from hundreds of researchers in dozens of disciplines was the kind of work that almost never occurs in government or universities because there is neither support nor reward for doing it. Nobody ever got tenure for analyzing other people's research. Billions are spent on individual scientific studies but virtually nothing to figure out what they collectively say about the state of the earth.

Colborn usually had little time to reflect on the implications of what she was doing. But once in a rare while, alone late at night, she would think about what all of the pieces might add up to. What were the long-term effects of these hormone-disrupting chemicals? Were we sabotaging our own fertility as well as that of wildlife? Was it possible that we were unknowingly and invisibly undermining the reproductive future of our children? On the face of it, the idea seemed preposterous. How could human fertility be in jeopardy when world population was soaring from 5 billion toward 10? Maybe she was chasing phantoms.

A few months later, any lingering doubts vanished. At a meeting in Ottawa in the summer of 1990, Colborn heard Richard Peterson of the University of Wisconsin talk about dioxin, a persistent contaminant produced in the manufacture of certain chlorine-containing chemicals such as pesticides and wood preservatives, as well as by bleaching paper with chlorine, incinerating trash containing plastics and paper, and burning fossil fuels. Peterson had given dioxin to pregnant rats to see how it would affect the development of their male offspring. As expected, the highly toxic dioxin damaged the male reproductive system if the pups were exposed during a critical period in their prenatal development. What surprised Peterson was how little dioxin it took to do the damage. He saw long-term effects on male pups even when the mother rats had ingested an astonishingly tiny amount of dioxin at a critical moment. These findings had direct and immediate relevance: the lowest doses given to the mother rats had been very near the levels of dioxin and related compounds reported in people in industrialized countries such as the United States, Japan, and in Europe.

Over the years, dozens of scientists had explored isolated pieces of the puzzle of hormone disruption, but the larger picture did not emerge until Colborn and John Peterson Myers, a zoologist who directs the W. Alton Jones Foundation, finally brought 21 key researchers together in July 1991. At the Wingspread Conference in Racine, Wisconsin, specialists from diverse disciplines ranging from anthropology to zoology shared what they knew about the role of hormones in normal development and about the devastating impacts of hormone-disrupting chemicals on wildlife, laboratory animals, and humans. As the evidence was laid out, the conclusion seemed inescapable: the hormone disrupters threatening the survival of animal populations are also jeopardizing the human future.

At levels typically found in the environment, these chemicals act like thugs on the biological information highway, sabotaging vital hormonal communication. They mug the messengers or impersonate them. They jam signals. They scramble messages. They sow disinformation. Because hormone messages orchestrate many critical aspects of development, from sexual differentiation to brain organization, hormone-disrupting chemicals pose a particular hazard before birth and early in life. The process that unfolds in the womb and creates a normal, healthy baby depends on getting the right hormone message to the fetus at the right time. Relatively low levels of contaminants that have no observable impact on adults can have a devastating impact on the unborn. The most important concept in thinking about this kind of toxic assault is not poisons, not carcinogens, but chemical messages.

At the end of the 1991 conference, the scientists issued the Wingspread Statement, an urgent warning that humans in many parts of the world are being exposed to chemicals that have disrupted development in wildlife and laboratory animals, and that unless these chemicals are controlled, we face the danger of widespread disruption of human embryonic development.

⋅⊙⋅

How many synthetic chemicals scramble the body's chemical messages? No one knows because no one has systematically screened the tens of thousands of chemicals created since World War II. Many of the known offenders have been discovered by accident.

To date, researchers have identified at least 51 synthetic chemicals—most of them ubiquitous in the environment—that disrupt hormones in one way or another. Some mimic estrogen, and others interfere with other parts of the body's control or "endocrine" system, such as testosterone and thyroid metabolism. This tally of hormone disrupters includes large chemical families such as the 209 compounds classified as PCBs, the 75 dioxins, and the 135 furans. It also includes the most heavily used pesticide in American agriculture, atrazine, and the infamous DDT, which, while banned in the United States, continues to circulate in the global environment. The PCBs, DDT, and dioxin get the lion's share of attention because they happen to be the only hormone-disrupting chemicals that scientists have studied in any depth. But the age of discovery is far from over.

With so many chemicals at issue, the situation confronting us does not lend itself to easy prescriptions. Our current economy and civilization are built on a foundation of fossil fuels and synthetic chemicals. It has taken fifty years to work our way into this dilemma; it will almost certainly take just as long or longer to find our way out of it.

As we work to create a future where children can be born free of chemical contamination, our scientific knowledge and technological expertise will be crucial. Nothing, however, will be more important to human well-being and survival than the wisdom to appreciate that however great our knowledge, our ignorance is also vast. In this ignorance we have taken huge risks, and inadvertently gambled with survival. Now that we know better, we must have the courage to be cautious, for the stakes are very high. We owe that much, and more, to our children.

NO ↵

Stephen H. Safe

Environmental and Dietary Estrogens and Human Health: Is There a Problem?

$\mathbf{R}$ecent reports have suggested that background levels of industrial chemicals and other environmental pollutants may play a role in development of breast cancer in women and decreased male reproductive success as well as the reproductive failures of some wildlife species. These suggestions have been supported by articles in the popular and scientific press and by a television documentary which have described the perils of exposure to endocrine-disrupting chemicals such as estrogenic organochlorine pesticides and pollutants. During the past two decades, environmental regulations regarding the manufacture, use, and disposal of chemicals have resulted in significantly reduced emissions of most industrial compounds and their by-products. Levels of the more environmentally stable organochlorine pesticides and pollutants are decreasing in most ecosystems including the industrialized areas around the Great Lakes in North America. Decreased levels of organochlorine compounds correlates with the improved reproductive success of highly susceptible fish-eating water birds in the Great Lakes region. This article reviews key papers that have been used to support the hypotheses that environmental estrogens play a role in the increased incidence of breast cancer in women and decreased sperm counts in males. Environmental/dietary estrogens and antiestrogens are identified and intakes of "estrogen equivalents" are estimated to compare the relative dietary impacts of various classes of estrogenic chemicals.

Role of Estrogens in Breast Cancer and Male Reproductive Problems

Concerns regarding the role of environmental and dietary estrogens as possible contributors to the increased incidence of breast cancer were fueled by several reports that showed elevated levels of organochlorine compounds in breast cancer patients.... Polychlorinated biphenyls (PCBs) and 1, 1-dichloro-2, 2-bis(p-chlorophenyl)ethylene (DDE) are the two most abundant oganochlorine

From Stephen H. Safe, "Environmental and Dietary Estrogens and Human Health: Is There a Problem?" *Environmental Health Perspectives,* vol. 103, no. 4 (April 1995). References omitted.

pollutants identified in all human tissues with high frequencies. In one Scandinavian study, levels of DDE or PCBs in adipose tissue from breast samples were not significantly different in breast cancer patients compared to controls. In another study in Finland, β-hexachlorocyclohexane levels were elevated in breast cancer patients; however, this compound was not detected in adipose tissue of some individuals in the patient and control groups and has a relatively low frequency of detection in human tissue samples. Falck and co-workers reported that PCB levels were elevated in mammary adipose tissue samples from breast cancer patients in Connecticut. In contrast, serum levels of DDE (but not PCBs) were significantly elevated in breast cancer patients enrolled in the New York University Women's Health Study. DDE (but not PCB) levels were also elevated in estrogen receptor (ER)-positive but not ER-negative breast cancer patients from Quebec compared to levels in women with benign breast disease. It was initially concluded by Wolff and co-workers that "these findings suggest that environmental chemical contamination with organochlorine residues may be an important etiologic factor in breast cancer." The correlations reported in the two U.S. studies heightened public and scientific concern regarding the potential role of these compounds in development of breast cancer. These observations undoubtedly reinforced advocacy by some groups for a ban on the use of all chlorine-containing chemicals. However, the proposed linkage between PCBs and/or DDE and breast cancer is questionable for the following reasons:

- Most studies with PCBs indicate that these mixtures are not estrogenic, and the weak estrogenic activity observed for lower chlorinated PCB mixtures may be due to their derived hydroxylated metabolites;
- p,p'-DDE, the dominant persistent metabolite of 1, 1, 1-trichloro-2, 2-bis(p-chlorophenyl)ethane (p,p'-DDT), is not estrogenic, and levels of $o,p' = $ DDT, the estrogenic member of the DDT family, are low to nondetectable in most environmental samples;
- Epidemiology studies of individuals occupationally exposed to relatively high levels of DDT or PCBs do not show a higher incidence of breast cancer; and
- No single class of organochlorine compounds was elevated in all studies, suggesting that other factors may be critical for development of breast cancer.

Krieger and co-workers recently reported results from a nested case–control study of women from the San Francisco area which showed that there were no differences in serum DDE or PCB levels between breast cancer patients and control subjects. The authors concluded that "the data do not support the hypothesis that exposure to DDE and PCBs increases risk of breast cancer." This was duly noted in *Time* magazine by a three-line statement in "The Good News" section. Moreover, combined analysis of the 6 studies which report PCB and DDE levels in 301 breast cancer patients and 412 control patients showed that there were no significant increases in either DDE or PCB levels in breast cancer patients versus controls.

The second major link between environmental/dietary estrogens and human disease was precipitated by an article published in the *Lancet*, in which Sharpe and Skakkebaek hypothesized that increased estrogen exposure may be responsible for falling sperm counts and disorders of the male reproductive tract. Unlike the proposed link between environmental estrogens and breast cancer, this hypothesis was not based on experimentally derived measurements of increased levels of any estrogenic compounds in males. Previous studies with diethylstilbestrol, a highly potent estrogenic drug, showed that *in utero* exposure results in adverse effects in male offspring, and the authors' hypothesized that *in utero* exposure to environmental/dietary estrogens may also result in adverse effects in male offspring. A critical experimental component supporting the authors' hypothesis was their analysis of data from several studies which indicated that male sperm counts had decreased by over 40% during the past 50 years. These observations, coupled with the hypothesis that environmental estrogens including organochlorine chemicals were possible etiologic agents, were reported with alarm in the popular and scientific press and in a BBC television program entitled "Assault on the Male: a Horizon Special." Subsequent and prior scientific studies have cast serious doubts on both the hypothesis and the observed decrease in male sperm counts. In 1979, Macleod and Wang reported that there had been no decline in sperm counts, and reanalysis of the data presented by Carlsen and co-workers showed that sperm counts had not decreased from 1960 to 1990. Thus, during the time in which environmental levels of organochlorine compounds were maximal, there was not a corresponding decrease in sperm counts. Moreover, a reevaluation of the sperm concentration data was recently reported by Brownwich et al. in the *British Medical Journal*, and their analysis suggested that the decline in sperm values in males was a function of the choice of the normal or reference value for sperm concentrations. The authors contend that their analysis of the data does "not support the hypothesis that the sperm count declined significantly between 1940 and 1990."

These results suggest that the increasing incidence of human breast cancer is not related to organochlorine environmental contaminants and that decreases in sperm counts is highly debatable. Nevertheless, human populations are continually exposed to a wide variety of environmental and dietary estrogens, and these compounds clearly fit into the category of "endocrine disrupters." The remainder of this article briefly describes the different structural classes of both environmental and dietary estrogens and quantitates human exposures to these compounds.

Synthetic Industrial Chemicals With Estrogenic Activity

The estrogenic activities of different structural classes of industrial chemicals were reported by several research groups in the late 1960s and 1970s in which *o,p'*-DDT and other diphenylmethane analogs and the insecticide kepone were characterized as estrogens. Subsequent studies have confirmed the estrogenic activity of *o,p'*-DDT and related compounds whereas the *p,p'*-substituted analogs were relatively inactive. In addition, p,p'-methoxychlor and its hydroxylated metabolites elicit estrogenic responses. Ecobichon and Comeau investigated the estrogenic activities of commercial PCB mixtures (Aroclors) and individual congeners in the female rat uterus and reported estrogenic responses for some Aroclors and individual congeners. Studies in this laboratory showed that a number of commercial PCBs did not significantly increase secretion of procathepsin D, an estrogen-regulated gene product, in MCF-7 human breast cancer cells. It should be noted that several hydroxylated PCBs bind to the ER, and it is possible that *para*-hydroxylated PCB metabolites may be the active estrogenic compounds associated with lower chlorinated PCBs. A recent study reported that several additional organochlorine pesticides including endosulfan, toxaphene, and dieldrin exhibit estrogenlike activity and induce proliferation of MCF-7 human breast cancer cells.

Other industrial chemicals or intermediates that have been identified as estrogenic compounds include bisphenol-A, a chemical used in the manufacture of polycarbonate-derived products; phenol red, a pH indicator used in cell culture media; and alkyl phenols and their derivatives, which are extensively used for preparation of polyethoxylates in detergents.

Natural Estrogenic Compounds

Human exposure to estrogenic chemicals is not confined to xenoestrogens derived from industrial compounds. Several different structural classes of naturally occurring estrogens have been identified, including plant bioflavonoids and various mycotoxins including zearalenone and related compounds. The plant bioflavonoids include different structural classes of compounds which contain a flavonoid backbone: flavones, flavanones, flavonols, isoflavones, and related condensation products (e.g., coumesterol). The estrogenic activities of diverse phytoestrogenic bioflavonoids and mycotoxins have been extensively investigated in *in vivo* models, *in vitro* cell culture systems, and in ER binding assays, and most of these compounds elicit multiple estrogenic responses in these assays. In addition, a number of plant foodstuffs contain 17 β-estradiol (E_2) and estrone.

Environmental and Dietary Antiestrogens

Several different structural classes of chemicals found in the human diet also exhibit antiestrogenic activity. 2, 3, 7, 8-Tetrachlorodibenzo-*p*-dioxin (TCDD) and related halogenated aromatics including polychlorinated dibenzo-*p*-dioxins (PCDDs), dibenzofurans (PCDFs), and PCBs are also an important class of organochlorine pollutants that elicit a diverse spectrum of biochemical and toxic responses. These chemicals act through the aryl hydrocarbon receptor (AhR)-mediated signal transduction pathway, which is thought to play a role in most of the responses elicited by these compounds. AhR agonists such as TCDD have been characterized as antiestrogens using rodent and cell models similar to those used for determining the estrogenic activity of dietary and environmental chemicals. In the rodent model, TCDD and related compounds inhibit several estrogen-induced uterine responses including increased uterine wet weight, peroxidase activity, cytosolic and nuclear progesterone receptor (PR) and ER binding, epidermal growth factor (EGF) receptor binding, EGF receptor mRNA, and c-*fos* mRNA levels. In parallel studies, the antiestrogenic activities of TCDD and related compounds have also been investigated in several human breast cancer cell lines. For example, structurally diverse AhR agonists inhibit the following E_2-induced responses in MCF-7 human breast cancer cells: post-confluent focus production, secretion of tissue plasminogen activator activity, procathepsin D (52-kDa protein), cathepsin D (34-kDa protein), a 160-kDa protein, PR binding sites, glucose-to-lactate metabolism, pS2 protein levels, and PR, cathepsin D, ER, and pS2 gene expression. Moreover, TCDD inhibits formation and/or growth of mammary tumors in athymic nude mice and female Sprague-Dawley rats after long-term feeding studies or initiation with 7,12-dimethylbenzanthracene. A recent epidemiology study on women exposed to TCDD after an industrial accident in Seveso reported that breast cancer incidence was decreased in areas with high levels of TCDD contamination (particularly in the age class 45 to 74) and among women living longest in an area of low TCDD contamination. Endometrial cancer showed a remarkable decrease, particularly in areas with medium and low TCDD contamination. Thus, TCDD and related compounds exhibit a broad spectrum of antiestrogenic activities and, not surprisingly, so do other AhR agonists such as the polynuclear aromatic hydrocarbons (PAHs), indole-3-carbinol (IC), and related compounds found in relatively high levels in foodstuffs. PAHs are found in cooked foods and are ubiquitous environmental contaminants. IC is a major component of cruciferous vegetables (e.g., brussels sprouts, cauliflower) and exhibits antiestrogenic and anticancer (mammary) activities.

Bioflavonoids have been extensively characterized as weak estrogens and therefore may also be active as antiestrogens at lower concentrations. The interaction between estrogenic bioflavonoids and E_2 depends on their relative doses or concentrations, the experimental model, and the specific estrogen-induced endpoint. Markaverich and co-workers reported that the estrogenic bioflavonoids quercetin and luteolin inhibited E_2-induced proliferation of MCF-7 human breast cancer cells and E_2-induced uterine wet weight increase in 21-day-old female rats. Similar results were also observed in this laboratory for quercetin, resperetin, and naringenin. For example, the bioflavonoid naringenin inhibited estrogen-induced uterine hyptertrophy in female rats and estrogen-induced luciferase activity in MCF-7 cells transfected with an E_2-responsive plasmid construct containing the 5'-promoter region of the pS2 gene and a luciferase reporter gene (unpublished results). In contrast, a recent study reported that coumestrol, genistein, and zearalenone were not antiestrogenic in human breast cancer cells. The antiestrogenic activities of weak dietary and environmental estrogens require further investigation; however, it is clear that at subestrogenic doses, some of these compounds exhibit antiestrogenic activities in both *in vivo* and *in vitro* models.

Mass/Potency Balance

The uptake of environmental or dietary chemicals that elicit common biochemical/toxic responses can be estimated by using an equivalency factor approach in which estrogen equivalents (EQs) in any mixture are equal to the sum of the concentration of the individual compounds (EC_i) times their potency (EP_i) relative to an assigned standard such as diethylstilbestrol (DES) or E_2. The total EQs in a mixture would be:

$$EQ = \Sigma \left([EC_i] \times EP_i \right)$$

A similar approach is being used to determine the TCDD equivalents (TEQs) of various mixtures containing halogenated hydrocarbons. Verdeal and Ryan have previously used this approach with DES equivalents assuming that the oral potency of E_2 is 15% that of DES. Winter has estimated the dietary intake of pesticides based on FDA's total diet study, which includes estimates of food intakes and pesticide residue levels in these foods. The results presented in Table 1 summarize the estimated exposure of different groups to estrogenic pesticides. For example, 14- to 16-year-old males were exposed to a total of 0.0416 [μ]g/kg/day of the estrogenic pesticides, DDT, dieldrin, endosulfan, and *p,p'*-methoxychlor (note: the DDT value represents *p,p'*-DDE and related metabolites, which are primarily nonestrogenic). Thus, the overall dietary intake of these compounds by this age group was 2.5[μ]g/day.

Table 1

Estimated Dietary Intake of Estrogenic Pesticides by Different Age Groups Based on Food Intakes and Pesticide Levels in These Foodstuffs

	Estimated exposure (μg/kg/day)		
	6–11	14–16[a]	60–65
Pesticide	months	years	years
DDT (total)	0.077	0.0260	0.0103
Dieldrin	0.0014	0.0016	0.0016
Endosulfan	0.0274	0.0135	0.0210
p,p'- Methoxychlor	0.0005	0.0005	0.0001

[a] Maximum exposure: 60 x 0.0416 = 2.5 μg/day.

The relative potencies of dietary and xenoestrogens are highly variable. The results of *in vitro* cell culture studies suggest that estrogenic potencies of bioflavonoids relative to E_2 are 0.001 to 0.0001 whereas Soto and co-workers have assigned an estrogen potency factor of 0.000001 for the estrogenic pesticides. These relative estrogen potency factors for bioflavonoids and pesticides may be lower when derived from *in vivo* studies since pharmacokinetic factors and metabolism may decrease bioavailability. Thus, a more accurate assessment of dietary/environmental EQs requires further data from dietary feeding studies that evaluate these compounds using the same experimental protocols.

The results in Table 2 summarize human exposure to dietary and environmental estrogens and the estimated daily dose in terms of EQs. The relative estrogenic intakes for various hormonal drug therapies were previously estimated by Verdeal and Ryan; the average estimated daily intake of all flavonoids in food products was 1020 and 1070 mg/day, (winter and summer, respectively). The results show that the estimated dietary EQ levels of estrogenic pesticides are 0.0000025 [μ]g/day, whereas the corresponding dietary EQ levels for the bioflavonoids are 102 [μ]g/day. Thus, the EQ values for the dietary intake of flavonoids was 4 × 10 times higher than the daily EQ intake of estrogenic pesticides, illustrating the minimal potential of these industrial estrogens to cause an adverse endocrine-related response in humans.

Table 2

Estimated Mass Balance of Human Exposures to Environmental and Dietary Estrogens and Antiestrogens

Source	Estrogen equivalents (μg/day)
Estrogens	
Morning after pill	333,500
Birth control pill	16,675
Post-menopausal therapy	3,350
Flavonoids in foods (1,020 mg/day x 0.0001)	102
Environmental organochlorine estrogens	
(2.5 x 0.000001)	0.0000025
Antiestrogens	TCDD antiestrogen equivalents (μg/day)
TCDD and organochlorines (80–120 pg/day)	0.000080–0.000120[a]
PAHs in food (1.2–5.0 x 106 pg/day;	
relative potency ~ 0.001)	0.001200–0.0050[b]
Indolo[3,2-*b*]carbazole in 100 g brussels sprouts	0.000250–0.00128[b]
(0.256–1.28 x 106 pg/day; relative potency	
~ 0.001)	

[a] In most studies, 1 nM TCDD inhibits 50–100% of 1 nM E_2-induced responses in MCF-7 cells; therefore, 1 estrogen equivalent ≅ 1 antiestrogen equivalent.
[b] The antiestrogenic potencies of PAHs and indolo[3,2-*b*]carbazole compared to E_2 were approximately 0.001.

Previous studies have also shown that AhR agonists, such as TCDD and related compounds, PAHs and IC and its most active derivative, indolo[3,2-b]carbazole (ICZ) all inhibit E_2-induced responses in MCF-7 cells. At a concentration of 10^{-9} M, TCDD inhibits 50–100% of most E_2-induced responses *in vitro* in which the concentration of E_2 is 10^{-9} M. Therefore, 1 TEQ is approximately equal to 1 EQ. The estimated daily intakes of TCDD and related compounds, PAHs, and ICZ (in 100 g brussels sprouts) are summarized in Table 2. The relative potencies of PAHs and ICZ as antiestrogens compared to TCDD are approximately 0.001 in MCF-7 cells. Thus, the TEQs or antiestrogen TEQs can be calculated for the dietary intakes of TCDD and related organochlorines and PAHs (in all foods). The antiestrogen TEQs for the three classes of dietary AhR agonists are orders of magnitude higher than the estimated dietary intakes of estrogenic pesticide EQs. Thus, the major human intake of endocrine disrupters associated with the estrogen-induced response pathways are naturally occurring estrogens found in foods. Relatively high serum levels of estrogenic bioflavonoids have also been detected in a Japanese male population, whereas lower levels were observed in a Finnish group, and this is consistent with their dietary intakes of these estrogenic compounds, *p,p'*-DDE is present in human serum; however, the estrogenic *o,p'*-DDE and *o,p'*-DDT analogs and other weakly estrogenic organochlorine compounds are not routinely detected in serum samples. A recent study identified several hydroxylated PCB congeners in human serum. All of the hydroxylated compounds were also substituted with chlorine groups at both adjacent meta positions. Based on results of previous structure-activity studies (43) for hydroxylated PCBs, these compounds would exhibit minimal estrogenic activity; however, further studies on the activity of hydroxylated PCBs are warranted.

Summary

The hypothesized linkage between dietary/environmental estrogens and the increased incidence of breast cancer is unproven; there is a lack of correlation between higher organochlorine levels in breast cancer patients compared to controls and the low levels of organochlorine EQs in the diet (Table 2). Higher levels of bioflavonoids are unlikely to contribute to increased breast cancer incidence because these compounds and the foods they are associated with tend to exhibit anticarcinogenic activity. The hypothesis that male reproductive problems and decreased sperm counts are related to increased exposure to environmental and dietary estrogens is also unproven. As noted above, dietary exposure to xenoestrogens derived from industrial chemical residues in foods is minimal compared to the daily intake of EQs from naturally occurring bioflavonoids. Moreover, there are serious questions regarding the decreased sperm counts reported by Carlsen and co-workers. Reanalysis of Carlsen et al.'s data suggests that there has not been a decrease in sperm counts in males over the past 30 years and possibly over the past 50 years. Thus, in response to articles in the popular and scientific press such as "The Estrogen Complex" and "Ecocancers: Do Environmental Factors Underlie a Breast Cancer Epidemic?", the results would suggest that the linkage between dietary or environmental estrogenic compounds and breast cancer has not been made, and further research is required to determine the factors associated with the increasing incidence of this disease.

Note Added in Proof: A recent study reported a 2.1% decrease in sperm concentrations in France from 1973 to 1979.

POSTSCRIPT

Do Environmental Hormone Mimics Pose a Potentially Serious Health Threat?

The statement by Safe that is most often quoted by authors of popular articles about the environmental estrogen controversy is, "The suggestion that industrial estrogenic chemicals contribute to an increased incidence of breast cancer in women and male reproductive problems is not plausible." This quote is taken from the abstract that was published along with his article in *Environmental Health Perspectives*. Note that in the article he draws the much more cautious conclusion that a link between these consequences and estrogenic compounds is "unproven." Drawing conclusions about estrogenic potency based upon the average person's low dietary exposure to a few common pesticides is suspect for several reasons. Individuals living near highly contaminated bodies of water and consuming fish and other foods from them may take in far higher levels of synthetic estrogens. The complex and variable manner by which different compounds with estrogenic properties affect organisms makes comparisons risky. Some researchers have suggested that humans and other animals may have evolved tolerances to naturally occurring estrogenic compounds but not to the synthetics.

Although Colborn says that she would like to be proven wrong, she is among those researchers who believe that there is already convincing evidence of extensive damage to wildlife caused by synthetic estrogenic chemicals and that the likelihood that humans are experiencing similar health problems due to these industrial pollutants is frighteningly high. This argument is amplified in *Our Stolen Future* (Dutton, 1996). Recent data reinforce her points; see Rebecca Renner, "Human Estrogens Linked to Endocrine Disruption," *Environmental Science and Technology* (January 1, 1998); J. Raloff, "Common Pollutants Undermine Masculinity," *Science News* (April 3, 1999); and T. Hesman, "DDT Treatment Turns Male Fish Into Mothers," *Science News* (February 5, 2000). See also Ted Schettler et al., *Generations at Risk: Reproductive Health and the Environment* (MIT Press, 1999).

There have been numerous articles about the environmental endocrine issue in both popular and technical journals and magazines. For three different perspectives on the controversy, see "Hormone Mimics Pose Challenge," *Chemistry and Industry* (May 20, 1996); "Environmental Estrogens," by J. A. McLachlan and S. F. Arnold, *American Scientist* (September/October 1996); and "Hormone Hell," by C. Dold, *Discover* (September 1996). The National Environmental Health Association has called for more research and product testing; see Ginger L. Gist, "National Environmental Health Association Position on Endocrine Disrupters," *Journal of Environmental Health* (January–February 1998).

ISSUE 12

Is the Environmental Protection Agency's Decision to Tighten Air Quality Standards for Ozone and Particulates Justified?

YES: Carol M. Browner, from Statement Before the Committee on Environment and Public Works, U.S. Senate (February 12, 1997)

NO: Daniel B. Menzel, from Statement Before the Subcommittee on Clean Air, Wetlands, Private Property and Nuclear Safety, Committee on Environment and Public Works, U.S. Senate (February 5, 1997)

ISSUE SUMMARY

YES: Carol M. Browner, administrator for the Environmental Protection Agency (EPA), summarizes the evidence and arguments that were the basis for the EPA's proposal for more stringent standards for ozone and particulates. She contends that research indicates that the present acceptable levels for these pollutants may be inadequate and that the new regulations are both necessary and sufficient to protect public health.

NO: Daniel B. Menzel, a professor of environmental medicine and a researcher on air pollution toxicology, agrees that ozone and particulates are serious health hazards. He argues, however, that adequate research has not been done to demonstrate that the new standards will result in the additional public health benefits that would justify the difficulty and expense associated with their implementation.

The fouling of air due to the burning of fuels has long plagued the inhabitants of populated areas. After the industrial revolution, the emissions from factory smokestacks were added to the pollution resulting from cooking and household heating. The increased use of coal combined with local meteorological conditions in London produced the dense, smoky, foggy condition first referred to as "smog" by Dr. H. A. Des Vouex in the early 1900s.

Dr. Des Vouex organized British smoke abatement societies. Despite the efforts of these organizations, the problem grew worse and spread to other industrial centers during the first half of the twentieth century. In December 1930 a dense smog in Belgium's Meuse Valley resulted in 60 deaths and an estimated 6,000 illnesses due to the combined effect of the dust and sulfur oxides from coal combustion. This event, and a similar one in Donora, Pennsylvania, received public attention but little response. Serious smog control efforts did not begin until a disastrous "killer smog" in London in December 1952 resulted in approximately 4,000 deaths.

The first major air pollution control victory was the regulation of high-sulfur coal burning in populated areas. The last life-threatening London smog incidents were recorded in the mid-1960s. Unfortunately, a new smog problem, linked to the rush hour traffic in large cities such as Los Angeles, was fast developing. This highly irritating, smelly, smoky haze—now referred to as photochemical smog—is caused by sunlight acting on air laden with nitrogen oxides, unburned hydrocarbons from automotive exhausts, and other sources. The resulting chemical reactions produce a variety of exotic chemicals that are very irritating to lungs and nasal passages—even at very low levels.

The Clean Air Act of 1963 was the first comprehensive U.S. legislation aimed at controlling air pollution from both stationary sources (factories and power plants) and mobile sources (cars and trucks). Under this law and its early amendments, regulations have been issued to establish maximum levels for both ambient air concentrations and emissions from tailpipes and smokestacks for common pollutants, such as sulfur dioxide, suspended particulates (dust), carbon monoxide, nitrogen dioxide, ozone, and airborne lead.

After years of debate Congress finally passed the 1990 Clean Air Act amendments. This complex piece of legislation includes more stringent standards and extends controls to many additional sources of air emissions. It is designed to improve air quality by the end of the century and to reduce the incidence of acid precipitation that occurs when nitrogen- and sulfur-oxide emissions return to earth hundreds of miles from their sources in the form of rain, snow, or fog laden with sulfuric and nitric acids. The 1990 amendments did not respond to concerns, expressed by many environmental health experts, that the standards for the maximum allowed ambient air levels of ground-level ozone and suspended particulates are too high.

Based on a controversial series of research reports, demands have been made for these levels to be made more stringent to protect the health of sensitive segments of the population, including children, asthmatics, and the elderly. In July 1997, with the support of the Clinton administration, the EPA announced its decision to significantly lower permissible ozone levels and to phase in a new, more restrictive standard for suspended particulate matter aimed at controlling smaller, presumably more toxic particles.

The following selections are excerpts from statements by Carol M. Browner and Daniel B. Menzel, which were included in extensive, contentious testimony at the series of congressional hearings held prior to the implementation of the new standards.

Carol M. Browner  **YES**

Statement of Carol M. Browner

Mr. Chairman, members of the Committee, I want to thank you for inviting me to discuss the Environmental Protection Agency's [EPA] proposed revisions to the national ambient air quality standards for particulate matter and ozone.

On these two pollutants, over the past three and a half years, EPA has conducted one of its most thorough and extensive scientific reviews ever. That review is the basis for the new, more stringent standards for particulate matter and ozone that we have proposed in order to fulfill the mandate of the Clean Air Act.

On average, an adult breathes in about 13,000 liters of air each day. Children breathe in 50 percent more air per pound of body weight than do adults.

For 26 years, the Clean Air Act has promised American adults and American children that they will be protected from the harmful effects of dirty air—based on the best available science. Thus far, when you consider how the country has grown since the Act was first passed, it has been a tremendous success. Since 1970, while the U.S. population is up 28 percent, vehicle miles travelled are up 116 percent and the gross domestic product has expanded by 99 percent, emissions of the six major pollutants or their precursors have *dropped* by 29 percent.

The Clinton Administration views protecting public health and the environment as one of its highest priorities. We have prided ourselves on protecting the most vulnerable among us—especially our children—from the harmful effects of pollution. When it comes to the Clean Air Act, I take very seriously the responsibility the Congress gave me to set air quality standards that "protect public health with an adequate margin of safety"—based on the best science available.

Mr. Chairman, the best available, current science tells me that the current standards for particulate matter and ozone are not adequate, and I have therefore proposed new standards that I believe, based on our assessment of the science, are required to protect the health of the American people.

The standard-setting process includes extensive scientific peer review from experts outside of EPA and the Federal Government. Under the law, we are not to take costs into consideration when setting these standards. This has been the case through six Presidential administrations and 14 Congresses, and has

From U.S. Senate. Committee on Environment and Public Works. Subcommittee on Clean Air, Wetlands, Private Property and Nuclear Safety. *Clean Air Act: Ozone and Particulate Matter Standards.* Hearing, February 12, 1997. Washington, D.C.: Government Printing Office, 1997. (S. Hrg. 105-50, Part 1.)

been reviewed by the courts. We believe that approach remains appropriate. However, once we revise any given air quality standard, it is both appropriate and, indeed, critical that we work with states, local governments, industry and others to develop the most cost-effective, common-sense strategies and programs possible to meet those new standards....

Background

The Clean Air Act directs EPA to identify and set national standards for certain air pollutants that cause adverse effects to public health and the environment. EPA has set national air quality standards for six common air pollutants —ground-level ozone (smog), particulate matter (measured as PM_{10}, or particles 10 micrometers or smaller in size), carbon monoxide, lead, sulfur dioxide, and nitrogen dioxide.

For each of these pollutants, EPA sets what are known as "primary standards" to protect public health and "secondary standards" to protect the public welfare, including the environment, crops, vegetation, wildlife, buildings and monuments, visibility, etc.

Under the Clean Air Act, Congress directs EPA to review these standards for each of the six pollutants every 5 years. The purpose of these reviews is to determine whether the scientific research available since the last review of a standard indicates a need to revise that standard. The ultimate purpose is to ensure that we are continuing to provide adequate protection of public health and the environment. Since EPA originally set the national air quality standards (most were set in 1971), only two of EPA's reviews of these standards have resulted in revised primary standards—in 1979, EPA revised the ozone standard to be less stringent; and in 1987, EPA revised the particulate matter standard to focus on smaller particles (those less than 10 micrometers in diameter), instead of all sizes of suspended particles.

By the early 1990's, thousands of new studies had been published on the effects of ozone and there was an emerging body of epidemiological studies showing significant health effects associated with particulate matter. EPA was sued by the American Lung Association to review and make decisions on both the ozone and particulate matter standards. I directed my staff to conduct accelerated reviews of both standards....

Rationale for EPA's Proposed Revision of the Ozone Standards

Since the mid-1980's, there have been more than 3,000 scientific studies published that are relevant to our understanding of the health and environmental effects associated with ground-level ozone. These peer-reviewed studies were published in independent scientific journals and included controlled human exposure studies, epidemiological field studies involving millions of people (including studies tracking children in summer camps), and animal toxicological studies. Taken as a whole, the evidence indicates that, at levels below the current standard, ozone affects not only people with impaired respiratory systems, such as asthmatics, but healthy children and adults as well. Indeed, one of the groups most exposed to ozone are children who play outdoors during the summer ozone season.

Certain key studies, for example, showed that some moderately exercising individuals exposed for 6 to 8 hours at levels as low as 0.08 parts per million (ppm) (the current ozone standard is set at 0.12 ppm and focuses on 1-hour exposures) experienced serious health effects such as decreased lung function, respiratory symptoms, and lung inflammation. Other recent studies also provide evidence of an association between elevated ozone levels and increases in hospital admissions. Animal studies demonstrate impairment of lung defense mechanisms and suggest that repeated exposure to ozone over time might lead to permanent structural damage in the lungs, though these effects have not been corroborated in humans.

As a result of these and other studies, EPA's staff paper recommended that the current ozone standard be revised from the current 1-hour form (that focuses on the highest "peak" hour in a given day) to an 8-hour standard (that focuses on the highest 8 hours in a given day). It also recommended setting an 8-hour standard in the range of 0.07 ppm to 0.09 ppm, with multiple exceedances (between one and five per year).

The CASAC [Clean Air Scientific Advisory Committee] panel reviewed the scientific evidence and the EPA staff paper and was unanimous in its support of eliminating the 1-hour standard and replacing it with an 8-hour standard. While I do not base my decisions on the views of any individual CASAC member (as a group they bring a range of expertise to the process), it is instructive to note the views of the individual members on these matters. While ten of the 16 CASAC members who reviewed the ozone staff paper expressed their preferences as to the level of the standard, all believe it is ultimately a policy decision for EPA to make. All ten favored a multiple exceedance form. Three favored a level of 0.08 ppm; one favored a level of either 0.08 or 0.09 ppm; three favored the upper end of the range (0.09 ppm); one favored a 0.09–0.10 range with health advisories when a 0.07 level was forecast to be exceeded; and two just endorsed the range presented by EPA as appropriate.

Consistent with the advice of the CASAC scientists and the EPA staff paper, we proposed a new eight-hour standard at 0.08 ppm, with a form that allows for multiple exceedances, by taking the third highest reading each year and averaging those readings over three years. We are asking for comments on a number of alternative options, ranging from eight-hour levels of 0.07 to 0.09 ppm to an option that would retain the existing standard. Just as a point of reference, based on our most recent analysis of children outdoors, when measuring the exposures and risks of concern, as well as the number of areas of the country that would be in "nonattainment" status, the current 1-hour ozone standard of 0.12 ppm is roughly equivalent to a 0.09 ppm 8-hour standard with approximately two to three exceedances.

We considered a number of complex public health factors in reaching the decision on the level and form proposed. The quantitative risk assessments that we performed indicated differences in risk to the public among the various levels within the recommended ranges, but they did not by themselves provide a clear break point for a decision.[1] The risk assessments did, however, point to clear differences among the various standard levels under consideration. These differences indicate that hundreds of thousands of children are not protected under the current standard but would be under EPA's ozone proposal.

Also, consistent with EPA's prior decisions over the years, it was my view that setting an appropriate air quality standard for a pollutant for which there is no discernible threshold means that factors such as the nature and severity of the health effects involved, and the nature and size of the sensitive populations exposed are very important. As a result, I paid particular attention to the health-based concerns reflected in the independent scientific advice and gave great weight to the advice of the health professionals on the CASAC. To me, this is particularly important given the fact that one of the key sensitive populations being protected would be children. The decision to propose at the 0.08 ppm level reflects this, because, though it is in the middle of the range recommended for consideration by CASAC and the EPA staff paper, as a policy choice it reflects the lowest level recommended by individual CASAC panel members and it is the lowest level tested and shown to cause effects in controlled human-exposure health studies.

Finally, air quality comparisons have indicated that meeting a 0.08 ppm, third highest concentration, 8-hour standard (as proposed by EPA) would also likely result in nearly all areas not experiencing days with peak 8-hour concentrations above the upper end of the range (0.09 ppm) referred to in the CASAC and the EPA staff paper. Given the uncertainties associated with this kind of complex health decision, we believe that an appropriate goal is to reduce the number of people exposed to ozone concentrations that are above the highest level recommended by any of the members of the CASAC panel. The form of the standard we proposed (third highest daily maximum 8-hour average) appears to do the best job of meeting that goal, while staying consistent with the advice of the CASAC as a group, as well as the personal views of individual members.

It is also important to note that ozone causes damage to vegetation including:

- interfering with the ability of plants to produce and store food, so that growth, reproduction and overall plant growth are compromised;
- weakening sensitive vegetation, making plants more susceptible to disease, pests, and environmental stresses; and
- reducing yields of economically important crops like soybeans, kidney beans, wheat and cotton.

Nitrogen oxides is one of the key pollutants that causes ozone. Controlling these pollutants also reduces the formation of nitrates that contributes to fish kills and algae blooms in sensitive waterways, such as the Chesapeake Bay.

As part of its review of the ozone science, the CASAC panel unanimously advised that EPA set a secondary standard more stringent than the current standard in order to protect vegetation from the effects of ozone. However, agreement on the level and form of the secondary standard was not reached.

Rationale for EPA's Proposed Revision to the Particulate Matter Standards

For particulate matter standard review, EPA assessed hundreds of peer reviewed scientific research studies, including numerous community-based epidemiological studies. Many of these community-based health studies show associations between particulate matter (known as PM) and serious health effects. These include premature death of tens of thousands of elderly people or others with heart and/or respiratory problems each year. Other health effects associated with exposure to particles include aggravation of respiratory and cardiovascular disease, including more frequent and serious attacks of asthma in children. The results of these health effects have been significantly increased numbers of missed work and school days, as well as increased hospital visits, illnesses, and other respiratory problems.

The recent health studies and a large body of atmospheric chemistry and exposure data have focused attention on the need to address the two major subfractions of PM_{10}—"fine" and "coarse" fraction particles—with separate programs to protect public health. The health studies have indicated a need to continue to stay focused on the relatively larger particles or "coarse" fraction that are a significant component of PM_{10} and are controlled under the current standards. We continue to see adverse health effects from exposures to such coarse particles above the levels of the current standards. As a result, CASAC scientists were unanimous that existing PM_{10} standards be maintained for the purpose of continuing to control the effects of exposure to coarse particles.

However, a number of the new health and atmospheric science studies have highlighted significant health concerns with regard to the smaller "fine" particles, those at or below 2.5 micrometers in diameter. These particles are so small that several thousand of them could fit on the type-written period at the end of a sentence. In the simplest of terms, fine particles are of health concern because they can remain in the air for long periods both indoors and outdoors contributing to exposures and can easily penetrate and be absorbed in the deepest recesses of the lungs. These fine particles can be formed in the air from sulfur or nitrogen gases that result from fuel combustion and can be transported many hundreds of miles. They can also be emitted directly into the air from sources such as diesel buses and some industrial processes. These fine particles not only cause serious health effects, but they also are a major reason for visibility impairment in the United States in places such as national parks that are valued for their scenic views and recreational opportunities. For example, visibility in the eastern United States should naturally be about 90 miles, but has been reduced to under 25 miles.

EPA analyzed peer-reviewed studies involving more than five and a half million people that directly related effects of "fine" particle concentrations to human health. For example, one study of premature mortality tracked almost 300,000 people over the age of 30 in 50 U.S. cities.

Based on the health evidence reviewed, the EPA staff paper recommended that EPA consider adding "fine particle" or $PM_{2.5}$ standards, measured both annually and over 24 hours. The staff paper also recommended maintaining the current annual and/or 24-hour PM_{10} standards to protect against coarse fraction exposures, but in a more stable form for the 24-hour standard. This more stable form would be less sensitive to extreme weather conditions.

When CASAC reviewed the staff paper, 19 out of 21 panel members recommended establishment of new standards (daily and/or annual) for $PM_{2.5}$. They also agreed with the retention of the current annual PM_{10} standards and consideration of retention of the 24-hour PM_{10} standard in a more stable form.

Regarding the appropriate levels for $PM_{2.5}$, staff recommended consideration of a range for the 24-hour standard of between 20 and 65 micrograms per cubic meter (ug/m3) and an annual standard to range from 12.5 to 20 ug/m^3. Individual members of CASAC expressed a range of opinions about the levels and averaging times for the standards based on a variety of reasons. Four panel members supported specific ranges or levels within or toward the lower end of the ranges recommended in the EPA staff paper. Seven panel members recommended ranges or levels near, at or above the upper end of the ranges specified in the EPA staff paper. Eight other panel members declined to select a specific range or level.

Consistent with the advice of the EPA staff paper and CASAC scientists, in November [1996] I proposed adding new standards for $PM_{2.5}$. Specifically, based on public health considerations, I proposed an annual standard of 15 ug/m^3 and a 24-hour standard of 50 ug/m^3. In terms of the relative protection afforded, this proposal is approximately in the lower portion of the ranges or options recommended by those CASAC panel members who chose to express their opinions on specific levels. However, taking into account the form of the standard proposed by EPA, we understand that the proposal would fall into the lower to middle portion of the ranges or options. In order to ensure the broadest possible consideration of alternatives, I also asked for comment on options both more and less protective than the levels I proposed.

Also consistent with the advice of the EPA staff paper and CASAC scientists, I proposed to retain the current annual PM_{10} standard and to retain the current 24-hour PM_{10} standard, but with a more stable form. I also requested comment on whether the addition of a fine particle standard and the maintenance of an annual PM_{10} standard means that we should revoke the current 24-hour PM_{10} standard.

As has been the case throughout the 25-year history of environmental standard setting, uncertainty has played an important role in decisionmaking on the particulate matter standards. Specifically, the uncertainty about the exact mechanism causing the observed health effects has led some to argue that not enough is known to set new or revised standards. In this case, however, because of the strong consistency and coherence across the large number of epidemiological studies conducted in many different locations, the seriousness and magnitude of the health risks, and/or the fundamental differences between "fine" and "coarse" fraction particles, the CASAC scientists and the experts in my Agency clearly believed that "no action" was an inappropriate response. The question then became one of how best to deal with uncertainty—that is, how best to balance the uncertainties with the need to protect public health.

Given the nature and severity of the adverse health effects, I chose to meet the Congressional requirement of providing the public with an "adequate margin of safety," by proposing $PM_{2.5}$ standards within the ranges recommended in the EPA staff paper and commented upon in the CASAC closure letter. I believe the levels chosen reflect the independent, scientific advice given me about the relationship between the observed adverse health effects and high levels of fine particle pollution. That advice led to a proposed decision toward the lower end of the range of levels for the annual standard which is designed to address widespread exposures and toward the middle of the range for the 24-hour standard, which would serve as a backstop for seasonal or localized effects.

One final note on particulate matter. Some have suggested we need more research before decisions are made about these standards. I strongly support the need for continued scientific research on this and other air pollutants as a high priority. However, as we pursue this research, we must simultaneously take all appropriate steps to protect public health. We believe that tens of thousands of people each year are at risk from fine particles and I believe we need to move ahead with strategies to control these pollutants.

Finding Common Sense, Cost-Effective Strategies for Implementing a Revised Ozone or PM Standard

Throughout the 25-year history of the Clean Air Act and air quality management in the United States, national ambient air quality standards have been established based on an assessment of the science concerning the effects of air pollution on public health and welfare. Costs of meeting the standards and related factors have never been considered in setting the national ambient air quality standards themselves. As you can see from the description of the process I went through to choose a proposed level on ozone and particulate matter, the focus has been entirely on health, risk, exposure and damage to the environment.

I continue to believe that this is entirely appropriate. Sensitive populations like children, the elderly and asthmatics deserve to be protected from the harmful effects of air pollution. And the American public deserves to know whether the air in its cities and counties is unsafe or not; that question should never be confused with the separate issues of how long it may take or how much it may cost to reduce pollution to safe levels. Indeed, to allow costs and related factors to influence the determination of what levels protect public health would be to mislead the American public in a very fundamental way.

While cost-benefit analysis is a tool that can be helpful in developing strategies to *implement* our nation's air quality standards, we believe it is inappropriate for use to *set* the standards themselves. In many cases, cost-benefit analysis has overstated costs. In addition, many kinds of benefits are virtually impossible to quantify—how do I put a dollar value on reductions in a child's lung function or the premature aging of lungs or increased susceptibility to respiratory infection? Very often I cannot set a value and these types of health benefits are, in effect, counted as zero.

At the same time, both EPA and industry have historically tended to over-state costs of air pollution control programs. In many cases, industry finds cheaper, more innovative ways of meeting standards than anything EPA esti-mates. For example, during the 1990 debates on the Clean Air Act's acid rain program, industry initially projected the costs of an emission allowance (the au-thorization to emit one ton of sulfur dioxide) to be approximately $1,500, while EPA projected those same costs to be $450 to $600. Today those allowances are selling for less than $100....

On the other hand, the Clean Air Act has always allowed that costs and feasibility of meeting standards be taken into account in devising effective emission control strategies and in setting deadlines for cities and counties to comply with air quality standards. This is certainly the case for any revision we might make to either the ozone or the particulate matter standards. This pro-cess has worked well. In fact, our preliminary studies indicate that from 1970 to 1990 implementation of the Act's requirements has resulted in significant monetizable benefits many times the direct costs for that same period.

If we ultimately determine that public health is better served by revising one or both of these standards, the Clean Air Act gives us the responsibility to devise new strategies and deadlines for attaining the revised standards. In doing so, we are determined to develop the most cost-effective, innovative imple-mentation strategies possible, and to ensure a smooth transition from current efforts.

To meet this goal, we have used the Federal Advisory Committee Act to establish a Subcommittee for Ozone, Particulate Matter and Regional Haze Implementation Programs. It is composed of almost sixty members of state and local agencies, industry, small business, environmental groups, other Fed-eral agencies and other groups and includes five working groups comprised of another 100 or so members of these same kinds of organizations.

The Subcommittee and the various workgroups have been meeting regu-larly for well over a year working to hammer out innovative strategies for EPA to consider in implementing any revised standards. Members from industry, state governments and others are putting forward position papers advocating inno-vative ways to meet air quality standards. It is our belief that results from this Subcommittee process will lead us to propose innovative approaches for im-plementing any new standards. The Subcommittee will continue to meet over the next year to help develop cost-effective, common-sense implementation programs.

The issues being addressed by the Subcommittee include:

- What will be the new deadlines for meeting any new standards? [If EPA tightens a standard, it has the authority to establish deadlines of up to ten years—with the possibility of additional extensions—beyond the date an area is designated "nonattainment."]
- What will be the size of the area considered "nonattainment"? If it revises an air quality standard, EPA has the ability to change the size

of the affected nonattainment areas and focus control efforts on those areas that are causing the pollution problems, not just the downwind areas that are monitoring unhealthy air.

- How do we address the problem of the pollutants that form ozone and/ or fine particles being transported hundreds of miles and contributing to nonattainment problems in downwind areas?
- What kinds of control strategies are appropriate for various nonattainment areas? Can we use the experience of the past several years to target those control strategies that are the most cost-effective?
- How can we promote innovative, market-based air pollution control strategies?

The implementation of these new standards is likely to focus on sources like trucks, buses, power plants and cleaner fuels. In some areas, as with the current standards, our analysis shows that reaching the standards will present substantial challenges. All of the air pollution control programs we are pursuing to meet the current ozone and particulate matter standards, as well as programs to implement other sections of the Clean Air Act, will help meet any revised standards. For example, the sulfur dioxide reductions achieved by the acid rain program will greatly help reduce levels of fine particles, particularly in the eastern United States. Cleaner technology in power plants would also greatly reduce the nitrogen oxides that help form ozone across the eastern United States. In fact, we believe that under certain comprehensive control strategies, more than 70 percent of the counties that could become nonattainment areas under a new ozone standard would be brought back into attainment as a result of a program to reduce nitrogen oxides from power plants and a large number of other sources. Programs underway to reduce emissions from cars, trucks, and buses will also help meet a revised particulate matter or ozone standard.

I intend to announce our proposals on implementation of the proposed new standards in phases that correspond to the Federal Advisory Committee Act Subcommittee's schedule for deliberating on various aspects of the program. I expect to propose the first phase of that program at the same time that I announce our final decision on revisions to the ozone and particulate matter standards.

In announcing the proposed ozone and particulate matter standards last November, I directed my Office of Air and Radiation to further expand the membership of the Federal Advisory Subcommittee to include more representation from small business and local governments. Also, in conjunction with the Small Business Administration and the Office of Management and Budget, we are holding meetings with representatives of small businesses and small governments to obtain their input and views on our proposed standards.

There is one last point I would like to make on this matter. Critics of the proposals have been saying that meeting these proposed standards means widespread carpooling and the elimination of backyard barbecues, among other lifestyle changes. The broad national strategy is being developed by EPA, as I have described, with extensive input from industry, small business, state and local governments and others. While the ultimate decisions as to what programs are needed to meet air quality standards are up to the state and local governments, I would like to state categorically that there will *not* be any new federal mandates eliminating backyard barbecues or requiring carpooling. These kinds of claims are merely scare tactics designed to shift the debate away from the critical, complex public health issues we are attempting to address.

Conclusions

Mr. Chairman, I commend you for holding these hearings. The issues we are discussing today are critical to the state of the Nation's public health and environment. It is imperative that the American public understand these important issues. In that regard, I am disappointed that some have chosen to distort this important discussion by raising distracting and misleading pseudo issues like "junk science" and "banning backyard barbecues." I am hopeful that this and other hearings and public forums will help focus the national debate on the real health and environmental policy implications of these national air quality standards.

In the Clean Air Act, the Congress has given me the responsibility to review every 5 years the most recent science to determine whether revisions to national air quality standards are warranted. In doing so, the law tells me to protect the public health with an adequate margin of safety.

We are constantly reviewing the science associated with these standards, but we do not often propose revisions to them. I have done so in the case of ozone and particulate matter because of compelling new scientific evidence. For the past three and a half years we have targeted our resources to conduct a thorough, intensive review of this scientific evidence. The scope and depth of this review process has been based on unprecedented external peer review activities.

Given the sensitive populations affected by these pollutants—children, asthmatics, the elderly—as well as possible effects on outdoor workers and other healthy adults, it was my judgment that it was appropriate to propose standards that tended to fall in the lower end of the range of protection supported by my independent science advisors and recommended by experts in my technical offices. Based on the record before the Agency at the time of proposal, including the advice and recommendations of the CASAC panels, I concluded—subject to further consideration based on public comments—that the proposed standards were both necessary and sufficient to protect the public health, including sensitive populations, with an adequate margin of safety.

At the same time, I recognize that the proposed standards involve issues of great complexity and I look forward to receiving a broad range of comments from all affected and interested parties. As I have described, we have gone to unprecedented lengths to provide the public with opportunities to express their views on the proposed standards. We have also expressly requested comments on options (including alternative levels and forms of the standards) that are both more protective and less protective than the levels we proposed.

Note

1. CASAC itself agreed that there are a continuum of effects—even down to background—and that there is no "bright line" distinguishing any of the proposed standards as being significantly more protective of public health.

NO ⟵

Statement of Daniel B. Menzel

My name is Daniel B. Menzel. I am Professor and Chair of the Department of Community and Environmental Medicine, University of California at Irvine, Irvine, California. I have had more than 30 years' experience in research in air pollution and toxicology. My expertise centers in two areas: mechanisms of air pollution toxicity and mathematical modeling of toxicology, particularly deposition of air pollutants in the respiratory tract. I have served as a senior author on multiple EPA Criteria Documents and recently as a Consultant to the Clean Air Scientific Advisory Committee examining the Particulate Matter Criteria Document and proposed standard.

The Committee has requested that I provide my views on the ozone and particulate matter standards, which EPA has published in the Federal Register and intends to implement under the Clean Air Act. I am pleased to do that and would also like to extend my testimony to include the research effort of EPA because it directly affects the standard-setting process. I understand that the two standards present different problems in terms of the form of the standard, the scientific data supporting each standard and the process by which the standard was promulgated. In my view, however, there are similarities between the two standards that reflect a major deficiency in EPA's efforts. The common deficiency is the lack of solid scientific data. EPA is a grossly underfunded Agency given the scope of its responsibilities. EPA has not done well with its resources by not sustaining research to meet the long-term goals of the Agency. Thus, I hope that the committee will allow me to express my concerns about the research planning at EPA.

Air Pollution Is a Major Long Term Public Health Problem

Air pollution is a worldwide problem. In the United States air pollution is of such public health importance that it is critical that a national debate be undertaken on the future directions of air pollution research and regulation. This committee is providing a very valuable forum to the people so that they may learn more about the scientific controversy surrounding these two air pollutants and the alternative views that exist concerning the future of air pollution

From U.S. Senate. Committee on Environment and Public Works. Subcommittee on Clean Air, Wetlands, Private Property and Nuclear Safety. *Clean Air Act: Ozone and Particulate Matter Standards.* Hearing, February 5, 1997. Washington, D.C.: Government Printing Office, 1997. (S. Hrg. 105-50, Part 1.)

remediation efforts. I am at the moment writing a review of the toxicology of ozone[1]. This will be the third review of ozone that I have written for the scientific literature. Almost ten years have elapsed since my last effort, and I was surprised and saddened to note on examining the literature that questions which we raised in the review in 1988 still remain unresolved. Much new human data has become available on ozone supporting a lower standard and shorter averaging time, but the book is far from closed on ozone. I also wrote the first part of the health section of the SO_x (sulfur oxides) Particulate Matter Criteria Document for EPA in 1980. Many of the questions raised in that document also remain unanswered. As a consultant to the Clean Air Scientific Advisory Committee I assisted in the review of the current Particulate Matter Criteria Document. Not only were the fundamental questions raised in the original SO_x Particulate Matter Criteria Document still existent, but new important questions arose for which we have no answer. All of these experiences suggest to me that a greatly enhanced and invigorated research effort in air pollution is needed if we are to make sound, reasonable and rational decisions on the implementation of clean air standards. If anything, air pollution research is now more important to the national public health than ever before.

Both the ozone and particulate matter standards have vast implications for the quality of life and the economy of the United States. It is my opinion that the vast majority of Americans support improving and enhancing the quality of their life by eliminating or decreasing air pollution. Americans are quite willing to shoulder the burden of cleaner air, cleaner water, and cleaner food if they can understand clearly the benefits to be gained by these activities. The confidence of the American people in the decisions being made on environmental issues is critical to the ability of this government to govern and implement these decisions. If ever the public loses confidence in the environmental strategies promulgated by the Federal Government then it will be impossible to carry out large national programs designed to eliminate or at least ameliorate the adverse effects of air pollution. I am very concerned that the Environmental Protection Agency and the Congress maintain the confidence of the U.S. public and demonstrate to the public their vigorous support for a better quality of life and clean air. Scientific truth is the only lasting commodity upon which decisions can be based.

Generic Issues

From my view the difficulties that we face with both the ozone and the particulate matter standard stem from generic issues in toxicology which must be addressed in a sound scientific manner. The first of these generic issues is a plausible biological mechanism of action for the particular pollutant. The second is the nature of the dose-response relationship. I will address each of these and give examples of how they impinge upon the two standards that we are discussing today.

Plausible Biological Mechanisms

What is a plausible mechanism? We have learned a great deal about the quantitative nature of toxic reactions in the last 40 years. It is now possible to divide biological reactions to toxicants into several categories under which plausible mechanisms have been elucidated. A plausible mechanism of action for a toxin places the toxin within the context of our knowledge of disease processes. Having a plausible mechanism of action increases our confidence that health effects observed in animals will occur in humans. Understanding a mechanism of action also makes experiments more meaningful and relevant. In this forum it is not possible for me to elaborate in greater technical detail on how a plausible mechanism influences the experimental design and interpretation of the results of experiments. Experimental design and the concept of plausible mechanism of action are dealt with in standard textbooks of toxicology, such as "Casserett and Doull's Fundamentals of Toxicology."

A plausible mechanism of action is critically essential to controlled human exposure studies. The extrapolation from animal experiments to human exposures as they occur in nature, that is with free-living people, depends upon an intermediate link of controlled exposures of human volunteers to the toxin. We must have a clear idea of a plausible mechanism so that human studies can be developed with due care that no harm will ever result to the volunteers who courageously commit themselves to these kinds of experiments. In air pollution many of the human studies have been very limited because of the lack of a clear understanding of a plausible mechanism. Investigators have been very reluctant to engage in high level exposures of human subjects because they fear that some long-term harm will result from their experiments. Clearly, we cannot and will not tolerate human experimental studies that result in harm to the volunteer. This is simply not ethically acceptable.

Plausible Mechanism of Ozone Toxicity

One plausible mechanism of action of ozone is the production of free radicals by the reaction of ozone with cellular constituents. The free radical theory is that which we proposed in 1971.[2] It is now clear that this mechanism of action is too naive and simplistic and clearly does not explain the consequences of chronic exposure to ozone. Studies with experimental animals clearly show that the results of a continuous or intermittent lifetime exposure to ozone are highly complex and are not predictable from the free radical hypothesis alone. Further experiments are needed with life-term exposures of experimental animals using the most modern molecular biology techniques. The complex pattern of lifetime ozone exposure must involve multiple signal transduction pathways. Simply put, the adverse health effects of chronic exposure to ozone are complex and beyond the free radical theory which we now recognize as accounting for the brief initial contact of ozone with the lung.

Chronic exposure is the critical issue in ozone exposure. EPA initiated and was carrying out an excellently conceived and implemented research program on the chronic effects of ozone in support of the current ozone standard. But this research has stopped and support for ozone research by other Federal agencies has stalled. Basic research support for ozone by the National Institutes of Health and particularly the National Institute of Environmental Health Sciences (NIEHS), has fallen away. The scientific community is in error in allowing this to have happened.

Very compelling controlled human exposure experiments suggest that the current ozone standard (0.12 ppm) may be toxic. The short term exposures under which humans can be safely exposed do not allow us to study the chronic effects of ozone exposure. Epidemiologic studies are underway in the South Coast Air Basin, particularly those by Professor John Peters of the University of Southern California but this study is hampered because no quantitative biomarker of ozone health effects has been developed.

We would not be... engaging in this discussion if EPA's chronic ozone study in experimental animals had been carried out. Nor would we still have doubts about the ozone standard if ozone research had received a high priority in research support by the other Federal research agencies such as NIH and NSF.

In summary, there is a preliminary biologically plausible mechanism of action for ozone. The free radical theory is not comprehensive and does not explain all of the effects of chronic exposure to ozone. Much additional work is needed to understand the chronic effects of ozone.

Particulate Matter

In contrast to the ozone problem, no plausible biological mechanism of action has so far been proposed for particulate matter. It has been very difficult to demonstrate toxicity for particulate matter in experimental animals. In my laboratory and that of my colleagues at UCI we have not been able to show major toxicity with particulate matter at potencies approaching the levels reported from epidemologic studies.[3,4]

To place this problem in a more global context, urban particulate matter is a universal problem. Particulate matter seems to be a common result of human concentration in urban areas. To eliminate all of the particulate matter in our cities would, in my view, be only possible by the elimination of all human activity. Clearly this Draconian approach is not reasonable.

The studies of Schwartz and his colleagues[3,4] have challenged our conclusions from experimental animal studies. These studies indicate that all particles regardless of their geographic origin have the same toxicity. It is well known that the chemical composition of the urban particles differ widely between geographic areas. For example, in the western US, especially in the South Coast Air Basin of Los Angeles and its environs, the chemical processes responsible for the formation of particulate matter depend on photochemical reactions. Nitric acid is the dominant end product. There are very few oxides of sulfur present because of the nature of the fossil fuels used in California. On the other hand, in the East Coast Corridor the consumption of sulfur-containing fuels is much greater, and the chemistry of the reactions leading to the formation of particulate matter is not as dependent upon photochemistry as it is upon chemical reactions. Sulfuric acid, not nitric acid, is the dominant end product present in particulate matter. The chemical nature of the particles formed in California are quite different [from] those of the East Coast Corridor. Yet the health effects measured by epidemiologic techniques suggest that all particles have the same effect despite the differences in chemical composition. This is a very troublesome problem. One of the basic tenets of toxicology is that the toxicity occurs via chemical reaction. How then can the same effect result from very different kinds of chemistries? We must conclude that there is no plausible mechanism now available for particulate matter which can account for the reported results.

Particle Size and Site of Action of Respirable Urban Particles

The toxicity of particles also depends on the site within the respiratory tract where they are deposited. A major advance has been the recognition of the dependence of toxicity on the site of deposition. The site of deposition in the respiratory tract depends, in turn, on the physical size of the particle. By measuring the amount of particles within the size range which can be deposited in the human lung, EPA adopted a biologically based criterion for its standard setting. This concept of defining particulate air pollution in terms of the size of particles most likely to be responsible for the adverse health effects is referred to as PM_{10} where 10 refers to particles of 10 micrometers aerodynamic mass median diameter or less. PM_{10} is a fairly good surrogate measurement for the amount of material that would actually be inhaled and deposited in the human respiratory tract. Schwartz and his colleagues extrapolated from measured PM_{10} values. PM_{10} is a major advance in public health policy pioneered by EPA. The PM_{10} concept shifts emphasis to particles of that size which are likely to be the most harmful to people. A network of PM_{10} monitors has been constructed in the US and large amounts of data have been accumulated.

Schwartz and his colleagues went beyond PM_{10} and extrapolated from a very limited set of measurements of $PM_{2.5}$ and PM_{10} to estimate $PM_{2.5}$ values and to relate mortality and morbidity to particulate matter exposure smaller than PM_{10} or particles less than 2.5 micrometers mass median aerodynamic diameter. Only a few data exist on the $PM_{2.5}$ exposure in our major cities. By shifting from PM_{10} to $PM_{2.5}$ values, a major difference in the regional deposition within the lung of these particles is suggested as the site of action. The smaller the particle the more deposition occurs in the deeper parts of the lung. By assigning toxicity to particles in the $PM_{2.5}$ range the site of action is also assigned to the thoracic region of the lung. Because these $PM_{2.5}$ values are calculated and not measured, it is very difficult to place the heavy weight of evidence on this ultrafine particle range as EPA has done in its criteria document. Even with a shift in attention to particles of this size range, there is still is no plausible mechanism for toxicity. Further, some of the CASAC members questioned the potency of the particles calculated from the mortality and morbidity data. All of this underscores the importance of the research program reviewed by CASAC as part of the particulate matter standard setting process.

Dose-Response Relationship

The dose-response relationship is a curve that relates the number of individuals responding with an adverse reaction (mortality, morbidity or the like) to a certain exposure concentration of the chemical. The shape of the dose-response curve is important when setting standards. All theories of the dose-response relationship so far indicate that these curves will be non-linear; that is, there will be a point at which the probability that a response would occur is very unlikely. To put it another way, all theories suggest that there is a concentration at which nothing will occur while above that concentration adverse effects will occur. The point at which there is nothing detectable is the threshold. The dose-response relationship is at the heart of the risk assessment. In both the particulate matter and ozone standard the dose-response relationship is only poorly understood. Consequently, estimates of risk are also uncertain. Examples for ozone and particulate matter follow.

The Particulate Matter Dose-Response Curve Is Linear Not Curved

The current assumption of epidemiologic studies is that the mortality or morbidity is a linear function passing through zero at zero concentration of particles. The dose-response function has no point at which no adverse effects occur. The linear dose-response curve is in opposition to all of the theories and experimental data derived for a host of chemicals acting by a variety of different mechanisms of action.

The epidemiologic basis for a linear relationship between effect and dose is very poor. The data are not supported by any kind of a generalized theory and are in many cases a default assumption coming about because the epidemiologic data are weak. It is very difficult for epidemiologists to relate exposure to effect. The methodologies of epidemiology at present are insensitive to the concentration or exposure effect. This is especially true in ecological studies where indirect evidence is used for adverse health effect.

For example, the epidemiologic studies of particulate matter health effects depend upon death certificates and the coincidence of an increase in death with an increase in particulate matter exposure. These studies again provide no indication of how a person might have died from the exposure to particulate matter. The studies only associate the death with the exposure to particulate matter. Nonetheless, the increases in mortality associated with particulate matter are troublesome. If the magnitude of mortality suggested by these studies is correct, then we are faced with a major public health problem that demands immediate attention.

Time and Intensity Relationships in Ozone Health Effects

EPA initiated a time and intensity study in cooperation with the USSR. This program was well thought out and attacked the question of which variable is most important in determining the health effects of ozone. From the data that were generated by this study it appears that the intensity is the most critical factor rather than the duration of exposure for ozone toxicity. These studies of the time and concentration effects on ozone toxicity led to the current hypothesis upon which the proposed ozone standard is based. If it is correct that the magnitude of the exposure is more important, then extremes of exposure should be reduced. One strategy to reduce exposure to extreme concentrations of ozone is to change the averaging time for the standard, making implementation plans stricter for short-term excursions. The US-USSR research program to study the time and concentration dependency of ozone adverse health effects was very productive and was progressing along a track which would, if continued, have provided us a great deal of information at this time. Unfortunately, EPA chose to reduce and essentially eliminate this line of study. Extramural support for the program lagged and ozone in general has become an unpopular topic for support by other government agencies such as NIEHS.

Based on the fragmentary information that we have available, I feel that it is appropriate to support the EPA proposal of changing the averaging time for the ozone standard so that large excursions over short time periods will be eliminated or reduced. However, one should recognize that changing the averaging time will have a major impact on State implementation plans and will have major economic consequences. Clearly, understanding the nature of the dose-response relationship is very important and affects which alternatives we choose to reduce ozone health effects.

Time and Intensity Relationship for Particulate Matter Health Effects Are Unknown

As stated above, most time and intensity (dose and dose-rate) relationships for chemicals follow a simple relationship that the product of the dose rate and the time of exposure form a constant. This constant is arbitrary and unique for each chemical. Epidemiologic studies of the increases in mortality associated with increases in particulate matter are strictly linear with the amount of particulate matter. One reason why this assumption occurs is that a lag period has been assumed. The lag period means that the increases in mortality occurring 2 to 3 days after an exposure are related to the exposure to particulate matter, not earlier or later. The underlying hypothesis is that particulate matter toxicity is not immediately evident but occurs after this lag period. This very short acting time raises the question as to what happens when people are exposed to concentrations of particulate matter over the long term. We really have no data on the chronic effects in humans of exposure to particulate matter. Chronic exposure studies are very difficult to achieve using epidemiologic data.

To my knowledge there are no experimental animal data or controlled human studies which relate this kind of lag time to exposure to the toxicity of particulate matter. In my laboratory and that of my colleagues at UCI we have found that experimental animals such as the rat are very insensitive to particulate matter exposures. We have never observed potencies equivalent to that proposed for humans based on the epidemiologic data. This again raises the question of a plausible biological mechanism of action. . . .

Conclusions

The Proposed Ozone Standard

It is my opinion that we will have achieved only marginal effects by decreasing the current ambient air quality standards for ozone from 120 parts per billion to 90 parts per billion. The nature of the dose-response relationship is such that it may still be at a linear range and thus reduction to much lower levels may be necessary to result in the abolition of detectable health effects from ozone. My colleague, Robert Wolpert, and I published a simple analysis of different kinds of dose-response relationships for ozone looking toward this very issue. How much would one have to reduce the ozone concentration in the air in order to be able to find a detectable advance in public health? Because the data are so sparse, a multitude of different kinds of theoretical treatments are possible. None of them, however, are sufficiently sensitive that one could lead to a clear prediction of a health benefit. On the other hand, as I mentioned above, a change in the time constant alone is going to have a great benefit. I endorse EPA's analysis of the time constant and think that EPA's proposal to a change in the averaging time for ozone is likely to be of benefit to the public health.

Still, I think that translating these changes into new State implementation plans may be very difficult. To translate both a change in the concentration, that is the amount of ozone that is permissible in the air and the duration over which it is permissible, will be a very difficult task indeed to implement.

Continued research into the health effects of ozone are urgently needed. Further reductions in the ozone standard may be indicated in the near future. Because of the economic impact of ozone standards and strategies, the highest quality research is needed.

Particulate Matter Standard

As I have said previously, I do not doubt that the particulate matter problem is a very serious problem indeed. We need to place a very strong active and progressive research program into place in order for us to cope with this problem. It is my view that too little is known. In the report of the Clean Air Scientific Advisory Committee to Administrator Carol Browner, the committee pointed out that one of the areas in which additional research should be undertaken is chronic exposure.

I am not in favor of the use of a $PM_{2.5}$ standard. A viable network of monitoring instruments and sound research supports the PM_{10} standard. The $PM_{2.5}$ standard has no background. There is no existing research quality $PM_{2.5}$ network. Without a research quality $PM_{2.5}$ network it is not likely that we will make much progress towards the goal of a new particulate matter standard. We lack information on the actual $PM_{2.5}$ in the atmosphere of our cities. We do not know the duration of exposure of people to $PM_{2.5}$. The chemical nature of the $PM_{2.5}$ fraction is poorly known. We lack a plausible biological mechanism for particulate matter. We do not know if regulation of $PM_{2.5}$ will be of benefit. A strong aggressive long-term research program is essential to address the current data deficiencies if we are to convince people that this is a major problem.

Notes

1. Shoaf, C.R. and Menzel, D.B. Oxidative damage and toxicity of environmental pollutants. In: Cellular Antioxidant Defense Mechanisms. (ed., C. K. Chow) CRC Press, Inc, Vol. 1:197–213, 1988.

2. Roehm, J.N., Hadley, J.G. and Menzel, D.B. Oxidation of Unsaturated Fatty Acids by Ozone and Nitrogen Dioxide: A Common Mechanism of Action. *Arch. Environ. Health* 23:142–148, 1971.

3. Saldiva, P. H., Pope, C. A., Schwartz, J., Dockery, D. W., Lichtenfels, A. J., Salge, J. M., Barone, I. & Bohm, G. M. (1995) Air pollution and mortality in elderly people: a time-series study in Sao Paulo, Brazil. *Arch. Environ. Health* 50:159–163.

4. Schwartz, J. (1995) Short term fluctuations in air pollution and hospital admissions of the elderly for respiratory disease. *Thorax* 50: 531–538.

POSTSCRIPT

Is the Environmental Protection Agency's Decision to Tighten Air Quality Standards for Ozone and Particulates Justified?

The analysis of the options available for reducing the health effects of air pollution is complex, and, like most technological problems related to environmental protection, it requires the use of many unprovable assumptions. A specific proposal for a new, costly set of regulatory standards requires a judgment about whether or not the analysis is based on an adequate set of valid research results and whether or not the interpretation of the results provides sufficient confidence that the benefits achieved will justify the expense. It is on this point—rather than the question of whether or not suspended particulate matter and ozone are serious health threats—that Browner, Menzel, and many other experts disagree. Menzel points out one serious problem with the proposal to base particulate matter standards on particles smaller than 2.5 microns rather than 10 microns: no system for monitoring the smaller particles exists. In announcing the actual schedule for implementing the new standards, the EPA acknowledged this problem and established a timetable associated with the development of the needed monitoring network.

Many alternative strategies were evaluated by the EPA and its consultants during the decade of political struggle that preceded the enactment of the 1990 Clean Air Act amendments. The January/February 1991 issue of *EPA Journal* is devoted to a discussion of the new law. In this issue, the agency presents technical details about the amendments and a justification of the regulatory strategy that they embody.

The EPA first announced in late November 1996 that it planned to seek more stringent regulations to control fine particulates and ozone levels. The background to this plan is discussed in a feature article by Catherine M. Cooney in the July 30, 1996, issue of *Environmental Science and Technology*.

One of the key research efforts that the EPA used in formulating the new particulate standard is the Harvard School of Public Health's "Six Cities" study. For a report on epidemiologist Joel Schwartz's role in directing that research, see Renée Skelton's article "Clearing the Air" in the Summer 1997 issue of *The Amicus Journal*. Another debate about the new standards featuring Lester B. Lave and Robert W. Crandall was published in the Summer 1997 issue of *The Brookings Review* under the title "EPA's Proposed Air Quality Standards." For a lengthy denunciation of the new particulate and ozone standards and the scientific research on which they are based, see "Polluted Science," by Michael Fumento,

Reason (August/September 1997). For a summary of the Health Effects Institute's study released in July 2000, see Jocelyn Kaiser, "Evidence Mounts That Tiny Particles Can Kill," *Science* (July 7, 2000).

In "Who Will Be Protected by EPA's New Ozone and Particulate Matter Standards?" *Environmental Science and Technology* (January 1, 1998), Feng Liu reports that some moderation would occur under the new regulations in the disproportionate effect of air pollution on Hispanics, Asians/Pacific Islanders, and African Americans. In the June 1, 1998, issue of the same journal, see Allen S. Lefohn, Douglas S. Shadwick, and Stephen D. Ziman, "The Difficult Challenge of Attaining EPA's New Ozone Standard."

On October 15, 1997, the EPA issued a report entitled, *The Benefits and Costs of the Clean Air Act, 1970 to 1990.* On November 15, 1999, a second report was issued entitled, *The Benefits and Costs of the Clean Air Act, 1990 to 2010.* Both concluded that the benefits of the programs and standards required by the 1990 Clean Air Act Amendments significantly exceed costs. See http://www.epa.gov/airlinks. However, on October 29, 1999, the U.S. Court of Appeals for the District of Columbia upheld a May 1999 court decision calling the EPA's 1997 National Ambient Air Quality Standards (NAAQS) for ozone and particulates unconstitutional. See April Reese, "Bad Air Days," *E: The Environmental Magazine* (November–December 1999) and Richard J. Pierce, Jr., "The Inherent Limits on Judicial Control of Agency Discretion: The D.C. Circuit and the Nondelegation Doctrine," *Administrative Law Review* (Winter 2000) for criticism of the decision. In January 2000 New Jersey and Massachusetts asked the U.S. Supreme Court to review that decision. In May 2000 the Supreme Court agreed to hear the case, with a decision expected "early in 2001."

Much of the debate has centered on air quality standards, especially for particulates. Regulations to control downwind air pollution (mentioned by Browner) have also come under fire. Eastern states have demanded that the EPA enforce agreements that would reduce the amount of air pollutants emitted by Midwestern power plants, which travel on the wind and add to air quality problems in the East. Industry groups and states such as Michigan filed suit to block such reductions, but in June 2000 the U.S. Court of Appeals for the District of Columbia chose to allow the EPA to implement its plans.

On the Internet ...

DUSHKIN ONLINE

Committee for the National Institute for the Environment

This site contains links to information about many environmental issues. Informative reports to Congress are available on numerous topics, including all aspects of waste disposal.

http://www.cnie.org

Superfund

The EPA's Superfund information site also contains information about other aspects of the hazardous waste disposal problem.

http://www.epa.gov/superfund/

Waste Prevention/Resource Conservation Publications

This site offers publications and other information resources about waste prevention, recycling, and waste management from the Local Government Commission.

http://www.lgc.org/wasteprevn/welcome.html

Yucca Mountain

This is the EPA's information site on the proposed Yucca Mountain permanent nuclear waste repository.

http://www.epa.gov/radiation/yucca/

PART 3

Disposing of Wastes

*M*odern industrial societies generate many types of waste. Manufacturing and construction activities yield hazardous liquid and solid
residues; the most common methods of treating raw sewage produce
large quantities of sludge; mining operations generate mountains of tail-
ings; and radioactive waste results from the use of nuclear technology in
medicine, electric power production, and the weapons industry. Each of
these forms of waste, as well as ordinary household garbage, contains
toxins and pathogens that are major sources of air, water, and land
pollution if they are not disposed of properly. We must now deal with
the legacy of years of neglect and inappropriate waste disposal methods.
This section exposes some of the major controversies concerning proposed
solutions to three important waste categories.

- Hazardous Waste: Should the "Polluter Pays" Provision of Superfund Be Weakened?

- Municipal Waste: Is Recycling an Environmentally and Economically Sound Waste Management Strategy?

- Nuclear Waste: Should the United States Continue to Focus Plans for Permanent Nuclear Waste Disposal Exclusively at Yucca Mountain?

ISSUE 13

Hazardous Waste: Should the "Polluter Pays" Provision of Superfund Be Weakened?

YES: Bernard J. Reilly, from "Stop Superfund Waste," *Issues in Science and Technology* (Spring 1993)

NO: Ted Williams, from "The Sabotage of Superfund," *Audubon* (July/August 1993)

ISSUE SUMMARY

YES: DuPont corporate counsel Bernard J. Reilly argues that in defining standards and assigning costs related to waste cleanup, "Congress should focus the program on reducing real risk, not on seeking unattainable purity."

NO: *Audubon* contributing editor Ted Williams contends that insurers and polluters are lobbying to change the financial liability provisions of Superfund, and he warns against turning it into a public welfare program.

T he potentially disastrous consequences of improper hazardous waste disposal burst upon the consciousness of the American public in the late 1970s. The problem was dramatized by the evacuation of dozens of residents of Niagara Falls, New York, whose health was being threatened by chemicals leaking from the abandoned Love Canal, which was used for many years as an industrial waste dump. Awakened to the dangers posed by chemical dumping, numerous communities bordering on industrial manufacturing areas across the country began to discover and report local sites where chemicals had been disposed of in open lagoons or were leaking from disintegrating steel drums. Such esoteric chemical names as dioxins and PCBs have become part of the common lexicon, and numerous local citizens' groups have been mobilized to prevent human exposure to these and other toxins.

The expansion of the industrial use of synthetic chemicals following World War II resulted in the need to dispose of vast quantities of wastes laden with organic and inorganic chemical toxins. For the most part, industry adopted a casual attitude toward this problem and, in the absence of regulatory restraint, chose the least expensive means available. Little attention was paid to the ultimate fate of chemicals that could seep into surface water or groundwater. Scientists have estimated that less than 10 percent of the waste was disposed of in an environmentally sound manner.

The magnitude of the problem is truly mind-boggling: Over 275 million tons of hazardous waste is produced in the United States each year; as many as 10,000 dump sites may pose a serious threat to public health, according to the federal Office of Technology Assessment; and other government estimates indicate that more than 350,000 waste sites may ultimately require corrective action at a cost that could easily exceed $500 billion.

Congressional response to the hazardous waste threat is embodied in two complex legislative initiatives. The Resource Conservation and Recovery Act (RCRA) of 1976 mandated action by the Environmental Protection Agency (EPA) to create "cradle to grave" oversight of newly generated waste, and the Comprehensive Environmental Response, Compensation, and Liability Act of 1980, commonly called "Superfund," gave the EPA broad authority to clean up existing hazardous waste sites. The implementation of this legislation has been severely criticized by environmental organizations, citizens' groups, and members of Congress who have accused the EPA of foot-dragging and a variety of politically motivated improprieties. Less than 20 percent of the original $1.6 billion Superfund allocation was actually spent on waste cleanup.

Amendments designed to close RCRA loopholes were enacted in 1984, and the Superfund Amendments and Reauthorization Act (SARA) added $8.6 billion to a strengthened cleanup effort in 1986 and an additional $5.1 billion in 1990. While acknowledging some improvement, both environmental and industrial policy analysts remain very critical about the way that both RCRA and Superfund/SARA are being implemented. Efforts to reauthorize and modify both of these hazardous waste laws have been stalled in Congress since the early 1990s.

Although the following selections were written in 1993, they still accurately reflect the current controversy concerning the Superfund Act. Bernard J. Reilly argues that the legislation has turned into an "unjustifiable waste of the nation's resources at the expense of other critical society needs." He calls for major changes, which would "focus the program on practical risk reduction." One specific change he advocates is in the "polluter pays" provision of the law, which he argues holds companies liable for more than their fair share of the costs. Ted Williams acknowledges that the Superfund program has been very costly and has cleaned up little waste. But he blames this on sabotage of the law by the Bush and Reagan administrations. He specifically warns against recommendations to abandon the policy of holding polluters strictly liable for the damage they have caused, which he fears would "turn it into a public works program" whereby "citizens will pay twice, once with their environment and once with their tax money."

Bernard J. Reilly  **YES**

Stop Superfund Waste

President Clinton's economic plan is a clear attempt to reorder federal spending priorities by putting more money into "investments" that will spur economic growth and increase national wealth, while cutting unproductive activities. One important way he could further his agenda would be to push for reform of one of today's most misguided efforts: the Superfund hazardous waste cleanup program. The President has already paid lip service to this goal, telling business leaders in a February 11 speech at the White House that, "We all know it doesn't work—the Superfund has been a disaster."

Superfund, created by the Comprehensive Environmental Response, Compensation, and Liability Act of 1980 (CERCLA) in the wake of the emergency at the Love Canal landfill in Niagara Falls, New York, was designed as a $1.6-billion program to contain the damage from and eventually clean up a limited number of the nation's most dangerous abandoned toxic waste sites. But in short order it has evolved into an open-ended and costly crusade to return potentially thousands of sites to a near-pristine condition. The result is a large and unjustifiable waste of our nation's resources at the expense of other critical societal needs.

No one questions that the nation has a major responsibility to deal with hazardous waste sites that pose a serious risk to public health and the environment. It is the manner and means by which the federal government has pursued this task, however, that are wasteful. Superfund legislation has given the U.S. Environmental Protection Agency powerful incentives and great clout to seek the most comprehensive, "permanent" cleanup remedies possible—without regard to cost or even the degree to which public health is at risk. Although the EPA does not always choose the most expensive remedial solution, there is strong evidence that, in many cases, waste sites can be cleaned up or sufficiently contained or isolated for a fraction of the cost, while still protecting the public and the environment. Further, EPA's selection of "priority" cleanup sites has been haphazard at best. Indeed, it has no system in place for determining which of those sites—or the many potential sites it has not yet characterized—pose the greatest dangers.

A 180-degree turn in policy is needed. When the Superfund program comes up for reauthorization next year, Congress should direct the EPA to abandon its pursuit of idealistic cleanup solutions and focus the program on

From Bernard J. Reilly, "Stop Superfund Waste," *Issues in Science and Technology* (Spring 1993), pp. 57–60, 62–64. Copyright © 1993 by The University of Texas at Dallas, Richardson, TX. Reprinted by permission.

practical risk reduction, targeting those sites that pose the greatest health risks and tying the level and cost of cleanup to the degree of actual risk. Only by making such a fundamental change can the nation maximize the benefits of its increasingly huge investment in the remediation of hazardous waste sites.

Costs Are Escalating

Estimates for cleaning up, under current practice, the more than 1,200 sites on the EPA's "national priority list" (NPL) range from $32 billion by EPA (based on a $27 million per-site cost) to $60 billion by researchers at the University of Tennessee (based on a $50 million per-site cleanup cost). These estimates are likely to be well below the ultimate cost, since EPA can add an unlimited number of sites to the list. The agency plans to add about 100 sites a year, bringing the total by the year 2000 to more than 2,100. But more than 30,000 inactive waste sites are being considered for cleanup and the universe of potential sites has been estimated at about 75,000. Most experts believe that far fewer—from 2,000 to 10,000—will eventually be cleaned up. The University of Tennessee researchers make a best guess of 3,000 sites, which would put the cost at $150 billion (in 1990 dollars) over 30 years, not including legal fees.

This $150 billion might be acceptable if the U.S. economy were buoyant and limitless funds existed for other needs. It most certainly would be justified if many sites posed unacceptable dangers to the public. But neither of these situations exists.

Skewed Priorities

A key flaw in Superfund is that most of its effort and money are directed to a relatively small number of "priority" sites, while thousands of others are ignored and, in most cases, not even sampled or studied. For this reason, it is doubtful that the NPL includes all the worst sites.

"Deadly" chemical landfills buried under residential neighborhoods have hardly been typical of the sites EPA has placed on the NPL. Indeed, EPA's efforts to create a system for ranking hazards have not been geared to actually finding the riskiest sites but to satisfying the letter of the CERCLA law. In the first ranking scheme, sites were evaluated for various threats and a score of 28.5 (on a scale of 100) was determined to be sufficient for an NPL listing. However, the listings were not necessarily based on an actual determination of the degree to which they posed threats to public health or the environment. Rather, the sites were included because Congress had determined that 400 sites must be on the NPL, and a score of 28.5 resulted in 413 listings.

Several years ago, the ranking scheme was made much more elaborate, with threats from contaminants in the air, water, and soil weighed differently. The same maximum score of 100 and listing score of 28.5 were used. Why? EPA said that it was "not because of any determination that the cutoff represented a threshold of unacceptable risk presented by the sites" but because the 28.5 score was "a convenient management tool." So much for the rigors of a system designed to cull the Love Canals from town dumps.

A 1991 report by a committee of the National Research Council (NRC) strongly faulted EPA's methods of selecting sites and setting priorities. The report said that EPA has no comprehensive inventory of waste sites, no program for discovering new sites, insufficient data for determining safe exposure levels, and an inadequate system for identifying sites that require immediate action to protect public health.

In a perfect world, every "dirty" site would be cleaned, regardless of the degree of risk it presented. In practice, this is impossible, so we should be spending more to prioritize in order to focus our limited resources on real risks.

Extreme Remedies

However it is accomplished, once a site makes the NPL, money is no object in the remediation process. This was not necessarily the case under the original 1980 Superfund law. CERCLA left some ambiguity about how extensive the cleanups had to be—whether only reasonable risks needed to be eliminated or whether the site had [to] be returned to a preindustrial condition. When it enacted the Superfund Amendments and Reauthorization Act (SARA) in 1986, however, Congress, motivated by a deep distrust of the Reagan-era EPA, took a hard-line stance. SARA, which increased funding for the program to $8.5 billion and ordered action to begin at ever more sites, directed EPA to give preference to cleanup remedies that "to the maximum extent practicable" lead to "permanent solutions." The emphasis on permanence was further reinforced by a requirement that cleanups must comply with any "applicable or relevant and appropriate requirement" (ARAR) in any other state or federal law relating to protection of public health and the environment.

SARA was deeply flawed. For one thing, it effectively forced EPA to continue remedial action even after all realistic risks at a site had been eliminated. One example is the Swope Superfund site, a former solvent reclamation facility in Pennsauken, New Jersey. Although all major sources of contamination had been removed from the site, EPA ordered the installation of a $5-million vapor extraction system to remove more contaminants. The purpose was to protect groundwater in case any private wells were sunk in the future. But EPA neglected to consider the fact that private wells had been banned in the area.

SARA's requirements also serve to exclude the use of other far less costly remedies that would give the public the same or at least acceptable protection from harm. For example, at the Bridgeport Rental and Oil Services Superfund site in Logan Township, New Jersey, EPA ordered the construction of an onsite, $100-million incinerator after PCBs were found in several sludge samples. In making its decision, EPA used the ARAR requirement to retroactively apply the federal Toxic Substances Control Act (TSCA), which requires incineration of currently generated wastes if samples indicate that PCBs in the soil exceed 500 parts per million.

The absurdity of the plan became apparent when EPA decided to create an on-site landfill to dispose of the heavy metal residues from the incineration. Given that a landfill was to be created anyway and that PCBs at the site were so scarce that EPA had to import them for trial burns of the incinerator, the agency could have opted to contain the sludge on site in the first place—using existing proven technologies—while more than adequately protecting the public at an estimated one-fifth of the cost of incineration.

A similar tale is unfolding at another Superfund site in Carlstadt, New Jersey, which is contaminated with solvents, PCBs, and heavy metals. A trench has been cut around the site to an underlying impervious soil layer and then filled with clay to prevent any migration of the contaminants. The site has also been pumped dry to protect groundwater and capped to keep out rain. Remediation work has cost about $7 million, and DuPont as well as other responsible parties have pledged to maintain these containment systems for as long as necessary. However, despite the absence of any current or reasonably foreseeable public exposures, EPA may decide to require incineration of the top 10 feet of soil at an estimated cost of several hundred million dollars. This would be a foolish waste of money.

EPA must also consider that extreme, costly remediation solutions often are not without costs of their own. Incineration, for instance, cannot destroy metals. Does the public really benefit when lead is released into the air as a byproduct? By the same token, when contaminated soil is ordered excavated and carted elsewhere, one neighborhood gets a "permanent" solution, whereas another gets a landfill with toxic residues.

Risks Exaggerated

Superfund legislation is not the only force driving EPA to seek "permanent" solutions. EPA decides on a remedy only after assessing the risks at a site. However, EPA often uses unrealistic assumptions that exaggerate the risks and lead to excessive actions. For example, according to the Hazardous Waste Cleanup Project, an industry group in which DuPont has been involved, EPA may make estimates of exposure based on a scenario in which an individual is assumed to reside near a site for 70 years, to consume two liters of water every day during those 70 years, and to obtain all of that water from groundwater at the site. It has even made exposure estimates based on the length of time a child will play (and eat dirt) on a site in the middle of an industrial location surrounded by a security fence. Each of these scenarios is highly improbable.

Questions involving risk assessment are, of course, going to be contentious ones for some time to come. Clouding the Superfund debate is the fact that there is no scientific consensus as to the precise magnitude of the dangers posed by chemicals typically found at Superfund sites.

The existence of toxic wastes at a site does not necessarily mean that they pose a threat to nearby residents. Epidemiologic studies of waste sites have severe technical limitations, and it is difficult at best to determine whether exposure to hazardous wastes can be blamed for medical problems when a long gap exists between exposure and disease. Even at such a well-known site as Times Beach, Missouri, where the entire community was evacuated, research in recent years has shown that the potential health risks were relatively small or even nonexistent.

The most comprehensive assessment of the risks from Superfund sites came in the 1991 NRC report, which concluded that "current health burdens from hazardous-waste sites appear to be small," but added that "until better evidence is developed, prudent public policy demands that a margin of safety be provided regarding public health risks from exposures to hazardous-waste sites."

No one can argue with a margin of safety. However, that is not the focus of the current Superfund program, which, far more than any other environmental program, makes no rational attempt to link costs with benefits. EPA's own Science Advisory Board, in a 1990 report that attempted to rank the environmental problems for which the agency is responsible, concluded that old toxic waste sites appeared to be "low to medium risk." Other hazards, such as radon gas in homes and cigarette smoke, were considered to pose much larger risks.

The Liability Mess

The bulk of the Superfund tab will be picked up by industry, through taxes imposed under CERCLA, out-of-pocket cleanup costs, or settlements with insurance companies. Industry recognizes that it must assume its fair share of the financial and operating burden of the cleanup effort, and it acknowledges that Superfund has compelled it to become exceptionally vigilant not only in disposing of toxic wastes but also in minimizing their generation in the first place. But it objects to a system in which EPA seemingly has put a higher priority on pinning the blame and the bill on companies than on ensuring the protection of public health.

CERCLA dictated a "polluter-pays" philosophy to deal with what had largely been lawful disposal of wastes. CERLCA and court interpretations of it also have created an extremely broad liability scheme. Virtually any company remotely involved in a site-waste generators, haulers, site owners or operators, and even, in some cases, the companies' bankers—could be held responsible. One or a few companies could be forced to pay the entire bill, even though they were only minor participants and other parties were involved—a provision called joint-and-several liability. No limits were imposed on the amount of money that could be extracted from "guilty" parties.

One problem with this liability system is that it completely lacks cost accountability. With industry paying for most of the cleanup, the funds are not in EPA's budget and thus do not have to compete in budget battles with other cash-starved federal programs. And given the strictness of the law, why should EPA regulators subject themselves to possible congressional criticism by selecting a less-than-perfect solution, especially if money is no object? But let us not kid ourselves. Although this money may seem "free" to Congress, EPA, and the public, companies must make up the difference by raising prices, cutting investment and jobs, or taking other undesirable actions.

An even more damning problem is that the liability provisions have spawned countless legal brouhahas that are consuming a large and increasing share of Superfund resources—even as the cleanup process itself has languished. (The average length of a site cleanup is 8 years, and fewer than 100 sites have been "permanently" remediated.) In the approximately 70 percent of Superfund sites that involve multiple parties, companies must fight with the EPA, among themselves, and with their insurance companies over who dumped what, when, and how much—questions extremely difficult to answer many years after the alleged "dumping" is thought to have occurred. Some experts believe that these "transaction costs" will eventually account for more than 20 percent of all Superfund expenditures. This is a boon for lawyers but a waste for the nation.

Legal costs—as well as burdensome technical and administrative expenses —could potentially be greatly reduced if Congress would allow EPA to take a more practical approach to risk reduction. Unlike other environmental laws, such as the Clean Air Act and Clean Water Act, which have sought to deal with problems in successive stages, Superfund's emphasis on finding a one-time, complete, and permanent solution magnifies the stakes to all parties, prolonging disputes and greatly increasing the costs. If companies could count on a more realistic remediation approach, they might be more willing to compromise, which could lead to faster cleanups.

The liability mess could get completely out of hand if Congress goes along with a patently unfair proposal to exempt municipalities from liability at closed municipal landfills, which account for about 20 percent of NPL sites. Municipal governments argue that most of these landfills largely contain household wastes not covered by Superfund and thus they should not be billed for the cleanup. But in many cases this is not true. For example, at the Kramer Superfund site in Mantua, New Jersey, municipal governments contributed the greatest share of hazardous substances. Despite this, EPA is no longer even naming municipalities in cost-recovery suits. (EPA's tendency to selectively enforce the law has been increasing. At Kramer, EPA sued 25 parties even though hundreds were potentially responsible.)

Industry recognizes that many municipal governments are severely strapped for revenues. Yet companies, which provide jobs and help create the tax base needed to support municipal services, should not be milked to pay for Superfund shares properly owed by others.

One last concern with the liability provisions is that they may be having a chilling effect on new investment at sites in older urban areas—areas that sorely need such investment. The reason is that any party that buys such a property would be caught in Superfund's liability web. For example, investors seeking to build a coal-fired power plant in an area with a projected need for such a use recently approached DuPont about buying a property that had been used for manufacturing for more than 100 years and clearly contains some contaminated soil. Virgin land is not needed for a site to burn coal, and risk assessments indicated that workers could be protected with commonsense steps such as paving. But efforts to get reasonable compromises from regulators on containing the site proved fruitless, and now the investment will not be made, at least in this area.

Steps to Reform

It is time for a major redirection of the Superfund program. Congress should tell EPA to abandon its focus on idealistic cleanup remedies and emphasize practical risk reduction. Instead of continuing its haphazard site selection and unjustifiably costly cleanup remedies, EPA should first define the universe of sites that may present real health risks and then take steps to deal with the most immediate dangers, taking costs into consideration. Once a national inventory has been established, extensive site evaluations can be undertaken, with the purpose of setting priorities for cleanup. Only after these actions are taken will we be able to make non-hysterical decisions as to how much we should invest in cleaning these sites, balancing such factors as risks, costs, and other societal needs.

It is particularly crucial that remedy decisions be based on the expected future use of the land and the costs and practicality of the proposed solution. If residential development is planned near the site, the cleanup may need to be extensive. In many cases, however, especially when another industrial use is planned on or near an old waste site, use of containment technologies may be sufficient to protect against risk of exposure. In the most troublesome cases, where major remediation is necessary, costs are high, and existing technology has limitations, it makes much more sense to isolate the site until more cost-effective treatment techniques are developed or increased land values justify a large investment.

In making these decisions, it would be helpful if EPA had much better information on the benefits and costs of different levels of cleanup. Currently, less than 1 percent of EPA's Superfund budget goes for research on the scientific basis for evaluating Superfund sites. Much more should be spent. EPA also should increase its research on the environmental consequences of different types of remedial actions, such as whether incineration actually increases risk by transferring hazardous substances from the ground or water into the air.

The liability provisions of the Superfund program also need to be changed. DuPont and companies in the chemical, petroleum, and other industries favor replacing the very unfair joint-and-several liability provision (making one or a few companies liable for all the costs, even though many others, often defunct, were also responsible) with proportional liability. In other words, responsible parties would pay only in proportion to the share of the cleanup costs associated with the wastes that they contributed at a site. EPA would then be forced to either find and sue all responsible parties or pay for the remainder of the cleanup costs itself. EPA is already authorized to pay for cleanup costs in cases where parties cannot be found or cannot afford to pay—shares which are often sizable. But in practice it has sought to recoup all cleanup costs under the joint-and-several provision. Proportional liability would inject more fairness into the process, and since the polluter-pays principle would be retained, it would continue to encourage responsible parties to pressure EPA to pursue the most cost-effective cleanup remedy. Most important, proportional liability would impose much-needed financial discipline on EPA, since it would be forced to pay for more of the cleanups out of its own budget. For the first time, EPA would have to consider whether the benefits were worth the costs.

Proportional liability would not, of course, solve the problem of how to divide up responsibility in the first place. One possible way out of this morass is to formalize in the law an alternative dispute resolution process in which any or all potentially responsible parties could participate. It would be chaired by neutral parties satisfactory to all. Its findings on shares could be appealed to the courts, but any party that concurred with the decision would be authorized to pay its share and exit the process. This solution would help cut site contention, reward cooperative parties, and leave messy litigation to those unwilling to pay their fair shares. It would also diminish Superfund's luster as a federally mandated entitlements program for lawyers.

More extensive reform of the liability provisions has been proposed by the insurance industry, which wants to eliminate all liability at sites in which more than one party is involved and in which waste disposal occurred prior to enactment of either CERCLA in 1980 or SARA in 1986. Site cleanup would then be paid out of the Superfund budget, financed by increased taxes on industry, including insurers. Although this proposal would eliminate contentious fights over specific site responsibility, substantially cut transaction costs, and possibly speed up site cleanups, it would be unacceptable to DuPont and other parties at Superfund sites if the new taxes were unfairly levied on the same companies already paying disproportionately large shares of the current Superfund cleanups.

Finally, the liability scheme must be changed so that prospective owners of older urban sites are not deterred from making new investments in them. New owners should certainly not be held responsible for contamination that they did not cause. One approach would be for current owners to demonstrate, before sale, that their sites, while not pristine, are adequately contained and do not pose unacceptable risks to the public. The new owner would be expected to maintain or monitor whatever containment system was developed. If EPA later did a more extensive site evaluation and determined that greater threats existed, the new owner would not have to pay. In addition, current owners should be able to make new investments in their property if they demonstrate that the sites are adequately contained.

✿

The limits of our national wealth have not been so obvious since the 1930s. More than ever, we must make choices among competing, compelling demands for scarce resources. We recognize that a dollar spent on defense cannot be spent on health care. We must also recognize that a dollar spent on hazardous waste cleanup is similarly unavailable. As with other federal programs, Superfund spending must be balanced and managed. This can be done if we refocus our Superfund investment on real risks, give EPA a stake in doing its job cost-effectively, and bring more fairness into the process.

Ted Williams

The Sabotage of Superfund

T he setting was perfect: Cold rain and gull-filled mist blowing in from Buzzards Bay. Litter clinging to the bare ribs of dead brush like shards of rotten um-
brella silk. Derelict, graffiti-streaked trailers stuffed to overflowing with bald
truck tires. Ratty mattresses and broken easy chairs strewn about the cratered
parking lot. Glass from the abandoned mill crunching under my boots and
snatches of Eliot's *The Waste Land* resounding in my brain as I trudged along
Wet Weather Sewage Discharge Outfall No. 022: "Sweet Thames run softly, till I
end my song..."

Until this day, March 28, 1993, I had avoided Superfund sites. So this
was my first visit to the waterfront of New Bedford—an impoverished, predominantly Hispanic seaport in southern Massachusetts, now as famous for the
polychlorinated biphenyls (PCBs) on the bottom of its harbor as for its whaling
history. PCBs, widely used in the manufacture of electrical components until
banned in 1978, do hideous things to creatures that come in contact with them
—such as causing their cells to proliferate wildly and warping their embryos. I
wasn't about to touch anything without my rubber ice-fishing gloves.

Where the sewage dribbled into the dark Acushnet River I jumped down
onto gray silt and, breathing through my teeth, scooped up five handfuls of
muck. The Environmental Protection Agency's guideline for protecting marine
life from chronic toxic effects of PCBs is 30 parts per trillion. I cannot accurately report the PCB content of my amateur sample (taken illegally, the EPA
later informed me), but the greasy globules that floated up through the surface scum were likely very rich. Had I been able to get out into the river and
upstream to the old Aerovox Inc. discharge pipe, I could have found concentrations of at least 200,000 parts per *million,* or 20 percent, among the highest ever
recorded. That means that with a similar test dredging I'd have retrieved one
handful of pure PCBs, along with a tangle of wriggling sludge worms, about
the only creatures that can live in such habitat.

I was rinsing my gloves and boots in a rain-filled pothole when *The Waste
Land's* Fisher King materialized through the gloom—a wispy, gray-haired figure
in a red plaid jacket, toting a stout spinning rod. He had parked next to a sign
that read in Spanish, Portuguese, and English: "Warning. Hazardous Waste. No
wading, fishing, shellfishing. Per order of U.S. EPA." His name, which he printed

From Ted Williams, "The Sabotage of Superfund," *Audubon* (July/August 1993). Copyright © 1993
by Ted Williams. Reprinted by permission.

with his forefinger on the wet trunk of his car, was Robin Rivera; he knew only enough English to make me understand that he and his family eat the fish he catches here.

Not until 1978 did the nation get angry about the indiscriminate disposal of poisonous chemicals. In that year people who lived near Hooker Chemical Company's Love Canal dump in Niagara Falls, New York, were distressed to smell vile chemical odors in their basements and observe a malevolent secretion bubbling out of the ground at a local school yard. County health officials and Hooker reps tried vainly to contain the alarm, but their assurances that all was well sounded as wrong as the uncontained leachate looked and smelled.

Eventually the citizens took their case to the young commissioner of the state Department of Environmental Conservation—Peter A. A. Berle, now president of the National Audubon Society. Although Berle had no authority to act on public health issues, he sent his people out to test houses on the strength of his environmental mandate. The benzene levels they found were, in his words, "right off the chart." Eventually 600 homes were abandoned and 2,500 residents relocated.

In response to the Love Canal horror show Congress enacted the Comprehensive Environmental Response, Compensation, and Liability Act of 1980, better known as Superfund. Amended in 1986, the law uses taxes on crude oil and 42 commercial chemicals to maintain a fund with which the EPA may, as it likes to say, "remediate" hazardous-waste sites. If perpetrators can be found and are still in business, the EPA may require one or all to clean up the entire site. This essential principle of Superfund is called joint and several liability.

For an idea of the pace at which cleanup proceeds, consider that the EPA and its contractors have been studying and planning what to do about New Bedford's harbor ever since it was declared a "National Priority" Superfund site 11 years ago. Nationwide, the EPA has spent $7.5 billion on its Superfund program, with pitiful results. In some cases remediation has created more problems than it has solved by stirring up contaminants that had been dormant. In other cases vast sums have been squandered at sites that posed little threat to the public, while deadly brews seethed nearby. Superfund contractors have consistently ripped off the EPA, billing it for everything from office parties to work they were supposed to do and didn't.

At this writing only 163 of 1,204 sites have been remediated, and in many cases polluters have been granted what the EPA calls the "containment" option —a feline approach to toxic-waste management in which they just cover their messes and walk away. The average cleanup has cost about $25 million and taken 7 to 10 years to complete.

No one remotely connected with Superfund is happy about the way it has functioned. Polluters identified by the EPA have been madly rummaging through dumps, trying to identify other polluters by their trash and so spread liability. In the process small towns, businesses, and individuals that contributed legally and insignificantly to landfills have been intimidated and assessed for cleanup costs in a fashion utterly inconsistent with the intent of Congress. Environmentalists are at the throats of insurance companies who want to do away with the polluter-pay tenet. The insurance companies are warring in court with industries to whom they have rashly sold pollution-liability policies. People who live atop and beside toxic waste claim—often correctly —that they have been ignored and lied to by the EPA, and as a result, they sometimes oppose well-advised cleanup plans.

Hearings for Superfund's 1994 reauthorization are already under way. "We all know it doesn't work," says President Bill Clinton. "Superfund has been a disaster."

⋅⦿⋅

Even as I wished the Fisher King good luck, I found myself greeting the first of 94 demonstrators from the New Bedford–area citizen's group Hands Across the River. We stood in the rain, listening to fiery speeches amplified by bullhorn about the EPA's plan to dredge the five acres that contain roughly 45 percent of the PCBs extant in the 28-square-mile site, then cleanse the spoil by fire in portable incinerators set up on the downstream side of Sewage Outfall 022.

"You and I will be breathing their mistakes," bellowed rally leader Richard Wickenden. "They made their decision to incinerate behind closed doors with a total disregard for the local citizenry."

He spoke the truth. The New Bedford City Council had found out about the plan not from the EPA but from Hands Across the River, which had found out about it from federal documents at the library. "State-of-the-art incinerators" have a long history of malfunction, even when run by the most conscientious contractors. This one will be operated by Roy F. Weston, a large environmental-consulting firm based in West Chester, Pennsylvania, which in 1990 agreed to pay $750,000 to settle charges that it had defrauded the EPA by backdating data and submitting a bill for work it never did. Finally, people downwind—mostly people like Robin Rivera—are likely to be breathing mistakes, along with all manner of toxic PICs (products of incomplete combustion) that won't be monitored or even identified.

In attempting to clean up one point of pollution the EPA and its contractors will be creating others, asserted New Bedford City Councilman George Rogers. They'll be unleashing PCBs and heavy metals on moving seawater, hauling them onto the bank, then casting them to the four winds during dewatering and combustion. "They're doing this because New Bedford is a poor community; we don't have clout. They wouldn't do it in Miami Beach." He, too, spoke the truth. A study released last September by *The National Law Journal* reveals that the EPA is lenient in penalizing polluters of minority and low-income communities and that cleanups in such areas are slower and less thorough. Toxic racism, activists call it.

Equally veracious were the allegations of David Hammond, president and founder of Hands Across the River, that polluters love incineration because their liability goes up the stack along with the toxic PICs and that the EPA has undermined Superfund's effectiveness and its own credibility by ordering remediation studies from companies that make their money remediating. In particular, Hammond is upset that Weston was hired to do the New Bedford Remedial Action Master Plan, then wound up with the $19.4 million incineration contract.

Both the city council and Hands Across the River hasten to point out that they are not against remediation. But instead of incineration they favor the "Eco Logic" process—a relatively contained heat treatment developed in Rockwood, Ontario, which combines hydrogen with PCBs to form methane and hydrogen chloride and which has been getting rave reviews in the press. "Stunning New Method Zaps Toxic Chemicals Efficiently," shouted a headline in *The Toronto Star* on January 30, 1993.

After the New Bedford speechmaking the congregation marched back and forth over the Acushnet River bridge, waving placards, obstructing traffic (much of which honked in sympathy), and chanting, "No way, EPA," and "Hey, Carol [Browner], if you please, don't you burn those PCBs. Not New Bedford, not the nation. We don't want incineration."

The citizens could scarcely have done a better, more honest job of drawing public attention to the perils of dredging and incinerating PCBs. But this doesn't mean that the EPA ought not to press ahead with its plan for New Bedford. When PCB concentrations are this high, the perils of doing something else or nothing at all are probably greater. One day the Eco Logic process may indeed be a "stunning method" of remediation. Now, despite the effusions of *The Toronto Star,* it's largely an experimental technology and therefore fraught with risk. Meanwhile, the PCBs are spreading out into the Atlantic with every tide and every storm. Humans and marine ecosystems—including half the North American population of endangered roseate terns, which nests on a single island in Buzzards Bay—don't have another decade to wait while the EPA collects data and shuffles papers.

❧

In other contract deals the EPA has paid the New England office of Roy F. Weston, which it has criticized for poor performance, $635,000 to administer fieldwork that cost $340,000. But Weston looks like a model contractor when compared with some of the others.

Take, for example, consulting-engineering colossus CH2M Hill, which has worked on 275 major sites, including Love Canal, and which holds $1.4 billion worth of Superfund contracts. An inquiry by the House Subcommittee on Oversight and Investigations revealed that as part of alleged Superfund work, CH2M Hill billed the EPA $4,100 for tickets to basketball, baseball, and football games; $167,900 for employee parties and picnics, including the cost of reindeer suits, magicians, and a rent-a-clown; $15,000 for an office bash at a place called His Lordship; "thousands of dollars' worth" of chocolates stamped with the company logo; $63,000 for general advertising; $10,000 for a catered lobbying cruise on the Potomac; and $100 for a Christmas-party dance instructor. "I am all for rocking around the Christmas tree," commented Congressman Thomas J. Bliley Jr. (D-VA) at the hearing, "but does it have to be at the taxpayers' expense?"

Apparently yes, according to the testimony of CH2M Hill's president, Lyle Hassebroek. "No matter what differences of opinion exist on the manner in which we allocate costs," he explained, "CH2M Hill's charges to the government are fair to the taxpayer."

By no means is CH2M Hill aberrant. Last summer EPA investigators found that 23 companies hired for hazardous-waste cleanup in 1988 and 1989 spent 28 percent of their $265 million budget on wasteful administrative costs. Such inefficiency is cited by polluters and their insurance companies as a reason to "overhaul" Superfund—i.e., turn it into a public-works program whereby Uncle Sam would bail them out by picking up toxic litter (provided the offense preceded some stipulated date—1987, according to one proposal) and citizens will pay twice, once with their environment and once with their tax money.

Major polluters further foment discontent with Superfund by attempting to squeeze alleged shares of cleanup costs from everyone who might ever have sent a can of shoe polish to a landfill. The EPA and the courts don't want a nickel a day for 1,000 years and so avoid going after mom-and-pop polluters. But Mom and Pop don't know this, and technically they are liable. The real motive, charge environmental leaders, is not so much to collect money as to contrive broad support for Superfund "reform."

When Ford, Chrysler, General Motors, BASF Corporation, and Sea Ray Boats were fingered by the EPA for fouling the Metamora, Michigan, landfill with arsenic, lead, vinyl chloride, and the like, they proclaimed that 382 towns, businesses, and individuals were copolluters and tried to assess them $50 million to settle alleged liability quickly, including any unforeseen costs. Even the local Girl Scout troop was assessed $100,000. "That's a lot of cookies," declared a troop spokesperson.

In another case Doreen Merlino, the 25-year-old proprietor of a two-table pizzeria in Chadwicks, New York, offered the court officer the following plea when he served her with a two-inch-thick lawsuit in October 1990: "Aren't you at least gonna buy a pizza?" He kindly complied, but she didn't feel much better. In fact, she felt terrified. Cosmetics giant Chesebrough-Pond's and Special Metals Corporation were trying to extract $3,000 from her for helping them poison the local landfill. They weren't sure just what the trashman might have collected from her during the seven months she'd been in business—maybe pesticide containers or empty cleanser cans, they opined. But Merlino tells me she's never used pesticides and that she has always rinsed out her cleanser containers. A cover letter advised her that if she settled fast, she'd only have to pay $1,500. As it turned out, she had to pay no one save her lawyer, and none of her anger is directed at Superfund. Not all the 603 defendants were so philosophical.

When the EPA hits up polluters for toxic-waste-cleanup costs, polluters, naturally enough, hit up the insurance companies from which they have purchased pollution-liability coverage. Now the insurance companies would like to hit up the public—that is, rewrite Superfund so bygones can be bygones and taxpayers can spring for cleaning up old sites.

"Superfund's mission should be protecting human health and the environment, not fund-raising," contends the American International Group, an insurance company marshaling support for what it calls a National Environmental

Trust Fund, by which Superfund money could be raised "from all economic sectors without regard to site-specific liability" via a surcharge on commercial- and industrial-insurance premiums.

Generating pity for the insurance industry requires a greater heart than beats in the breast of environmental consultant Curtis Moore. As an aide to former Republican senator Robert Stafford of Vermont, Moore was instrumental in writing both the original Superfund law and the amended 1986 version. Insurance companies, he points out, tend not to cover purposeful acts by anyone, including God; so the policies were restricted to "sudden and accidental" pollution. "You dump crude oil on the ground for fifteen years," he says, "and over a twenty-five-year period it migrates to the water table. Would you consider that sudden? Accidental? No? Well, I got news for you: The courts do. There was a string of decisions that construed the terms *sudden* and *accidental* as covering groundwater contamination. This trend started a long time ago—in the 1970s or earlier. It was clearly discernible. Any insurance lawyer with manure for brains could see it happening. Notwithstanding, the insurance industry continued to use the terms *sudden* and *accidental* in its policies.

"So here comes Superfund, and the chemical companies start casting around, trying to figure out how they can get someone else to pick up the tab. They file suits against their insurance companies and win. Well, there's only one way to fix the insurance industry's problem. You can either shift liability or, failing that, repeal Superfund."

Presuming to speak for the insurance industry, the American International Group complains that Superfund is "bogged down in a morass of legal warfare that delays cleanup and wastes enormous financial and human resources." True enough, but what it doesn't mention is that the insurance industry has been responsible for a great deal of this legal warfare. A Rand Corporation study reveals that between 1986 and 1989 insurers spent $1.3 billion on Superfund. Of this, $1 billion went to defending themselves against their policyholders or defending their policyholders against the EPA. One leading attorney for the policyholders—Eugene Anderson, of the New York City firm of Anderson, Kill, Olick & Oshinsky—has gone so far as to suggest publicly that refusing all large claims is now seen as smart business procedure by insurers: Half the policyholders get scared away, and most of the others will settle out of court for less than full coverage.

⋅⟨◉⟩⋅

Superfund has bombed, as the President, environmentalists, inhabitants of toxic neighborhoods, brewers of toxic waste, and especially the insurance industry have observed. But it is essential to remember the difference between Superfund the law and Superfund the program.

"The law was a creation of people like Bob Stafford, Ed Muskie, Jennings Randolph, John Chafee, Jim Florio," remarks Moore. "The program was the creation of Ronald Reagan; the people who were put in charge of implementing the law six weeks after it was enacted were people who six weeks earlier had been lobbying against it. They set out with the intent of making it unworkable, and they succeeded."

There is nothing in the statute that directs the EPA's contractors to dress up like reindeer or distribute customized confections at government expense. They engage in such excess because the EPA lacks the personnel to keep them honest. Nor is there anything in the statute that mandates stonewalling and procrastination on the part of polluters. But they have learned that endless negotiation is profitable because the EPA lacks the personnel to haul more than a few of them into court. Mr. Clinton, who proposes to trim $76 million from Superfund the program, appears not to understand this.

Certainly, the statute could stand repair. It needs to define how clean is clean, provide a better, more flexible means of selecting remedies, ensure state and local participation, create incentives for companies to take voluntary action instead of suing everyone in sight. But the fact is that Superfund the law isn't broken.

Even Superfund the program, disastrous though it has been, has produced some splendid if accidental results. "Joint and several liability," says Peter Berle, "has put the fear of God into everybody, which means they are careful in ways they never were before in what they do with their waste. I also think the cost risk of inappropriate toxic-waste disposal has been the major impetus toward waste minimization. When it gets too expensive to deal with it, then you make less."

Rick Hind, toxics director for Greenpeace, agrees. "It doesn't cost you and me anything if a big company wants to spend ten million dollars on lawyers to avoid an eleven-million-dollar cleanup," he offers. "That costs the company. Good! So it costs them twice what it should. That will teach them a lesson. When a Colombian drug cartel is in court nobody cares what their legal expenses are. Nor should we care about polluters."

Insurers and polluters—not environmentalists—are the ones driving for major surgery on Superfund the law. If they are permitted to degrade it from a dedicated fund to a public-works program whereby big government passes around public-generated revenues, the hemorrhage of federal pork will make the EPA nostalgic for the days when it used taxes on crude oil and chemicals to rent clowns for CH2M Hill. If they are permitted to do away with Superfund's liability provisions and weaken its polluter-pay principle, the United States will be poisoned on a scale unimagined even in New Bedford, Massachusetts.

It may be that Superfund is mortally wounded from a dozen years of sabotage. But it also may be that it can be salvaged and made to work. We need to try. Vendors of insurance and chemicals will shriek and sob, but the law wasn't written for them. It was written for Love Canal couples forced to watch as bulldozers razed their homes, for Robin Rivera and his family, for roseate terns, for sick and deformed children, for children yet unborn.

POSTSCRIPT

Hazardous Waste: Should the "Polluter Pays" Provision of Superfund Be Weakened?

Note that although Reilly does not question the need to "deal with hazardous waste sites that pose a serious risk to public health and the environment," he later qualifies this responsibility by claiming that "the existence of toxic waste at a site does not necessarily mean that they pose a threat to nearby residents." As citizens' groups—such as the New Bedford activists, whose concerns are described by Williams—have made clear, "nearby residents" reject Reilly's qualification and invariably demand remediation of toxic waste that has been identified in their neighborhoods. The "joint-and-several liability" provision of Superfund that Reilly thinks is unfair is a common provision of tort law that is considered necessary by many legal experts to enable courts to apportion penalties when there are several disputing liable parties.

W. Kip Viscusi and James T. Hamilton share many of Reilly's views. In their essay in the summer 1996 issue of *The Public Interest*, they argue that accurate risk assessment, determining the extent of exposed populations, and appropriately balancing benefits and costs are the principles that should be used in reforming the Superfund. Some of this has been accomplished by EPA administrative reforms, the results of which are discussed by Linda Raber in "Superfund Reforms Helpful," *Chemical and Engineering News* (January 26, 1998). However, the title of Lani Sinclair 's "Lack of Support Slows Superfund Reform," *Safety & Health* (January 1997) remains apt.

The National Commission on the Superfund—established in 1992 as a joint project of the Environmental Law Center, Vermont Law School, and the Keystone Center—has issued its comprehensive *Final Consensus Report* (March 1, 1994). For a copy of this panel's report and recommendations, contact the Keystone Center, P.O. Box 8606, Keystone, Colorado 80435.

One controversial response to hazardous waste problems is the practice of buying out neighboring communities by the companies that are responsible for the problem. The pros and cons of this practice are discussed in "A Town Called Morrisonville," by John Bowermaster, *Audubon* (July/August 1993).

Another serious dimension of the hazardous waste problem is the growing use of developing nations as the dumping grounds for waste from the United States and other wealthier nations. "The Basel Convention: A Global Approach for the Management of Hazardous Wastes," by Iwonna Rummel-Bulska, *Environmental Policy and Law* (vol. 24, no. 1, 1994) is an assessment of the international treaty designed to prevent such waste dumping.

ISSUE 14

Municipal Waste: Is Recycling an Environmentally and Economically Sound Waste Management Strategy?

YES: Richard A. Denison and John F. Ruston, from "Recycling Is Not Garbage," *Technology Review* (October 1997)

NO: Chris Hendrickson, Lester Lave, and Francis McMichael, from "Time to Dump Recycling?" *Issues in Science and Technology* (Spring 1995)

ISSUE SUMMARY

YES: Environmental Defense Fund scientist Richard A. Denison and economic analyst John F. Ruston rebut a series of myths that they say have been promoted by industrial opponents in an effort to undermine the environmentally valuable and successful recycling movement.

NO: Engineering and economics researchers Chris Hendrickson, Lester Lave, and Francis McMichael assert that ambitious recycling programs are often too costly and are of dubious environmental value.

Since prehistoric times, the predominant method of dealing with refuse has been to simply dump it in some out-of-the-way spot. Worldwide, land disposal still accommodates the overwhelming majority of domestic waste. In the United States roughly 90 percent of residential and commercial waste is disposed of in some type of landfill, ranging from a simple open pit to so-called sanitary landfills where the waste is compacted and covered with a layer of clean soil. In a small, but increasing, percentage of cases, landfills may have clay or plastic liners to reduce leaching of toxins into groundwater.

By the last quarter of the nineteenth century, odoriferous, vermin-infested garbage dumps in increasingly congested urban areas were identified as a public health threat. Large-scale incineration of municipal waste was introduced at that time in both Europe and the United States as an alternative disposal method. By 1970 more than 300 such central garbage incinerators existed in U.S. cities, in addition to the thousands of waste incinerators that had been built into large apartment buildings.

Virtually all of these early garbage furnaces were built without devices to control air pollution. During the period of heightened consciousness about urban air quality following World War II, restrictions began to be imposed on garbage burning. By 1980 the new national and local air pollution regulations had reduced the number of large U.S. municipal waste incinerators to fewer than 80. Better designed and more efficiently operated landfills took up the slack.

During the past two decades, an increasing number of U.S. cities have been unable to find suitable, accessible locations to build new landfills. This has coincided with growing concern about the threat to both groundwater and surface water from toxic chemicals in leachate and runoff from dump sites. Legislative restrictions in many parts of the country now mandate costly design and testing criteria for landfills. In many cases, communities have been forced to shut down their local landfills (some of which had grown into small mountains) and to ship their wastes tens or even hundreds of miles to disposal sites.

The lack of long-range planning coupled with skyrocketing disposal costs created a crisis situation in municipal waste management in the 1980s. Energetic entrepreneurs seized upon this situation to promote European-developed incineration technology with improved air pollution controls as the panacea for the garbage problem. Ironically, the proliferation of these new waste incinerators in the United States coincided with increasing concern in Europe about their efficiency in containing the toxic air pollutants produced by burning modern waste. Citizen groups became aware of this concern and organized opposition to incinerator construction. The industry countered with more sophisticated air pollution controls, but these trapped the toxins in the incinerator ash, which presents a troublesome and expensive disposal problem. The result has been a rapid decrease in the number of municipalities that are choosing to rely on modern incineration to solve their waste disposal problems.

Recycling, which until recently has been dismissed as a minor waste disposal alternative, is being encouraged as a major option. The Environmental Protection Agency (EPA) and several states have established hierarchies of waste disposal technologies with the goal of using waste reduction and recycling for as much as 50 percent of the material in the waste stream. Several environmental groups are urging even greater reliance on recycling, citing studies that show that more than 90 percent of municipal waste can theoretically be put to productive use if large-scale composting is included as a component of recycling.

In the following selection, Richard A. Denison and John F. Ruston maintain that most arguments against recycling, which they believe is environmentally valuable and economically viable, are self-serving attempts by the organizations that create municipal waste problems to avoid scrutiny. In the second selection, generalizing from an analysis of the recycling program of their home city, Chris Hendrickson, Lester Lave, and Francis McMichael conclude that recycling is neither cost-effective nor environmentally advantageous.

Richard A. Denison
and John F. Ruston

 **YES**

Recycling Is Not Garbage

Ever since the inception of recycling, opponents have insisted that ordinary citizens would never take the time to sort recyclable items from their trash. But despite such dour predictions, household recycling has flourished. From 1988 to 1996, the number of municipal curbside recycling collection programs climbed from about 1,000 to 8,817, according to BioCycle magazine. Such programs now serve 51 percent of the population. Facilities for composting yard trimmings grew from about 700 to 3,260 over the same period. These efforts complement more than 9,000 recycling drop-off centers and tens of thousands of workplace collection programs. According to the EPA [Environmental Protection Agency], the nation recycled or composted 27 percent of its municipal solid waste in 1995, up from 9.6 percent in 1980.

Despite these trends, a number of think tanks, including the Competitive Enterprise Institute and the Cato Institute (both in Washington, D.C.), the Reason Foundation (in Santa Monica, Calif.), and the Waste Policy Center (in Leesburg, Va.), have jumped on the anti-recycling bandwagon. These organizations are funded in part by companies in the packaging, consumer products, and waste-management industries, who fear consumers' scrutiny of the environmental impacts of their products. The anti-recyclers maintain that government bureaucrats have imposed recycling on people against their will—conjuring up an image of Big Brother hiding behind every recycling bin. Yet several consumer researchers, such as the Rowland Company in New York, have found that recycling enjoys strong support because people believe it is good for the environment and conserves resources, not because of government edict.

Alas, the debate over recycling rages on. The most prominent example was an article that appeared... in the *New York Times Magazine,* titled "Recycling Is Garbage," whose author, John Tierney, relied primarily on information supplied by groups ideologically opposed to recycling. Here we address the myths he and other recycling opponents promote.

Myth: The modern recycling movement is the product of a false crisis in landfill space created by the media and environmentalists. There is no shortage of places to put our trash.

From Richard A. Denison and John F. Ruston, "Recycling Is Not Garbage," *Technology Review* (October 1997). Copyright © 1998 by *Technology Review,* published by The Association of Alumni and Alumnae of MIT. Reprinted by permission of *Technology Review*; permission conveyed through Copyright Clearance Center, Inc.

Fact: Recycling is much more than an alternative to landfills. The so-called landfill crisis of the late 1980s undoubtedly lent some impetus to the recycling movement (although in many cities around the country, recycling gained momentum as an alternative to incineration, not landfills). The issues underlying the landfill crisis, however, were more about cost than space.

Landfill space is a commodity whose price varies from time to time and from place to place. Not surprisingly, prices tend to be highest in areas where population density is high and land is expensive. In the second half of the 1980s, as environmental regulations became more stringent, large numbers of old landfills began to close, and many simply filled up, particularly in the Northeast. New landfills had to meet the tougher standards; as a result, landfill prices in these regions escalated dramatically. In parts of northern New Jersey, for example, towns that shifted their garbage disposal from local dumps to out-of-state landfills found that disposal costs shot from $15–20 per ton of garbage to more than $100 per ton in a single year. Although the number of open landfills in the United States declined dramatically—according to *BioCycle* magazine, from about 8,000 in 1988 to fewer than 3,100 in 1995—huge, regional landfills located in areas where land is cheap ultimately replaced many small, unregulated town dumps. Landfill fees declined somewhat and the predicted crisis was averted. Nonetheless, the high costs of waste disposal in the Northeast and, to a lesser extent, the West Coast, have spurred local interest in recycling: two-thirds of the nation's curbside recycling programs operate in these regions.

But landfills are only part of the picture. The more important goals of recycling are to reduce environmental damage from activities such as strip mining and clearcutting (used to extract virgin raw materials) and to conserve energy, reduce pollution, and minimize solid waste in manufacturing new products. Several recent major studies have compared the lifecycle environmental impacts of the recycled materials system (collecting and processing recyclable materials and manufacturing them into usable form) with that of the virgin materials system (extracting virgin resources, refining and manufacturing them into usable materials, and disposing of waste through landfills or incineration). Materials included in the studies are those typically collected in curbside programs (newspaper, corrugated cardboard, office paper, magazines, paper packaging, aluminum and steel cans, glass bottles, and certain types of plastic bottles). The studies were conducted by Argonne National Labs, the Department of Energy and Stanford Research Institute, the Sound Resource Management Group, Franklin Associates, Ltd., and the Tellus Institute. All of the studies found that recycling-based systems provide substantial environmental advantages over virgin materials systems: because material collected for recycling has already been refined and processed, it requires less energy, produces fewer common air and water pollutants, and generates substantially less solid waste. In all, these studies confirm what advocates of recycling have long claimed: that recycling is an environmentally beneficial alternative to the extraction and manufacture of virgin materials, not just an alternative to landfills.

Myth: Recycling is not necessary because landfilling trash is environmentally safe.

Fact: Landfills are major sources of air and water pollution, including greenhouse gas emissions.

According to "Recycling Is Garbage," municipal solid waste landfills contain small amounts of hazardous lead and mercury, but studies have found that these poisons stay trapped inside the mass of garbage even in the old unlined dumps that were built before today's stringent regulations. But this statement is simply wrong. In fact, 250 out of 1,204 toxic waste sites on the Environmental Protection Agency's Superfund National Priority List are former municipal solid waste landfills. And a lot more than just lead and mercury goes into—and comes out of—ordinary landfills. The leachate that drains from municipal landfills is remarkably similar to that draining from hazardous waste landfills in both composition and concentration of pollutants. While most modern landfills include systems that collect some or all of this leachate, these systems are absent from older facilities that are still operating. Moreover, even when landfill design prevents leachate from escaping and contaminating groundwater, the collected leachate must be treated and then discharged. This imposes a major expense and burden on already encumbered plants that also treat municipal sewage.

What's more, decomposing paper, yard waste, and other materials in landfills produce a variety of harmful gaseous emissions, including volatile organic chemicals, which add to urban smog, and methane, a greenhouse gas that contributes to global warming. Only a small minority of landfills operating today collect these gases; as of 1995, the EPA estimates, only 17 percent of trash was disposed of in landfills equipped with gas-collection systems. According to a 1996 study by the EPA, landfills give off an estimated 36 percent of all methane emissions in the United States. We estimate that methane emissions from landfills in the United States are 24 percent lower than they would be if recycling were discontinued.

Myth: Recycling is not cost effective. It should pay for itself.

Fact: We do not expect landfills or incinerators to pay for themselves, nor should we expect this of recycling. No other form of waste disposal, or even waste collection, pays for itself. Waste management is simply a cost society must bear.

Unlike the alternatives, recycling is much more than just another form of solid waste management. Nonetheless, setting aside the environmental benefits, let's approach the issue as accountants. The real question communities must face is whether adding recycling to a traditional waste-management system will increase the overall cost of the system over the long term. The answer, in large part, depends on the design and maturity of the recycling program and the rate of participation within the community.

Taking a snapshot of recycling costs at a single moment early in the life of community programs is misleading. For one thing, prices of recyclable materials fluctuate, so that an accurate estimate of revenues emerges only over time. For another, costs tend to decline as programs mature and expand. Most early curbside recycling collection programs were inherently inefficient because they duplicated existing trash-collection systems. Often two trucks and crews drove down the same streets every week to collect the same amount of material that one truck used to handle. Many U.S. cities have since made their recycling collection systems more cost-effective by changing truck designs, collection schedules, and truck routes in response to the fact that picking up recyclable refuse and yard trimmings leaves less trash for garbage trucks to collect. For example, Visalia, Calif., has developed a truck that collects refuse and recyclable materials simultaneously. And Fayetteville, Ark., added curbside recycling with no increase in residential bills by cutting back waste collection from twice weekly to once.

Several major cities—Seattle, San Jose, Austin, Cincinnati, Green Bay, and Portland, Ore.—have calculated that their per-ton recycling costs are lower than per-ton garbage collection and disposal. In part, these results may reflect the overall rate of recycling: a study of recycling costs in 60 randomly selected U.S. cities by the Ecodata consulting firm in Westport, Conn., found that in cities with comparatively high levels of recycling, per-ton recycling collection costs were much lower than in cities with low recycling rates. A similar survey of 15 North Carolina cities and counties conducted by the North Carolina Department of Environment, Health, and Natural Resources found that in municipalities with recycling rates greater than 12 percent, the per-ton cost of recycling was lower than that for trash disposal. Higher rates allow cities to use equipment more efficiently and generate greater revenues to offset collection costs. If we factor in increased sales of recyclable materials and reductions in landfill disposal costs, many of these high-recycling cities may break even or make money from recycling, especially in years when prices are high.

Seattle, for example, has achieved a 39 percent recycling/composting collection rate in its residential curbside program and a 44 percent collection rate citywide. Analysis of nine years of detailed data collected by the Seattle Solid Waste Utility shows that, after a two-year startup period, recycling services saved the city's solid waste management program $1.7 to $2.8 million per year. These savings occurred during a period of reduced market prices for recyclable materials; the city's landfill fees, meanwhile, are slightly above the national average. In 1995, when prices for recyclable materials were higher, Seattle's recycling program generated savings of approximately $7 million in a total budget of $29 million for all residential solid waste management services.

To reduce the cost of recycling programs, U.S. communities need to boost recycling rates. A study of 500 towns and cities by Skumatz Economic Research Associates in Seattle, Wash., found that the single most powerful tool in boosting recycling is to charge households for the trash they don't recycle. This step raised recycling levels by 8 to 10 percent on average. These kinds of variable-rate programs are now in place in more than 2,800 communities, compared with virtually none a decade ago.

Myth: Recycled materials are worthless; there is no viable market for them.
Fact: While the prices of recycled materials fluctuate over time like those of any other commodity, the volume of major scrap materials sold in domestic and global markets is growing steadily. Moreover, many robust manufacturing industries in the United States already rely on recycled materials. These businesses are an important part of our economy and provide the market foundation for the entire recycling process.

In paper manufacturing, for example, new mills that recycle paper to make corrugated boxes, newsprint, commercial tissue products, and folding cartons generally have lower capital and operating costs than new mills using virgin wood, because the work of separating cellulose fibers from wood has already occurred. Manufacturers of office paper may also face favorable economics when using recycling to expand their mills. Overall, since 1989, the use of recycled fiber by U.S. paper manufacturers has been growing faster than the use of virgin fiber. By 1995, 34 percent of the fiber used by U.S. papermakers was recycled, compared with 23 percent a decade earlier. During the 1990s, U.S. pulp and paper manufacturers began to build or expand more than 50 recycled paper mills, at a projected cost of more than $10 billion.

Recycling has long been the lower-cost manufacturing option for aluminum smelters; and it is essential to the scrap-fired steel "mini-mills" that are part of the rebirth of a competitive U.S. steel industry. The plastics industry, however, continues to invest in virgin petrochemical plants rather than recycling infrastructure—one of several reasons why the market for recycled plastics remains limited. Another factor not addressed by the plastics industry is that many consumer products come in different types of plastic that look alike but are more difficult to recycle when mixed together. Makers and users of plastic—unlike those of glass, aluminum, steel, and paper—have yet to work together to design for recyclability.

Myth: Recycling doesn't "save trees" because we are growing at least as many trees as we cut to make paper.
Fact: Growing trees on plantations has contributed to a severe and continuing loss of natural forests.

In the southern United States, for example, where most of the trees used to make paper are grown, the proportion of pine forest in plantations has risen from 2.5 percent in 1950 to more than 40 percent in 1990, with a concomitant loss of natural pine forest. At this rate, the acreage of pine plantations will overtake the area of natural pine forests in the South during this decade, and is projected to approach 70 percent of all pine forests in the country during the next few decades. While pine plantations are excellent for growing wood, they are far less suited than natural forests to providing animal habitat and preserving biodiversity. By extending the overall fiber supply, paper recycling can help reduce the pressure to convert remaining natural forests to tree farms.

Recycling becomes even more important when we view paper consumption and wood-fiber supply from a global perspective. Since 1982, the majority of the growth in worldwide paper production has been supported by recycled fiber, much of it from the United States. According to one projection, demand for paper in Asia, which does not have the extensive wood resources of North America or northern Europe, will grow from 60 million tons in 1990 to 107 million tons in 2000. To forestall intense pressures on forests in areas such as Indonesia and Malaysia, industry analysts say that recycling will have to increase, a prediction that concurs with U.S. Forest Service projections.

> *Myth: Consumers needn't be concerned about recycling when they make purchasing decisions, since stringent U.S. regulations ensure that products' prices incorporate the costs of the environmental harms they may cause. Buying the lowest-priced products, rather than recycling, is the best way to reduce environmental impacts.*
> *Fact: Even the most regulated industries generate a range of environmental damages, or "externalities," that are not reflected in market prices.*

When a coastal wetland in the Carolinas is converted to a pine plantation, estuarine fish hatcheries and water quality may decline but the market price of wood will not reflect this hidden cost. Similarly, a can of motor oil does not cost more to a buyer who plans to dispose of it by pouring it into the gutter, potentially contaminating groundwater or surface water, than to a buyer who plans to dispose of it properly. And there is simply no way to assign a meaningful economic value to rare animal or plant species, such as those endangered by clearcutting or strip mining to extract virgin resources. While many products made from recycled materials are competitive in price and function with virgin products, buying the cheapest products available does not provide an environmental substitute for waste reduction and recycling.

> *Myth: Recycling imposes a time-consuming burden on the American public.*
> *Fact: Convenient, well-designed recycling programs allow Americans to take simple actions in their daily lives to reduce the environmental impact of the products they consume.*

In a bizarre example of research, the author of "Recycling Is Garbage" asked a college student in New York City to measure the time he spent separating materials for recycling during one week. The total came to eight minutes. The author calculated that participation in recycling cost the student $2,000 per ton of recyclable trash by factoring in janitors' wages and the rent for a square foot of kitchen space, as if dropping the newspapers on the way out the door could be equated with going to work as a janitor, or as if New Yorkers had the means to turn small, unused increments of apartment floor space into tradable commodities.

Using this logic, the author might have taken the next step of calculating the economic cost to society when the college student makes his bed and does his dishes every day. The only difference between recycling and other routine housework, like taking out the trash, is that one makes your immediate environment cleaner while the other does the same for the broader environment.

Sorting trash does take some extra effort, although most people find it less of a hassle than sorting mail, according to one consumer survey. More important, it provides a simple, inexpensive way for people to reduce the environmental impact of the products they consume.

If we are serious about lowering the costs of recycling, the best approach is to study carefully how different communities improve efficiency and increase participation rates—not to engage in debating-club arguments with little relevance to the real-world problems these communities face. By boosting the efficiency of municipal recycling, establishing clear price incentives where we can, and capitalizing on the full range of environmental and industrial benefits of recycling, we can bring recycling much closer to its full potential.

NO ⬅

Chris Hendrickson, Lester Lave, and
Francis McMichael

Time to Dump Recycling?

After decades of lobbying by environmentalists and extensive experience with voluntary programs, municipal solid waste recycling has recently received widespread official acceptance. The U.S. Environmental Protection Agency (EPA) has set a national goal that 25 percent of municipal solid waste (MSW) be recycled. Forty-one states plus the District of Columbia have set recycling goals that range up to 70 percent. Twenty-nine states require municipalities or counties to enact recycling ordinances or develop recycling programs. Before celebrating this achievement, however, we need to take a hard look at the price of victory and the value of the spoils.

No one seeing the overflowing trash containers in front of each house on collection day can deny that MSW is a serious concern. Valuable resources are apparently being squandered with potentially serious environmental consequences. The popular media have carried numerous warnings that landfills are close to capacity, and we expect to find vehement local opposition to the siting of any new landfills. At first glance, recycling seems to be the perfect antidote, and it does have widespread public support.

Because it seemed to be the right thing to do, we have tolerated numerous glitches in establishing recycling programs. The supply of recycled material has grown much faster than the capacity for converting them to useful products. Prices for materials have fluctuated wildly, making planning difficult. It takes time to develop efficient collection and processing systems. But the public and policymakers have been willing to be patient as the kinks in the system are worked out. The self-evident wisdom of recycling reassured everyone that all these problems could be solved.

But as these difficulties are being resolved, we are developing a much clearer picture of the economics of recycling. Beneath the debates about markets and infrastructure lurk two fundamental questions: Is it cost effective? Does it actually preserve resources and benefit the environment? What "obviously" makes sense sometimes does not stand up to careful scrutiny.

From Chris Hendrickson, Lester Lave, and Francis McMichael, "Time to Dump Recycling?" *Issues in Science and Technology* (Spring 1995). Copyright © 1995 by The University of Texas at Dallas, Richardson, TX. Reprinted by permission.

Understanding the Problem

The U.S. gross domestic product (GDP) of $6 trillion entails a lot of "getting and spending." From short-lived items such as food and newspapers to clothing, computers, cars, household furnishings, and the buildings we live and work in, everything eventually becomes municipal solid waste. The average U.S. citizen produces 1,600 pounds of solid waste a year.

For most of our history, waste was carted to an open site outside of town and dumped there. When the public became unhappy with the smell, the appearance, and the threat to public health of these traditional dumps, EPA ruled that waste would have to be placed in engineered landfills. These sophisticated capital-intensive facilities must have liners to keep the leachate from spreading, collection and treatment systems for leachate, and covers to keep away pests and to inhibit blowing dust and debris. EPA's regulations resulted in the closure of most dumps and the elimination of the most serious environmental problems caused by MSW. Still, most people objected if their neighborhood was picked as the site of a landfill. Some analysts erred in interpreting the closure of dumps and siting difficulties as signs that the country was running short of landfills. Although a few cities, notably New York and Philadelphia, are indeed having trouble finding nearby landfills, there is no national shortage of landfills. Thus, lack of space for disposing of waste is not a rationale for recycling.

But even without a pressing need to find a new way to manage MSW, many people would promote recycling as an economically and environmentally superior strategy. Recycling is portrayed as a public-spirited activity that will generate income and conserve valuable resources. These claims need to be examined critically.

The Pittsburgh Story

To get a detailed picture of how current recycling programs work, we focus on Pittsburgh—an example of an older Northeastern city where one would expect waste disposal to be an expensive problem. In response to a state mandate, Pittsburgh introduced MSW recycling in selected districts in 1990 and gradually increased coverage of the municipality and the number of products accepted for recycling.

After studying numerous alternatives, Pittsburgh implemented a system by which recyclable trash was commingled in distinctive blue bags, separately collected at curbside, and delivered to a privately operated municipal recovery facility (MRF) for separation and eventual marketing to recyclers. The contract for operating the facility is awarded on the basis of competitive bidding. Recyclable trash is collected weekly by municipal employees using standard MSW trucks and equipment owned by the city. In addition, special leaf collections are made in the fall for composting purposes.

In 1991, the last year for which complete data are available, Pittsburgh collected 167,000 tons of curbside MSW. This represents roughly two-thirds of the city's total MSW; the other third included retail, industrial, office, and park wastes. Curbside pickup of glass, plastic, and metal produced 5,100 tons

(3.1 percent of curbside MSW) for recycling. In 1993, newsprint collection was added, and the total curbside pickup of recyclable material was 6,700 tons of newsprint and 5,300 tons of glass, plastic, and metal.

When Pittsburgh started its recycling program in 1989, it sought bids from MRF operators. The best bid was an offer to pay the city $2.18 per ton of glass, metal, and plastic delivered to the MRF facility and to charge the city $8.39 per ton to take the material if newsprint was included. The tipping fee at the landfill at the time was $24 per ton. Either option was therefore less expensive than landfilling if—and as we will see, this is a very big "if"—one does not take collection costs into account.

In the second round of bidding in 1992, the city was committed to recycling newsprint, so it solicited only bids that included newsprint. The best bid was a cost to the city of $31.60 per ton. Meanwhile, the fee for landfilling had fallen to $16.15 per ton. The city therefore had to pay almost twice as much per ton to get rid of its recyclable MSW—again, without accounting for collection costs.

The increased tipping fees for recyclable materials reflects recognition of the sorting costs associated with the Pittsburgh blue bags and the difficulties of marketing MSW recyclables. A study by Waste Management, Inc. found that the price of a typical set of recyclable MSW materials had fallen from $107 per ton (in 1992 dollars) in 1988 to $44 per ton in 1992. Prices have continued to fluctuate widely since then. Although they are high at the moment, there is no guarantee about the future.

Collector's Item

The price instability of recycled material has darkened the economic prospects for recycling and received extensive public attention. But an even more troubling problem—the cost of collecting recyclable material—has been largely overlooked. Pittsburgh's experience is particularly eye-opening. The city uses the same employees and type of equipment as it uses for regular MSW, but the trucks on the recycling collection routes use a crew of two instead of three. Using the city's own accounting figures and dividing the costs between recycling and regular collection in proportion to employee hours worked and time of truck use, we calculated total collection costs. In 1991, it cost Pittsburgh $94 per ton to collect regular MSW and $470 per ton to collect recyclable MSW. With tipping fees for recyclables now higher than those for regular MSW, the total cost of disposing of recyclable MSW is more than four times the cost for regular MSW.

Several factors account for the very large difference. First is the lower density of recyclables; a full truck will hold fewer tons of recyclables. A second reason is that the amount of material picked up at each house is much smaller (recyclable material is less than 10 percent of the total MSW in Pittsburgh) so that the truck has to travel farther and make more stops to collect each ton of recyclable MSW. Because the purpose of recycling is to preserve resources and protect the environment, it should be noted that collecting recyclable MSW results in a significant increase in fuel use and combustion emissions.

Care must be taken in generalizing from Pittsburgh, where the narrow streets and hilly terrain make collection difficult, to other cities. The cost of collecting recyclable MSW is not that high in most cities. Waste Management, Inc., reports an average collection and sorting cost of $175 per ton for recycled material, based on its experience with 5.2 million households in more than 600 communities. However, the cost of collecting regular MSW is also significantly lower elsewhere, so that the difference in the costs of collecting recyclable and regular MSW is very large everywhere. Data available for other municipalities suggests that Pittsburgh's experience is not atypical. For example, San Jose reports costs of $28 per ton to landfill versus $147 per ton to recycle.

Although the cost estimates cited above are a very rough estimate of actual costs, the difference between landfilling and recycling is so large that we are convinced that more finely tuned financial data would not have any significant effect on the bottom-line conclusion that most recycling is too expensive. City officials are apparently beginning to reach the same conclusion. After some years of experience, Pennsylvania's cities have begun to scale back their recycling program as a result of the unforeseen additional costs.

Disappointing Alternatives

Because collection accounts for such a large share of the cost of recycling, we need to look at alternatives to Pittsburgh's system of separate curbside pickup. One option would be to improve the efficiency of the current pickup system. In Pittsburgh, collection routes are determined by tradition, with little attention paid to minimizing cost. However, research by graduate students at Carnegie Mellon found that savings from improved routing and other improvements in the current collections system would be small. Although any reduction in cost would be desirable, the savings are available to regular as well as recyclable MSW collection so there should be no change in the relative costs.

A related strategy would be to decrease the frequency of collection. For example, under pressure from the city council to reduce costs, Pittsburgh adopted in mid-1994 a biweekly schedule for collecting recyclable MSW. By increasing the amount of recyclables at each residence, the density of collection has increased somewhat, but it still does not approach the density of regular MSW. Also, residences now have to store recyclable materials longer, which could weaken their willingness to participate. This might explain why Pittsburgh's 1994 collection of recyclables was 25 percent less by weight than it had been in 1993.

A second alternative for cost savings is to use a private firm for collection. This might result in marginal savings but could hardly be expected to make a significant difference. A third possibility is to use the same truck for collecting MSW and recyclables. The efficiency of this system depends on how much additional time is lost in collecting the recyclables and then dropping them off at the MRF on the way to emptying the MSW at a landfill. A few cities have adopted this approach, but no reliable economic evaluation has been done. For

Pittsburgh, we estimate that combined collection would actually increase costs by 10 percent.

Fourth, collection of recycled MSW might be abandoned in favor of disturbed dropoff stations. Households would reap the benefits of lower taxes at the expense of dropping off their recyclables. The efficiency of this system depends on the amount of recyclables to be dropped off and the number of additional miles driven. To obtain a rough estimate, assume that each household drives three extra miles (30 percent of an average shopping trip) every two weeks. The household generates 150 pounds of MSW every two weeks, of which 8 percent (12 pounds) is recyclable. Thus, the $0.90 additional driving cost amounts to $0.075 per pound or $150 per ton. Costs of dropoff center implementation and maintenance should also be added. In Wellesley, Massachusetts, the operating cost of a dropoff center is reported as $16 per ton of recycled material in 1988–1989 or roughly $18 in 1992 dollars. Thus, an estimate of the total direct cost of recycling in dropoff stations is $168 per ton. This does not include the value of volunteer labor such as sorting recyclable material and driving to the dropoff center. At $5 per hour, the labor cost is more than the vehicle costs, with a total of about $400 per ton. Having more dropoff centers would lower driving costs but add a neighborhood nuisance and increase center costs. Another consideration is that the total volume of recycled material might be much smaller because people would not want to do the extra work. Smaller volume would make it more difficult to establish a market for the recycled material and to benefit from economies of scale in processing the material.

Because the value of recycled material varies so much, efficiency might be increased by limiting collection to the most valuable materials. For example, assuming that typical MRF processing costs $150 per ton, that collecting recyclable MSW costs $75 per ton more than collecting regular MSW, and that tipping fees are about $35 per ton, the recyclable material would have to sell for at least $190 per ton to be worth separating from MSW. Only aluminum, which was selling for about $750 per ton in 1993, qualifies on this criteria. At that time, plastic was $100–$130 per ton, steel and bimetal from cans was $80 per ton, clear glass was $50 per ton, and newsprint was $30 per ton. By limiting collection to aluminum and other metal cans, plastic, and plastic containers, one could lower the separation costs at the MRF, but the unit collection cost would increase so much that it would probably dwarf the savings at the MRF. In addition, collecting only high-value materials contradicts the EPA goal of recycling 25 percent of all MSW.

A major problem with recycling is the low demand for recycled materials. For example, Germany instituted a packaging recycling program that collected essentially all used packaging, but now Europe is swamped with inexpensive (and subsidized) recycled plastic. One possible policy prescription for reducing the imposed costs of recycling is to stimulate the demand for recycled materials. For example, the federal government has changed its procurement policy to insure that 20 percent of paper purchases are of recycled pulp. In some cases, there is needless discrimination against recycled materials. However, at our estimated cost of $190 per ton for additional collection and separation costs, not many materials would be worth recycling even if demand for them surged.

Finally, we could move to a completely different arrangement such as the "take-back" system being tried in Germany in which the manufacturer is responsible for getting packaging material back from the consumer and recycling it. Germany is even considering legislation that would require manufacturers to take back and recycle their own products. In this system, firms would be required to arrange "reverse logistics" systems for collecting and eventually recycling their discarded products. For example, newspaper delivery services would have to collect used newspapers. The United States already has take-back regulation for a few particularly hazardous products such as the lead acid batteries used in automobiles. Although this approach creates strong incentives for manufacturers to reduce waste, the costs are likely to be much higher than those of the present system, because it will almost certainly require numerous collection systems.

What About the Environment?

Our analysis convinces us that recycling is substantially more expensive than landfilling MSW. But the primary motivation for recycling laws is not to save money; it is to save the environment. As it happens, saving the environment is not so different from saving money in this case. The greater costs stem from additional trucks, fuel, and sorting facilities. Every truck mile adds carcinogenic diesel particles, carbon monoxide, organic compounds, oxides of nitrogen, and rubber particles to the environment, just as building and maintaining each truck does. Collection in urban areas also increases traffic congestion and noise. Constructing, heating, and lighting for an MRF similarly use energy and other scarce resources. The variety of activities associated with the two- to four-fold increase in costs associated with recycling is almost certain to result in a net increase in resource use and environmental discharges.

For Pittsburgh and similar cities, the social cost of MSW recycling is far greater than the cost of placing the waste in landfills. No minor modifications in collection programs or prices of recycled materials are likely to change this conclusion. Approaches such as dropoff stations that attempt to hide the cost by removing it from the city ledger are likely to have the highest social cost.

Although many people object to landfill disposal, modern landfills are designed and operated to have minimal discharges to the environment. Current regulations are sufficient to minimize the environmental impacts of landfills for several decades. Nevertheless, landfills are unlikely to be the optimum long-term solutions.

The fundamental problem remains: A society in which each individual produces 1,600 pounds of MSW a year is consuming too much of our natural resources and is diminishing environmental quality. Today's MSW recycling systems are analogous to the "end-of-the-pipe" emission controls enacted 25 years ago. Air and water discharge standards were designed to stop pollution. They do so, but at a cost of about $150 billion per year. Recycling MSW lowers the amount going into landfills but at too high a cost.

EPA and some progressive companies have stressed "pollution prevention" and "green design" as the only real solution to pollution problems.

Just cleaning up Superfund waste sites has proven extraordinarily expensive. Less expensive but still inefficient is the cost of preventing environmental discharges through better management of hazardous waste. The ideal solution is to redesign production processes so that no hazardous waste is created in the first place and no money is needed for discharge control and remediation.

For MSW, this approach would mean designing consumer products to reduce waste and to facilitate recycling. The potential hazards associated with toxic materials in landfills could be reduced by eliminating the toxic components in many products. For example, stop adding cadmium to plastics to give them a shiny appearance and stop using lead pigments in paints and ink. Another example is choosing packaging to minimize the volume of waste. Finally, products can be designed so that at disposal time the high-value recyclable materials can be easily removed.

Producers and consumers don't have good information to help them make choices among materials. And even when they have the information, they are not sufficiently motivated to use it. Most consumers know that they shouldn't dump used motor oil down the drain and shouldn't put old smoke detectors or half-empty pesticide containers in their trash. If they were charged the social cost of these practices, they would find more environmentally satisfactory ways of handling these unwanted products. In some cases it may be cost effective for manufacturers to include prepaid shipping vouchers to encourage consumers to return highly toxic components such as radioactive materials in smoke detectors before disposing of a product.

The best way to inform consumers and producers and to motivate them to act in socially desirable ways is to establish a pricing mechanism for materials and products that reflects their full social cost, including resource depletion and environmental damage. Full-cost pricing of raw materials would lead producers to make more socially desirable choices of materials and lead them to designs that are easier to reuse or recycle. A major problem with the current system is that product wastes in MSW arrive at the MRF having been manufactured with little or no thought for making them easy to recycle. Full-cost pricing would change the choice of materials and design so that the MRF was an integrated part of a product's design.

Unfortunately, more research is required to determine the full cost of materials, and after that is done, it will be necessary to develop a means of implementing the concept. Neither task will be easy, but the alternative is to neglect environmental problems or to attempt to regulate every decision.

Even under the best of conditions, improved design and recycling will not eliminate the need for disposal. The waste stream will be smaller and less hazardous, but the total volume will still be daunting. We will have to come back to comparing the merits of landfills, recycling, and incineration. Changes in the waste stream will force us to examine each option with fresh eyes. At present, this might mean reserving recycling for metals, using the plastic and wood product portions of MSW as fuel for energy-producing incineration, and landfilling the rest.

MSW is a systems problem. Any one-dimensional solution, be it mandated recycling, incineration, or something else, is likely to do more harm than good. An assumed preference for recycling flies in the face of economic reality unless mechanisms can be found to greatly lower the costs of collection and sorting. The long-term answer to managing MSW is likely to include green design, materials choice, component reuse, and incineration, as well as recycling. Finding a way to use full-cost pricing so that decisions are decentralized and quickly adaptable will be the key to achieving thoughtful use of resources and improvements in environmental quality.

POSTSCRIPT

Municipal Waste: Is Recycling an Environmentally and Economically Sound Waste Management Strategy?

Denison and Ruston organize their selection around a series of claims denouncing the value of recycling that were included in "Recycling Is Garbage," the cover story by John Tierney in the June 30, 1996, issue of *The New York Times Magazine*. That story stimulated a torrent of heated responses from environmental organizations and activists who were outraged by its one-sided presentation and numerous examples of what they perceived to be misinformation. Hendrickson et al. focus on the area around Pittsburgh, Pennsylvania, where the greater availability of unpopulated areas suitable for landfills makes that option considerably more attractive and cheaper than in many other, more densely populated parts of the world. They argue that the pollution resulting from the collection and processing of recyclable materials exceeds the environmental benefits associated with recycling. Others who have examined this question maintain that the much lower pollution associated with the manufacture of paper, glass, aluminum, and steel from recycled feedstocks rather than from virgin raw materials makes the recycling of these waste-stream components highly beneficial from an environmental perspective.

At the same time that Hendrickson et al. were bemoaning the economic and environmental failings of recycling, waste management research experts William E. Franklin and Marjorie Franklin were presenting a description of a healthy, thriving recycling industry and making optimistic predictions about its future in their article in the March/April 1995 issue of the trade journal *MSW Management*. In the same issue of that journal, Delwin Biagi describes the progress that the sprawling city of Los Angeles, California, is making with its ambitious, new recycling program. In the May/June 1996 issue, Sue Eisenhold describes how the success of urban programs has inspired the development of creative rural recycling efforts throughout the United States and Canada. And in the May/June 1998 issue, Henry L. Henderson reports on the unorthodox but successful full-scale recycling program in Chicago, Illinois. Alice Horrigan and Jim Motavalli defend recycling in "Talking Trash," *E* (March–April 1997), and the continuing debate is reviewed by Mary H. Cooper in "The Economics of Recycling," *CQ Researcher* (March 27, 1998). Some, however, argue that the chief benefits of recycling are to our consciences; see Jesse Walker and Pierre Desrochers, "Recycling Is an Economic Activity, Not a Moral Imperative," *The American Enterprise* (January 1999).

ISSUE 15

Nuclear Waste: Should the United States Continue to Focus Plans for Permanent Nuclear Waste Disposal Exclusively at Yucca Mountain?

YES: Luther J. Carter and Thomas H. Pigford, from "Getting Yucca Mountain Right," *The Bulletin of the Atomic Scientists* (March/April 1998)

NO: D. Warner North, from "Unresolved Problems of Radioactive Waste: Motivation for a New Paradigm," *Physics Today* (June 1997)

ISSUE SUMMARY

YES: Science writer Luther J. Carter and nuclear engineer Thomas H. Pigford argue that although there are many unresolved issues related to nuclear waste disposal that need technical and congressional attention, these problems can be surmounted. They maintain that establishing clear goals that would culminate in a safe, permanent nuclear waste repository at Yucca Mountain is a realistic and sensible strategy.

NO: Risk assessment expert D. Warner North discusses the many formidable technical and political problems with the current strategy to site a permanent nuclear waste facility at Yucca Mountain and argues that a new, more open and flexible paradigm is needed to deal with disposal of radioactive materials.

The fission process by which the splitting of uranium and plutonium nuclei produces energy in commercial and military nuclear reactors generates a large inventory of radioactive waste. This waste includes both the "high-level" radioactive by-products of the fission reaction, which are contained in the spent fuel rods removed from the reactors, and a larger volume of "low-level" material, which has been rendered radioactive through bombardment with neutrons—neutral subnuclear particles—emitted during the reaction. "Low-level" waste also includes the refuse produced during medical and research uses of radioactive chemicals.

The amount of highly radioactive material that builds up in the core of a commercial nuclear power plant during its operation far exceeds the radioactive release that results from the explosion of a high-yield nuclear weapon. Because radioactive emissions are lethal to all biological organisms—causing severe illness and death at high doses and inducing cancer at any dose level—it is necessary to make sure that the radioactive wastes are kept isolated from the biosphere. Since some of the nuclear products remain radioactive for hundreds of thousands of years, this is a formidable task.

The early proponents of nuclear reactor development recognized the need to solve this problem. Confident that scientists and engineers would find the solution, a decision was made to proceed with a program, sponsored and funded by the U.S. government, to promote nuclear power before the serious issue of permanent waste disposal had been resolved.

Forty years later, with 100 commercial nuclear power plants licensed in the United States, more than 300 in other countries around the world, and hundreds of additional military nuclear reactors piling up lethal wastes in temporary storage facilities every day, the early confidence that the disposal problem could be solved has long since disappeared. The most recent Nuclear Waste Policy Act, legislated in 1982, set a step-by-step schedule to complete a permanent, operating "high-level" waste repository by 1998. This schedule has proven impossible to meet. In December 1987, recognizing that serious problems were again developing in implementing the new plan, Congress short-circuited the process by designating Yucca Mountain, Nevada, as the only site to be considered for the first high-level repository. Continuing technical and political problems made it impossible to meet the target date—the end of 1998—for the approval of the Yucca Mountain site, and there is considerable doubt that the new goal of opening a facility there by 2010 will be met. The continuing debate is summarized by Chuck McCutcheon in "High-Level Acrimony in Nuclear Storage Standoff," *Congressional Quarterly Weekly Report* (September 25, 1999) and Sean Paige in "The Fight at the End of the Tunnel," *Insight on the News* (November 15, 1999).

The history of the nuclear waste issue illustrates the folly of focusing on technological fixes without recognizing that solutions to real-world problems must meet political, socioeconomic, and ecological criteria that are not revealed by isolating the results of laboratory investigations from the other aspects of the issue. The evaluation of a proposed technological solution to a problem is related to social values, which in turn affect the political position of the participants in the process. Serious differences exist as to the degree of isolation and period of time necessary for "high-level" waste containment.

In the following selections, Luther J. Carter and Thomas H. Pigford maintain that the plan to build a permanent disposal facility at Yucca Mountain should proceed and that unresolved safety issues can be put to rest if Congress and the Department of Energy provide clear goals. D. Warner North asserts that complex political and technical problems will prevent solution of the nuclear waste problem by means of a single permanent waste disposal site at Yucca Mountain or anywhere else. He advocates a more open and flexible plan to deal with all aspects of the issue.

Luther J. Carter and
Thomas H. Pigford

 **YES**

Getting Yucca Mountain Right

The most commonly held image of the geologic disposal of nuclear waste is surely that of a deep underground labyrinth of tunnels for the emplacement of waste containers. Plans for the geologic repository proposed for the Yucca Mountain site in Nevada square well enough with that image: There would be more than 100 miles of tunnels, and into this extraordinary labyrinth would go about 12,000 very large, massively built containers of spent fuel from nuclear power reactors, together with about 4,500 smaller containers of high-level waste from the former nuclear weapons production complex.

But this may not be the most telling image for understanding what is afoot at Yucca Mountain. Rather, the mind's eye should focus on the great plume of contaminated ground water that would form over time beneath the repository and eventually extend out some 40 miles beyond it. This is the image that best reveals the important but obscure policy issues relevant to protecting future generations from harmful doses of radiation.

Calculations based on the Energy Department's latest published performance assessment indicate that the plume would begin to form within the first 5,000 years after the repository is sealed. Many of the waste canisters will have failed by then from the corrosive effects of water dripping from the ceiling of the waste emplacement tunnels, or "drifts."

Radionuclides will have started going into solution and contaminated water will have begun migrating hundreds of feet down through the rock to the water table and aquifer below. Yucca Mountain Project planners generally assume that the plume first takes on roughly the outline of the repository above it, measuring nearly two and a half miles wide and two miles long. But the plume, with an assumed depth of about 160 feet, becomes steadily more elongated, moving ever closer to the earth's surface as it follows the aquifer to the south.

Advancing at a rate of maybe 30 feet a year, after 7,000 years the plume will pass beneath U.S. 95, the highway from Las Vegas to Reno, at a point near the hamlet of Lathrop Wells. After 7,500 years the plume will reach the first farm well in the Amargosa Valley. After about 11,000 years it will reach its terminal point 40 miles from Yucca Mountain at a spot known variously as Franklin Lake Playa or Alkali Flat, where the aquifer nears the surface.

From Luther J. Carter and Thomas H. Pigford, "Getting Yucca Mountain Right," *The Bulletin of the Atomic Scientists* (March/April 1998). Copyright © 1998 by The Educational Foundation for Nuclear Science, 6042 South Kimbark, Chicago, IL 60637. A one-year subscription is $28. Reprinted by permission of *The Bulletin of the Atomic Scientists*.

Here the water will be drawn into the surface environment and atmosphere by capillary action, the roots of plants, and evaporation. Some radioactivity, principally the long-lived and highly mobile fission products iodine 129 and technetium 99, will begin to be deposited at or near the surface as solids, subject to dispersion by wind and water.

The plume, although undergoing some dissipation at its edges, will remain in place for hundreds of thousands of years, with the concentration of contaminants gradually increasing over its 40-mile length as more and more radionuclides migrate downward from the repository to the aquifer below. Concentrations will be greatest near the repository.

Calculations by the project show that in 10,000 years the annual dose from drinking contaminated water from wells three miles from the repository will be about 0.02 rem per year. When the dose from eating food contaminated by irrigation water from these same wells is added, the total dose will be about 0.13 rem. This is 13 times the annual dose limit established by the U.S. Nuclear Regulatory Commission (NRC) two decades ago for persons living near a nuclear power plant. It is five times the annual dose the NRC allows for persons making unrestricted use of a nuclear facility whose license has terminated. (The dose calculations allow a 5 percent probability of doses higher than those cited here.)

After 10,000 years, the calculated annual dose at a well three miles distant rises rapidly. Indeed, after 30,000 years the annual dose from iodine 129 and technetium 99 will have increased about 80-fold, to 10 rems. Then the longer-term annual dose from neptunium 237 appears and rises to about 50 rem by about 100,000 years, amounting in less than a decade to an exceedingly high, life-shortening cumulative dose.

The Energy Department recognizes that these doses exceed reasonable standards for public health protection—hence the pressing need for deeper analysis and a search for a more promising strategy.

Policy Questions

The projected plume brings a number of important policy matters into sharper focus. Some pertain to broader national and international considerations, others to technical strategies appropriate to the Yucca Mountain site.

The proposed repository should become the primary facility of a national center for nuclear waste storage and disposal. Permanent geologic disposal of spent fuel and high-level waste would take place inside the mountain following interim storage of these materials to the east of the repository at a place known as Jackass Flats, within the boundaries of the Nevada Test Site, the 1,300-square-mile reservation used from 1951 to the early 1990s for testing nuclear weapons.

Some 84,000 metric tons of spent fuel may eventually come to the Nevada center before the middle of the next century. That would impose two important responsibilities:

The first is to contain harmful radio-activity. The repository must be designed and built to safety standards robust enough to be convincing to the technical community and most opinion leaders. Indeed, with an estimated life-cycle cost of $33 billion (in 1994 dollars), the project will have to be convincing to receive the annual congressional appropriations necessary for sustaining it over the next several decades.

The second responsibility is to securely sequester the weapons-usable plutonium that the fuel contains. Every ton of fuel contains more plutonium than the 6.2 kilograms in the bomb that destroyed Nagasaki. Eighty-four thousand tons would contain about 800 tons of plutonium, or three times more than was present in the combined arsenals of the United States and the Soviet Union at the peak of the nuclear arms race. For security and accountability, as well as for long-term safety, central interim storage is far superior to continued storage at scores of nuclear power stations. With final disposal of spent fuel inside Yucca Mountain, security would be enhanced still further. Effective international oversight might have to continue as a safeguard against recovery of the plutonium by mining operations, but the sheer complexity and difficulty of such operations in the presence of the heat and radiation from radioactive decay will itself be a safeguard.

As for containing radioactivity, there is not at the moment a clear promise that the proposed repository can be made robustly and definitively safe. The project has been lumbering along, with its massive tunnel-boring machine completing a five-mile loop through the mountain [recently] and setting the stage for a deep in-situ investigation of the site. But defensible goals and scientific strategies for finishing the job are still lacking.

As yet, neither the Energy Department nor Congress is committed to providing a system of containment and isolation that can meet internationally recognized standards of safety and do so in a way that can be confirmed by reliable predictive models. These are the issues that cry out for a proper and timely response:

Stringently defining the "critical group" and the allowable annual radiation dose   Safety standards must protect the people most exposed, or the "critical group." Who makes up that group may vary from one era to the next, because the group is essentially self-selected by its choice of lifestyle and place of residence. But the group should include, for instance, a farm family whose drinking water comes from the wells near the repository (and down-gradient from it), and whose diet includes in substantial part vegetables watered from the same wells.

For repository performance standards to be convincing to a broad public, the critical group must be defined as the bounding case—that is, as the people who would receive the greatest exposures. The allowable annual radiation dose for that group should be no greater than the 0.01 to 0.025 rem permitted today for individuals living near nuclear power plants or freely using the site of a nuclear facility whose license has terminated.

If the critical group is protected, then all persons at lesser risk will be protected. Further, just as the critical-group concept works for people of a given time and generation, it also applies across great spans of time—protection of the critical group of a very distant epoch also protects people of times less distant.

This definition of the critical group is consistent with the one adopted by the International Commission on Radiological Protection in 1985, and since widely honored and respected by health physicists and radiation biologists worldwide. In contrast, the bill passed by the House of Representatives in October 1997 would have radiation safety assessed on the basis of a vaguely defined "vicinity average" annual dose not to exceed 0.1 rem.

That standard would permit an unconscionable leniency. A vicinity average dose could be kept low enough to meet virtually any health standard simply by manipulating population numbers, yet the critical group and many others in the vicinity could be harmfully exposed. Also unacceptable and susceptible to manipulation is the "probabilistic critical group," a concept proposed in 1995 by the National Research Council's Committee on Technical Bases for Yucca Mountain Standards. (Thomas Pigford, one of the authors of this essay, was a member of that committee; he set forth his dissenting views at the time.)

The Yucca Mountain Project assumes that a future critical group will obtain water from a well three miles from the repository, where the contaminants will be less concentrated than they will be at the repository boundary. Moreover, an attempt is now being made in some performance assessments to base compliance determinations on calculated radiation exposures from water extracted from a well 12 miles away, at Lathrop Wells, or even farther away, at Amargosa Farms.

Developing an adaptive strategy based on the particular strengths and weaknesses of the Yucca Mountain site The repository would be built high above the water table, which offers the important advantages of easy access and relatively dry rock. But there is a significant disadvantage: The presence of air in the repository complicates the hydrology and makes for an oxidizing atmosphere that promotes rapid deterioration of the spent fuel and early release of radionuclides from the waste containers.

An effective adaptive strategy would establish barriers to prevent water from ever reaching the waste containers. A two-layered capillary barrier might be the answer.

The Energy Department is now considering plans to slow the long-term release of radionuclides and to reduce the peak doses at 30,000 to 100,000 years by installing drip shields and by using a container material that will fail by pinhole penetrations rather than by general corrosion. But there is no data base —nor even adequate theories—for calculating the corrosion and mechanical failure of drip shields and the growth of pinholes over many tens of thousands of years. Reliable calculations of safety benefits cannot be made.

Adopting a carefully staged, iterative approach to developing and licensing the repository If geologic disposal is to succeed at Yucca Mountain, the design of the overall repository containment system must be rigorously and repeatedly tested against performance standards stringent enough to push the project to a higher level of resourcefulness and ingenuity. Licensing should come only at the end of a staged, iterative process. Instead, present plans call for the repository to be licensed and receiving waste by the end of the next decade.

Pending congressional legislation reaffirms the inflexible, counterproductive commitment to a fixed licensing schedule. The House bill would direct the Energy Department to have built and licensed a repository that would begin operations by January 17, 2010.

If the often-repeated cliché of the nuclear industry were true—that nuclear waste disposal is a political rather than a technical problem—then it might be possible to have a repository up and running by 2010. But this notion is wrong. Building a repository is much more than a matter of building tunnels and waste-emplacement drifts and installing the necessary waste handling equipment. Rather, the problem is to design and create—through iterative stages that could take decades—a system of containment capable of meeting rigorous standards of safety over many tens of thousands of years. Further, major new challenges often reveal themselves unexpectedly as site exploration and design progresses, independent of any scheduling constraints that Congress or project managers might choose to impose.

Recognizing the close and supportive relationship that an interim surface storage program can have to geologic disposal   Interim Storage in surface casks or monoliths relates to geologic disposal in two ways:

First, surface storage is the default solution to the nuclear waste problem —it is what will happen if geologic disposal fails or is abandoned.

Second, if a central interim storage facility is available, a repository can be developed and licensed in keeping with a staged, iterative process. Continued storage of spent fuel at the scores of widely dispersed reactor stations is not, however, likely to support such a patient, thorough-going process. This is what has led to the present intense lobbying by the nuclear industry, the state public utility commissions, and more than a dozen governors, for the present proposals for expedited, hurry-up solutions to the nuclear waste problem.

The Nevada Test Site offers special advantages for surface storage—isolation in a dry and almost uninhabited desert environment; the tight security of a former weapons-testing facility; and closed hydrologic basins from which no surface water flows.

The co-location of interim surface storage and the proposed repository project would offer yet another advantage. Jackass Flats, the area selected for surface storage in pending legislation, together with the repository site that is just to the west of it, is in a closed basin that discharges to Death Valley. Central storage there would be safer and more convenient than surface storage at virtually any other place in the country.

Sponsors of the House and Senate bills, strongly influenced by the nuclear industry, are solidly committed to providing surface storage at the Nevada Test Site. But the White House is not, objecting that to begin storing waste there might prejudice the outcome of the Yucca Mountain Project, either by tilting evaluations in the project's favor, or perhaps by lessening its urgency and political support.

Although these are not fanciful considerations, greater weight deserves to be given to the fact that central interim storage is an essential condition for the appropriate iterative approach to repository development. Central storage is, moreover, superior in its own right to continued storage at reactor sites. It affords greater long-term assurance of radiological safety. It also affords greater security and better accountability.

Improving Safety

Improving the repository waste isolation system, and doing it in ways that enhance the reliability of safety predictions, can be best accomplished through changes to the waste containers and their mode of emplacement.

The relevant maxim is that the farther the radioactivity migrates from its engineered emplacement, the greater the uncertainties in predicting the concentration of radioactivity in the ground water. The aquifer dilutes the radioactivity once it migrates down from the repository, but calculations of the dilution vary by factors of a few thousand to over 100,000, adding to uncertainties in radiation exposures for the critical group.

The calculation is subject to uncertainties in repository hydrology and in aquifer flow. In turn, the latter reflects uncertainties about hydraulic contours, rock porosity, and permeability. Some hydrologists suggest that, instead of the large elongated plume calculated by the project's idealized model, heterogeneities will cause several small plumes to form, each a few hundred feet across, creating some localized regions of higher concentrations and others of lower concentrations than those now predicted.

But if it is necessary to strengthen containment, how might it be done? And how can the reliability of our calculations be improved?

One possibility is to fill void spaces in the waste containers with a chemically reducing material such as pellets of depleted uranium. Experiments might show that such a reducing environment would lower the solubility and spent-fuel oxidation rate enough for better retention of the more troublesome radionuclides, especially isotopes of iodine, technetium, cesium, and neptunium. The fillers would also guard against the possibility of nuclear criticality.

Another possibility would be designing and installing barriers that divert the flow of water. This would be more than drip shields. The barrier would consist of a two-layered backfill covering the sides and tops of the waste containers with a course gravel layer next to the containers, then a layer of sand or finely ground volcanic tuff on top of that.

The outer layer would be a capillary barrier: the relatively large capillary forces of the sand layer would cause drip water to flow only through the sand and down into the porous rock below the emplacement tunnel. In the gravel layer, the spaces between the relatively large gravel particles would be too large to exert much capillary force.

Still to be determined is whether enough sand or ground tuff from the capillary layer would work its way down into the gravel, in a process called "fingering," to significantly reduce the barrier's effectiveness. Mock-up experiments conducted for the Electric Power Research Institute indicate that fingering is not a serious problem. However, the experiments should be extended using a shaking table to simulate the effects of earthquakes.

The properties that make such a diversion barrier promising can be measured in the laboratory. There is also a well-established mathematical theory for making reliable predictions of water flow and radionuclide release. In principle, capillary barriers could reduce the repository's peak annual dose from about 50 rem for the base case to a few microrem, a 10-million-fold reduction. Much greater predictive reliability also results because the capillary barrier's calculated performance appears insensitive enough to uncertainties in ground water flow for performance to remain unchanged even with a flow that is orders of magnitude greater than what is currently assumed.

Nevertheless, project managers seem reluctant to consider using capillary barriers because they would entail additional emplacement procedures and the redesign of equipment for waste emplacement. The use of any kind of backfill would be more complicated and costly than simply moving containers into their emplacement tunnels and leaving them, as in the present design. However, similar problems are being tackled in Europe, where project planners expect to install compacted clay backfill.

In any case, if the idea of capillary barriers is not carefully explored, their potential for insuring low doses and predictive reliability might be seized upon by skeptics of geologic disposal as a way to enhance the prospects for long-term surface storage.

Some 1,300 to 1,500 years ago, the Japanese protected burial vaults with capillary barriers. On the island of Honshu, near the town of Kumagaya along the Ara River, about a hundred of these vaults survive to this day. A wide variety of materials (steel, bronze, wood, cloth, and bone) have been well preserved because the barriers kept the moisture content extremely low inside the vaults and their surrounding gravel cocoons. These mounded structures, built on the surface, have remained undisturbed and functional despite Japan's damp climate and frequent earthquakes.

Potential Rewards

The rewards of a successful Yucca Mountain repository would be large. A fully developed site would accommodate all the commercial spent fuel and high-level nuclear weapons waste now in prospect, and then some. Under present law, the repository's capacity would be limited to 70,000 tons (with 7,000 reserved for weapons waste), but the actual potential is for some 140,000 tons. The U.S. spent fuel inventory at the middle of the next century is estimated at 84,000 tons, which could be high if no more reactors are ordered, for it assumes that all of about 100 existing reactors will remain on line for the rest of their 40-year operating permits.

Similarly, waste generation from the former U.S. nuclear weapons production complex is now well bounded. The Energy Department plans to dispose of 19,000 canisters of vitrified high-level waste from the Savannah River Plant in South Carolina, the Idaho National Engineering and Environmental Laboratory, and the Hanford site in the state of Washington. But that may be an overestimate. Much of this waste is expected to come from Hanford, where the special problems of recovery from the underground tanks there may well prove too difficult and costly to be achievable.

Consolidating spent fuel storage in Nevada before the repository is built would offer important advantages. If surface storage began early in the next decade, the Energy Department would be only a few years late in meeting its contractual and statutory obligations to the utilities, not to mention the rate payers, who have paid $12 billion into the Nuclear Waste Fund over the past 16 years.

Getting on with consolidated storage in Nevada will also be the key to the timely decommissioning of power reactors as they come to the end of their operating permits or are prematurely retired for economic reasons. It is important to prepare now for decommissioning, for if nuclear generation declines as steeply over the next 20 years as some studies predict, the user fees going into the Nuclear Waste Fund will decline correspondingly, perhaps causing a pinch with respect to building and operating the multi-billion-dollar spent-fuel transport and storage system that will be needed.

By consolidating spent-fuel storage in Nevada, the United States could provide a prototype for a global network of regional centers operating within a framework established by the International Atomic Energy Agency (IAEA). A network could make for greater nuclear materials accountability and control and thus a stronger nonproliferation regime.

The Nevada center could accept some fuel from other nations within the Western Hemisphere without taking on a heavy additional burden. While Canada has a substantial nuclear power program, it also has its own plans for spent-fuel management. The small tonnages of fuel that might come to Nevada from Mexico and Argentina, the only Latin countries that now have commercial nuclear programs, would add only marginally to the U.S. inventory.

Ultimately, with final disposal of U.S. and some foreign spent fuel deep inside Yucca Mountain, hundreds of tons of weapons-usable plutonium would be relatively inaccessible and under effective IAEA safeguards. The isolation and security of the spent fuel and plutonium would be enhanced by the installation of capillary barriers consisting of more than a million cubic yards of gravel and other material.

Here we come full circle to the importance of a Yucca Mountain repository as the centerpiece of a national center for nuclear waste storage and disposal in Nevada. If this facility is designed, licensed, and built to the appropriate standard, with radioactivity in the plume of contamination kept to innocuous levels, the United States will lead the world in finally coping with a nuclear waste problem which for far too long has gone without effective response.

D. Warner North

 **NO**

Unresolved Problems of Radioactive Waste

T he management of radioactive waste has become an increasingly intractable problem. US programs for the disposal of spent nuclear fuel and high-level waste, for disposal of transuranic waste, for clean-up of former nuclear weapons production facilities and even for disposal of low-level waste are far behind schedule and mired in public controversy. The US public has acquired a deep-seated fear of nuclear technology and its wastes, and mistrusts the institutions for managing these wastes. People's fear, mistrust and opposition are intensified when the proposed location for the wastes is nearby or is perceived to be a health or environmental hazard.

In the first decades of the nuclear age, the technical community viewed the problems posed by radioactive waste materials as modest and readily amenable to solution. Neither actinides, such as plutonium, nor the fission products from burning nuclear fuel were regarded as posing a serious environmental threat. Rather, they were regarded as dangerous materials that could be safely managed by using good engineering practices. The National Research Council of the National Academy of Sciences in 1957 first endorsed the concept of a repository in a deep geological formation that would effectively isolate waste from the biosphere for the time required for radioisotopes to decay (10^3 to 10^6 years). But finding a site where such isolation could be effectively guaranteed has proven to be an elusive goal.

The US government must deal with large inventories of high-level waste and transuranic waste from its weapons program, spent nuclear fuel and low-level waste. The government's policy is to work toward disposing of spent fuel and high-level waste in a geological repository at Yucca Mountain, Nevada; transuranic waste at the Waste Isolation Pilot Plant near Carlsbad, New Mexico; and low-level waste at a collection of sites licensed by states.

Proposed Yucca Mountain Repository

As envisioned in the 1982 Nuclear Waste Policy Act, a first geological repository was to be constructed and available to begin accepting spent nuclear fuel and high-level waste beginning in 1998. The Department of Energy has entered into

From D. Warner North, "Unresolved Problems of Radioactive Waste: Motivation for a New Paradigm," *Physics Today*, vol. 50, no. 6 (June 1997). Copyright © 1997 by The American Institute of Physics. Reprinted by permission. References omitted.

agreements with nuclear utilities to accept spent fuel beginning in 1998, and a recent court decision has held that DOE [Department of Energy] retains this obligation even if no facilities are available for disposal or for interim storage.

Spent nuclear fuel can be stored inexpensively in the fuel pools at operating reactors, provided storage capacity exists. But the pools at some reactors are reaching capacity, and maintaining pool storage becomes quite expensive after a reactor ceases operation and pool maintenance costs must be charged against storage rather than reactor operation. Storage of spent fuel in dry casks (large metal or concrete containers) has been implemented at a small number of US reactor sites, and the Nuclear Regulatory Commission has certified that such dry cask storage is acceptably safe for the order of a century. There has been extensive public opposition to the use of dry cask storage at some reactor sites, and nuclear utilities are eager for DOE to take over the problem of managing the spent nuclear fuel.

Proposals for a monitored retrievable storage facility, a centralized location for spent fuel storage, have failed to progress toward an operational facility. One of the most recent such proposals has involved the Mescalero Apaches and their reservation in New Mexico. A bill to establish an interim storage site for spent fuel and high-level waste at the Nevada Test Site, near Yucca Mountain, passed the Senate in 1996, but died due to inaction by the House of Representatives under a threatened presidential veto. Similar legislation, S. 104, has been passed by the Senate in the current session of Congress. Opposition from the Clinton Administration was reiterated by then–Undersecretary of Energy Thomas Grumbly. The Nuclear Waste Technical Review Board, an independent Federal oversight board established by Congress under the 1987 Nuclear Waste Policy Amendments Act, published a report in 1996 in which it concluded there was no compelling technical or safety reason to move spent fuel to a centralized storage facility in the next few years. However, the board recommended that efforts should commence now to plan for a storage facility that will be available once reactors start to shut down in large numbers beginning in about 2010. In the meantime, the board declared, DOE should use available funds to determine whether Yucca Mountain is suitable as the site for a geological repository.

Congress selected Yucca Mountain in the 1987 Nuclear Waste Policy Amendments Act as the single site for characterization to determine suitability, following an eruption of controversy over the selection of Yucca Mountain and two other sites as the candidates for detailed characterization under the provisions of the 1982 act. The congressional decision was decried in Nevada as imposition of a national problem on one western state with weak representation in Congress. Strong opposition to use of the Yucca Mountain site has been vigorously pursued ever since by Nevada's political leaders, supported by segments of the state's citizenry.

The Yucca Mountain site is on Federal land adjacent to the Nevada Test Site, which was formerly used for the testing of nuclear weapons. The mountain, in an uninhabited desert location, would allow waste to be emplaced in an unsaturated zone of tuffaceous rock about halfway between the land surface and the water table, 600 to 800 meters below the surface. The tuffaceous rock is relatively inexpensive to excavate—a significant factor in DOE's 1986 evaluation that Yucca Mountain was the best among the candidate sites for high-level waste. The relative absence of flowing groundwater could minimize corrosion of waste containers and the subsequent migration of radionuclides into locations where they could endanger human health and the environment. But the performance assessment of radionuclide releases and transport to the biosphere has proved more complex and controversial than expected.

The initial view of a geological repository was that waste isolation might be accomplished primarily by choosing an excellent geological setting. Subsequent repository performance assessment in the US and in other countries with nuclear waste programs has led to an alternative view—that waste isolation may be best accomplished by using a system of engineered barriers placed in a good geological setting.

The most critical aspect of the geological setting at many sites may be the extent to which the engineered barriers (the waste emplacement containers, the backfilling with appropriate materials, the sealing of excavated areas and other provisions for managing groundwater flow) will be able to limit radionuclide migration to acceptably low levels for thousands or millions of years. Developing and deploying such barriers successfully will require an understanding of basic geochemical phenomena, supported by geological evidence of the ability of certain metals (for example, copper) to withstand corrosion in favorable geochemical environments. Direct verification of corrosion resistance over time periods of millennia will not be possible for modern synthetic materials.

Human Intrusion

Human intrusion poses much more challenging problems for performance assessment than do materials corrosion and geological changes at a site. No one can reliably predict how human society and human behavior may change over time periods of centuries to millennia. The plutonium and uranium in spent fuel may become sufficiently valuable for energy production that, at some future time, humans will wish to retrieve spent fuel from a repository. Retrieving spent fuel buried 300 meters below the surface of Yucca Mountain would be a relatively straightforward task for a small group of individuals using even 19th-century mining technology. Retrieval of spent fuel from the seabed beneath 3000 meters of ocean is arguably marginally feasible for a large national or multinational effort using advanced late-20th-century technology. Deliberate intrusion to obtain spent fuel from dry cask storage might be as simple as overpowering a guard, breaching a chain-link fence and then transporting a container that could be moved by an oversized vehicle designed for heavy construction or electrical equipment.

Inadvertent human intrusion has been the main focal point of performance assessment for the Waste Isolation Pilot Plant, because, under undisturbed conditions, no releases are predicted to occur within the next 10,000 years. The regulatory criteria for Yucca Mountain suggested by the National Research Council [NRC] in its 1995 report in response to a congressional request in the Energy Policy Act of 1992 do not address deliberate intrusion, but do include inadvertent-intrusion scenarios such as penetration of a repository by a drilling rig. At this time, it is not yet known how the Environmental Protection Agency and, subsequently, the Nuclear Regulatory Commission will use the recommendations from the 1995 NRC report in writing revised standards for Yucca Mountain. It seems clear that contamination of groundwater by radionuclides and penetration of the repository in assumed inadvertent-intrusion scenarios will be addressed by these forthcoming revised standards.

Technical Issues

The most formidable problems associated with using the Yucca Mountain site are political ones. Many parties outside DOE have concluded that these problems are quite serious. Congress has grown frustrated and impatient, public trust and confidence are lacking and there are many technical problems that could lead to extensive additional technical debate and delay before a repository could be licensed by the Nuclear Regulatory Commission and construction could begin.

Such technical problems have already occurred. For example, a DOE scientist, Jerry Szymanski, hypothesized that seismic events could induce major changes in the water table (that is, the level of the saturated zone) such that the zone of waste emplacement could be immersed in groundwater. A National Research Council review found no significant support for this hypothesis. Another technical problem that has arisen stems from the claim by two scientists at Los Alamos National Laboratory that a critical mass of highly enriched uranium or plutonium-239 could be assembled by geological processes. Such a scenario is believed to be highly unlikely, especially with appropriate engineering of the waste form and other barriers as part of the engineering design.

A recent report by the General Accounting Office suggests another potentially controversial issue. Some scientists in the US Geological Survey are concerned that not enough is known about the hydrology of the saturated zone, particularly the steep hydraulic gradient at the northern end of Yucca Mountain. Another potential problem is the potential for climatic change leading to increased precipitation. Could such increased precipitation significantly increase groundwater flow or cause the water table to rise over the zone of waste emplacement? Such scenarios may be hard to rule out over a time period of tens of thousands or millions of years. The more issues of this type that are raised, the more complex and demanding will be the requirements for performance assessment, and the more difficult the task of assuring the concerned public that a repository at Yucca Mountain will be acceptably safe.

In the face of recent budget cuts imposed by Congress, DOE has cut back many planned scientific investigations at Yucca Mountain. Available funding has gone primarily to complete a five-mile tunnel adjacent to the planned zone of waste emplacement, and to begin a series of heater tests to simulate the effect of the heat added by radioactive decay on the rock in the emplacement zone. The density of loading of spent fuel and high-level waste will determine the elevation in temperature, which is expected to cause moisture in the unsaturated rock to migrate away from the waste. Although such migration may be helpful in achieving waste isolation and corrosion protection for containers in the first thousand years following closure of the repository, there are many open questions about adverse changes that might occur subsequently. For example, could such migration concentrate corrosive ions in groundwater that will migrate back in as the rock cools, leading to corrosive drips onto the containers and subsequent acceleration of the time at which the containers fail? Such issues pose challenges for performance assessment, but the problem for repository design is that large-scale heater tests may require decades to obtain needed data on geotechnical and geochemical rock behavior. In the meantime, significant uncertainties persist for repository design. The importance of the thermal loading issue has been stressed repeatedly in reports by the Nuclear Waste Technical Review Board.

Another technical issue involving repository design is the apparent existence of fast flow paths, the evidence for which is that water containing chlorine-36 from nuclear tests has been observed in the emplacement zone. Such flow must have occurred within the past 50 years, although the average flow through the unsaturated rock appears to be much slower. Is there a potential for significant water flow in these "fast paths" with increased precipitation? Will such flow accelerate the penetration of barriers to waste containment, or can the flow be diverted into "drains" such that barriers will remain dry and corrosion-resistant?

DOE has now acknowledged that it cannot proceed directly to a license application for Yucca Mountain but only toward a determination of site viability in 1998. The 1982 Nuclear Waste Policy Act provides that the secretary of energy can at any time find Yucca Mountain unsuitable as a repository site. DOE currently plans to make a recommendation to the President on the suitability of the site by 2001, and, assuming acceptance, to submit a license application to the Nuclear Regulatory Commission by 2002. The viability determination and the license application will require choosing a design for the repository. Because *in situ* tests of the effect of radioactive decay heat from the waste in the tuffaceous rock at repository depth at Yucca Mountain are just now getting under way, DOE will have difficulty supporting a choice of a thermal loading strategy, and this choice will determine how much spent fuel and high-level waste can be loaded into Yucca Mountain. Important choices must also be made on the container to be used for waste emplacement and on back-filling of the excavated emplacement areas.

In view of these difficulties and the history of previous missed project milestones, it seems unlikely that any waste will actually be emplaced until well after DOE's planned Yucca Mountain opening date of 2010. In the face of controversy, public opposition, continuing large expenditures, uncertainties and delays in the time for projected repository availability, continued congressional support for the Yucca Mountain project is uncertain.

A Call for a New Paradigm

The performance of a repository for spent fuel and high-level waste must be evaluated for time scales much longer than recorded human history. How much must we know about technical failures and human behavior in the distant future to make decisions now? By asking performance assessment if we can safely "set it and forget it" for ten thousand to one million years, our regulatory system poses an extreme challenge that available science may not meet. We should not restrict our use of science to Earth sciences and engineering. We should also ask social scientists for their help in communicating with the public about radioactive waste.

A major theme from a National Research Council review of Yucca Mountain conducted nearly a decade ago is that this repository should not be conceived of as a construction effort trying to adhere to an aggressive time schedule. Rather, it should be conceived of as a difficult, first-of-a-kind research effort to learn to do a complex task safely and to make adjustments appropriately as new information is obtained. The repository should not be closed soon after the waste is emplaced, but rather, albeit perhaps with some difficulty, it should be designed so that the waste remains retrievable for a time scale of at least 100 years.

In dealing with nuclear waste, especially spent fuel and high-level waste from its weapons program, the US needs to replace the previous goal in which the generation that created the waste would dispose of it once and for all (that is, "set it and forget it"). We must recognize that this goal is outdated and that we have failed to achieve it. Performance of a repository involves many potentially significant uncertainties. Ongoing surveillance and extended retrievability are appropriate for managing these dangerous materials. There are no clearly apparent practical alternatives for eliminating nuclear waste or for isolating the waste from the biosphere that are feasible both politically and technically. Nuclear waste remains a political as well as a technical challenge. A large part of the political problem is that we have set up unreasonable expectations with the American public and political leadership, that there is a simple technical solution available—geologic disposal.

As I argue elsewhere, a shift to a new paradigm may be needed, in which additional participation by stake-holder groups is solicited to get consensus on how to carry out the needed ongoing management of nuclear waste. It seems quite evident that we do not want a situation in which failure to accomplish a license application by 2002, or to construct a Yucca Mountain repository by 2010, places people at risk from waste in surface storage facilities that are not suitably constructed and maintained. The lessons of the Waste Isolation Pilot Plant and Ward Valley are that the much easier tasks of siting and then operating low-level waste and transuranic waste disposal facilities can take much longer and encounter much more determined opposition than was originally anticipated.

The US should not proceed recklessly and at high speed toward a geological repository for spent fuel and high-level waste. Rather, we should seek ongoing protection to safeguard human health and the environment from the dangerous radioactive materials that have been created in the first half-century of the nuclear age, and we should seek public understanding that such protection is being achieved.

POSTSCRIPT

Nuclear Waste: Should the United States Continue to Focus Plans for Permanent Nuclear Waste Disposal Exclusively at Yucca Mountain?

Carter and Pigford confine their attention to the need to protect a future "critical group" from excessive exposure to the radioactive plume that best technical estimates predict will ultimately develop if present plans for nuclear waste storage at Yucca Mountain are realized. They present some proposals for technological responses to this need. As North indicates, however, experts are not in agreement about the dimensions or timetable for this plume problem, nor about whether or not it is the most serious future hazard that should be anticipated.

Carter and Pigford reject the industry's assertion that radioactive waste disposal is only a political problem. North, on the other hand, states, "The most formidable problems associated with using the Yucca Mountain site are political ones." Since any final plan for the site will include the expensive and demanding requirement of maintaining its security and technological integrity for hundreds of thousands of years, the ultimate problem may be how this can be guaranteed, given the world's history of political instability over much shorter periods of time.

Carter and Pigford strongly support the construction of an interim, above-ground central storage facility for commercial nuclear waste at Yucca Mountain, primarily as part of a strategy to ensure continued development of a permanent facility there. This proposal is discussed and analyzed in detail in the January 1998 issue of *Congressional Digest*. The requirement for such an interim storage facility was removed from legislation passed by Congress in March 2000 to approve the Yucca Mountain site, but the bill was vetoed by President Clinton. See the summary of the bill's history prepared by the National Association of Regulatory Utility Commissioners at http://www.naruc.org/News/nuclear_waste_bil.htm.

The entire June 1997 issue of *Physics Today* (which includes North's article) is devoted to the radioactive waste issue. For an analysis of the issue that calls for an unequivocal halt to the Yucca Mountain project and, like North, proposes a rethinking of the entire waste disposal question, see "Overcoming Tunnel Vision," by James Flynn, Roger E. Kasperson, Howard Kunreuther, and Paul Slovic, *Environment* (April 1997). An even more provocative proposal is nuclear waste researcher Nicholas Lenssen's argument that all plans for permanent high-level nuclear waste storage be put on hold until the future of nuclear power is resolved. See "Facing Up to Nuclear Waste," *World Watch* (March/April 1992).

Nevada's former governor Richard Bryan's outspoken rejection of the congressional decision to designate Yucca Mountain as the permanent storage site set the stage for the state's continued opposition. For an elaboration of Bryan's arguments, read his article "The Politics and Promises of Nuclear Waste Disposal: The View from Nevada," *Environment* (October 1987). For a less politically partisan evaluation of the Yucca Mountain proposal, see Chris Whipple, "Can Nuclear Waste Be Stored Safely at Yucca Mountain?" *Scientific American* (June 1996).

The U.S. nuclear weapons program has created an ongoing disposal and cleanup controversy. For background information about this problem, see William F. Lawless's article in the November 1985 issue of *Bulletin of the Atomic Scientists*. In the October 1990 issue of the same journal, Scott Saleska and Arjun Makhijani present a frightening assessment of ongoing nuclear waste storage and treatment practices at the large military nuclear reservation in Hanford, Washington. Equally distressing is the history of incompetence in building the Waste Isolation Pilot Plant (WIPP) in New Mexico for permanent storage of military nuclear waste, as told by Keith Schneider in "Wasting Away," *The New York Times Magazine* (August 30, 1992). Debate continued for years, as recounted by Chris Hayhurst in "WIPP Lash: Doubts Linger About a Controversial Underground Nuclear Waste Storage Site," *E* (January–February 1998), but the WIPP opened for business in 1999.

For a chilling exposé of the hazardous nuclear waste disposal practices of the former Soviet Union, see William Broad's article, which begins on the front page of the November 21, 1994, issue of *The New York Times*. Thomas Orszag-Land describes plans to deal with the aftermath in "Removing Nuclear Waste From the Arctic," *Contemporary Review* (May 1997).

On the Internet ...

DUSHKIN ONLINE

National Councils for Sustainable Development

The National Councils for Sustainable Development (NCSD) is a program of the UN Earth Council devoted to exploring the implementation of sustainable development initiatives.

http://www.ncsdnetwork.org

The International Institute for Sustainable Development

The International Institute for Sustainable Development (IISD) encourages sustainable development through research partnerships and by aiding the dissemination of essential information.

http://iisd1.iisd.ca

Project: Environmental Regulation, Globalization of Production and Technological Change

This site offers a description of a United Nations University research project designed to investigate the implications of environmental regulation on the competitiveness of polluting industries.

http://www.intech.unu.edu/program/projects/proj223/index.htm

Union of Concerned Scientists

The Union of Concerned Scientists (UCS) describes itself as "citizens and scientists working together for a common goal: a healthy environment and a safe world, for today and for the next century." This site contains links to research reports on several current environmental issues, including global warming and resource depletion.

http://www.ucsusa.org

Worldwatch Institute

The Worldwatch Institute describes itself as "a nonprofit public policy research organization dedicated to informing policymakers and the public about emerging global problems and trends and the complex links between the world economy and its environmental support systems."

http://www.worldwatch.org

PART 4

The Environment and the Future

In addition to the many serious environmental problems of today, there are several potential crises that could be averted or diminished if preventive measures are taken now or in the near future. Continued pollution of the atmosphere could result in an increase in global temperature, which might have highly detrimental consequences. Also, the continued use of nonsustainable technologies could deplete vital resources, increase pollution, and exacerbate world hunger. How to improve efforts to initiate the necessary policy changes that could avert these crises is a basic, controversial question related to these issues.

- Is Sustainable Development Compatible With Human Welfare?

- Will Voluntary Action by Industry Reduce the Need for Future Environmental Regulation?

- Are Aggressive International Efforts Needed to Slow Global Warming?

- Are Major Changes Needed to Avert a Global Environmental Crisis?

ISSUE 16

Is Sustainable Development Compatible With Human Welfare?

YES: Julie L. Davidson, from "Sustainable Development: Business as Usual or a New Way of Living?" *Environmental Ethics* (Spring 2000)

NO: Jacqueline R. Kasun, from "Doomsday Every Day: Sustainable Economics, Sustainable Tyranny," *The Independent Review* (Summer 1999)

ISSUE SUMMARY

YES: Researcher Julie L. Davidson argues that a radical conversion to sustainable development offers a way to make future human freedom possible and consistent with the wider social and ecological good.

NO: Economics professor Jacqueline R. Kasun asserts that sustainable development poses threats to human freedom, dignity, and material welfare.

Over the last 30 years, many environmentalists, as well as other socioeconomic analysts have expressed concerns that humanity cannot continue indefinitely to increase population, industrial development, and consumption. The trends and their impacts on the environment are described in numerous books, including historian J. R. McNeill's *Something New Under the Sun: An Environmental History of the Twentieth-Century World* (W. W. Norton, 2000).

In the 1960s and 1970s sustainable development was expressed as the "Spaceship Earth" metaphor, which said that since we have limited supplies of energy, resources, and room, we must conserve and recycle. Otherwise, shortages will occur. In the early 1980s the United Nations' Secretary-General, Javier Perez de Cuellar, asked Gro Harlem Brundtland, a former prime minister and minister of environment in Norway, to organize and chair a World Commission on Environment and Development and to produce a "global agenda for change." The resulting report, *Our Common Future* (Oxford University Press, 1987), defines sustainable development as "development that meets the needs of the present without compromising the ability of future generations to meet their

294

own needs." The report acknowledges that limits on population size and resource use cannot be known precisely, that problems may arise gradually rather than suddenly and will be marked by rising costs, and that limits may be redefined by changes in technology. However, the report also recognizes that limits exist and must be taken into account when governments, corporations, and individuals plan for the future.

The report led to the United Nations Conference on Environment and Development held in Rio de Janeiro in 1992. This conference set sustainability firmly on the global agenda and made it a part of the efforts to deal with global environmental issues and to promote equitable economic development. In brief, sustainability means such things as cutting forests no faster than they can grow back, using ground water no faster than it is recharged by precipitation, using renewable energy sources rather than exhaustible fossil fuels, and farming in such a way that soil fertility does not decline. In addition, economics must be revamped to take account of environmental costs as well as capital, labor, raw materials, and energy costs. Many add that the distribution of the Earth's wealth must be made more equitable as well.

Given the continuous growth in population and the constant demand for resources, sustainable development is a difficult proposition. Some believe that it can be maintained, but others contend that for sustainability to work, either population or resource demand must be reduced. Many see sustainable development as in conflict with business and industrial activities, private property rights, and such human freedoms as the freedoms to procreate, to accumulate wealth, and to use the environment as one wishes.

In the following selections, Julie L. Davidson argues that the radical approach to sustainable development offers a necessary bridge from the traditional resource-greedy society to a more ecologically enlightened and equitable society. It also offers a way to make human freedom consistent with the wider social and ecological good. Jacqueline R. Kasun contends that traditional economics is not constrained by resource limitations and that there is no factual basis for concern about global environmental problems. She maintains that sustainable development will require sacrificing human freedom, dignity, and material welfare.

Julie L. Davidson

 **YES**

Sustainable Development: Business as Usual or a New Way of Living?

I. Introduction

Western societies are poised on a civilizational threshold similar to the one which marked the end of the feudal period when outmoded institutions, values, and systems of thought and their associated dogmas were ripe for transcendence by more relevant systems of organization and knowledge. As a consequence of the influential writings of Francis Bacon and René Descartes, it was thought that a rational social order founded on science and technology could overcome the miseries imposed by the vagaries of nature and the social-religious conflict that had consumed the sixteenth and seventeenth centuries. However, the success of modern industrial societies in providing for that material well-being and corporeal health thought by Bacon and Descartes to constitute the foundation for all the other goods of human existence now threatens life itself. As a result of the realization that there are biophysical limits to humankind's productive prowess, general agreement is developing that utilization of the world's material resources must be pursued without compromising the life chances of either present or future generations, albeit that there is much contestation about the strategies which can achieve this end.

At various strategic points in the history of modern societies, theorists have been compelled to reformulate their understanding of the systems of economic life—that is, the modes of production, consumptions, and distribution—in response to radical changes in their constraining factors, including availability of resources, distributional effects and, more recently, biophysical limits. My objective in this paper is to analyze sustainable development strategy as a contemporary reformulation of the economic problem and explore its capacity to address the problems of our historical era. Because knowledge of sustainability and its goals is now widespread, and if, as I conclude, problems related to it can only be addressed by radical societal transformation, the implementation of a sustainable development strategy has profound implications for contemporary liberal democratic understanding of the good life, the role of the state in its citizens' pursuit of their individual goods, and the free market's role

Adapted and excerpted from Julie L. Davidson, "Sustainable Development: Business as Usual or a New Way of Living?" *Environmental Ethics*, vol. 22, no. 1 (Spring 2000). Copyright © 2000 by Julie L. Davidson. Reprinted by permission of the author and *Environmental Ethics*, a quarterly publication of Environmental Philosophy, Inc. References omitted.

in determining what is a good life. It also has considerable import in contemporary understandings of the relationship between individual and community well-being and highlights the necessity of pursuing their reconciliation.

In this paper, I draw attention to the role of community in pursuing the mutuality of social and ecological goals, and the many negative effects of market relations on community relationships....

II. Defining the Economic Problem

In the late eighteenth century, when conditions prevailing at the time made a reformulation of the economic problem necessary, the change was achieved through a particular set of politico-economic components. Specifically, collective well-being was deemed to be had in each individual looking to his or her own interests, while maximizing individual freedom....

Limiting environmental conditions were not really an issue, and when they did appear, they were of local significance only. The limitations of resource availability were of little moment and, under the conditions of the time, the economic problem was one of maximizing output for any given level of resources. The issue for classical economists was how the wealth created by the new industrial capacity could be distributed through society. It was not until late in the nineteenth century, when it first appeared that industry might run out of coal, that Jevons redefined economics as the science of maximizing output under conditions of scarce resources. However, the problems of distribution and limits which should have been addressed along with the emerging scarcity of coal resources were postponed by the discovery of an alternative energy supply—petroleum. It has become obvious that the consequences of the era created by the "fossil fuel subsidy," the hydrocarbon age, including global environmental degradation and global inequalities of wealth, demand a new formulation of the economic problem. In the late twentieth century, the limiting conditions revolve around resource depletion, waste assimilation capacity, and intra and intertemporal inequalities, and they manifest themselves at local, national, and global levels.... That ecological and social sustainability should now be a cause for concern suggests that the sociability of human communities cannot have its basis in "individual human need and greed," which has been the underlying assumption of market societies.

... [A]t the end of the twentieth century, the economic solution is more than the achievement of controlled growth or a steady-state economy.... As in the late eighteenth century, it involves a politico-economic reformulation of what we conceive individual freedom to be under conditions of resource depletion, global ecological deterioration, and vast inequalities of wealth. It also involves the inherently existential question "How should we live?" and the moral question "How should we arrange our systems of production and consumption to ensure the sustainability of the Earth?" The question of how we should live was recognized by [Max] Weber to be an inherently political concern. For Weber, it was a question that distinguished politics from science and that therefore could not be resolved through applications of science and technology....

Further, those ancient philosophical questions demand a wisdom of the body politic that is not to be found in the doctrines of self-interest, nor in the parallel ethic of progress through economic growth. The dilemmas that sustainability concerns raise involve questions of the long-term and the collective interest. As [Daniel W.] Bromley argues, these are not problems that can be resolved by the spontaneous order of market processes. Within the context of societies dominated by possessive individualism, these dilemmas can only be resolved by the reconciliation of the interests of the individual with those of the collective.

III. The Discourse of Sustainable Development

Sustainable development has been proposed as the economic strategy which can provide the necessary response to the recognition of limits. It is, as [J. Ronald] Engel argues, "but the latest attempt to answer the perennial question of the purpose of human activity on the face of the earth—the elemental moral question of what way of life human beings ought to pursue." In the context of existential dilemmas, the questions to ask of sustainable development are: can it articulate an ethically sound understanding of what is a good life? Can it reconcile the age-old question of human autonomy and the demands of the common good?

Whether sustainable development is conceived as mere rhetoric, a cloak for business as usual, or as an ethic which forms the basis for restructuring human productive activity and its relationship to ecological integrity, its contestability as a political subject is involved. Michael Jacobs maintains that, now that the objective of sustainability has been generally accepted by radical greens, technocrats, and capitalists alike, the contestation revolves around how it should be interpreted and implemented in practice. This "contestation constitutes the political struggle over the direction of social and economic development." Given that such a struggle is currently being played out to a greater or lesser degree in most of the world's democracies, it is necessary to establish just which of the interpretations delineated by Jacobs—radical or conservative —has the capacity to meet the social and environmental requirements of our historical era. The "radical" and "conservative" orientations constitute competing conceptions about either pole of a continuum of understanding. Some writers make a distinction between sustainable development and sustainability, equating the former with the normatively weak conservative orientation and the latter with the normatively strong radical orientation.... I interpret sustainability as the normative goal which sets the parameters of sustainable economic development.

Jacobs has recognized in the sustainable development discourse four "fault lines" about which contestation occurs. These include the *degree of environmental protection* required, the *degree of intergenerational equity,* the approach to *participation,* and the *breadth of the subject area.* The "weak" version adopts the position that the benefits of environmental protection have to be balanced or traded off against those of economic growth, this is, *environmental conservation,* while the "strong" version holds that economic activity is subject to *environmental limits.* The latter is based on the notions of "carrying capacity" and "maximum sustainable yield," which find their expression in the general understanding that society should live within its limits. The notion of *equity,* which involves a commitment to ensure the basic needs of those living now and in the future (intra and intertemporal equity) has played a large part in the discourse about sustainability in the "developing" South, but is very largely ignored in the wealthier, "developed" North. The discourse in the South adopts an egalitarian stance concerning the distribution of global resources, a position which is less attractive in the industrialized North because of the fundamental challenge to levels of production and consumption and established patterns of global economic relations. The redistribution of global resources implied by the egalitarian interpretation means that Northern countries are open to the charge of confining their interpretation of *sustainable development* to an "environmental" one rather than also to a "developmental" one, and thus of demonstrating little commitment to the equity provisions inherent in the concept. Moreover, the constraints of the growth paradigm and of liberal representative democracy mean that governments are restricted to looking after the welfare of present generations and ignoring the preferences of future generations. It is easier for most liberal democratic governments to give the appearance of attending to equity concerns by such maneuvers as the designation of wilderness areas, rather than acknowledging this limitation and addressing the fundamental contradictions of the growth paradigm.

The "top-down" version of *participation* is that favored by most governments because, by limiting participation to major stakeholders, including business, local government, interest groups and other nongovernment organizations, they can retain control of the sustainable development agenda. It is a technocratic strategy in that objectives are set by governments using experts, with public participation limited to the implementation stage of policy formation. Thus, the reform strategies pursued by governments are more likely to be concerned with issues such as waste reduction, recycling, and energy conservation. By contrast, the "bottom-up" approach involves public participation in both the setting of objectives and implementation, since participation is held to be a good in itself—that is, it has intrinsic value. . . .

The interpretation of the *subject area* covered by the discourse varies from environmental protection as the dominant motivation to a much broader set of concerns including social and ecological well-being. This broader understanding flows from the notion that environmental protection is not possible without sound human development, a development which is not synonymous with income growth. The expansion of the "quality of life" criteria of sustainable development to include not just environmental quality but also basic human needs for self-fulfillment, equal opportunity, and access to education and information, participation, protection of local and indigenous culture, and human-scale development has enormous implications for all areas of human activity. . . .

The radical understanding of sustainable development tends to be held by many greens, environmental activists, and by some public sector bureaucrats, confined largely to environmental protection agencies, while the conservative view tends to be the preserve of governments and business....

IV. Sustainable Development and Well-Being

In this era of unprecedented ecological hazards and uncertainty (much of which is the result of human activities), the accumulating side effects of technological success demand a rethinking of what is a good human life and thus of what constitutes human well-being. Utilitarian notions of well-being, which predominate in modern pluralist democracies, emphasize the maximizing of happiness and/or preference satisfaction as the basis of well-being. Unfortunately, individual well-being has come to be equated with material comfort, with the result that the good of the individual in the liberty polity is a narrowly conceived one. However, if sustainability is to be the goal of human activity, well-being must be rethought with respect to a different set of goods for the making of good human lives. The dependence of liberal democracies on free or self-regulating markets, where self-interest theoretically acts as the main regulator, encourages a limited ethical stance and therefore a narrow view of what is a good life. If good living is not to be confined to physical comfort, then we must seek a broader understanding of the goods of life.

For such an enlarged meaning of the good life, we might recover the early Aristotelian understanding of success in life, where "doing well" meant being active in the pursuit and exercise of virtue, an objective which was achieved through active involvement in collective life. Contrast this objective with the accepted measure of success in modern Western democracies where the best individual life is measured by the degree of success in accruing material trappings while responsibility to collective existence is similarly confined to a narrow range of civic duties, to pay one's taxes, to vote for one's democratic representatives at intervals, and to avoid wrongdoing. Thus, in Aristotle's classical democracy the health of society depended not only on physical well-being but also on the moral health of citizens. This idea was to be echoed in later times by Rousseau, Jefferson, J. S. Mill and T. H. Green. However, the problems that beset us now are not only those of relationships between citizens. We also need to be good citizens in relation to nature.

A recent attempt to look at human/nature relationships in this sense has been that of John O'Neill, who makes reference to the Aristotelian account of a flourishing human life and reinterprets it in the light of these changed circumstances. The Aristotelian account embraces a much broader view of human well-being than the satisfaction of individual preferences.... The market ethic presupposes a particular conception of human well-being—a narrow, want-regarding one—which is "institutionally fostered by the market itself, the environmental problems engendered by the market stem[ming] in part from the forms of self-understanding it develops." The culture of self-interest is delusory and stunting to the development of human capacities and blinds the modern individual to another more satisfying, more fulfilling kind of richness in human existence.... As O'Neill remarks, "The ethical life is one that incorporates a far richer set of goods and relationships than egoism would allow."

Additionally, as technology and science have become the primary shapers of the human condition, they have also become problematic with regard to the development of human capacities. Through the modern period, technological progress has been seen as the key to moral progress, though, in fact, the hopes for moral betterment remain unfulfilled.

... [G]iven recent insights in evolutionary biology, which dispute the self-interested, competitive model of species behaviors ("survival of the fittest") underpinning the market society, ... it appears feasible to abandon this model, and to argue for an extension of the relational community and therefore of the sphere of moral obligation as O'Neill has done.

Thus, the quest for sustainable modes of being represents a challenge to liberal democratic notions of how the collective good is determined. At issue in liberal democracies is the role of the state in the determination of individual and collective well-beings. The resurrection of positive life values—the sanctity of human life, the universal right to self-actualization, and respect for all nonhuman agencies and beings—and the associated understanding of the good citizen living the good life as a prerequisite for the common weal flies in the face of what has been the dominant consensus on the role of the liberal state for some centuries. This is the tradition rooted in Hobbes that the function of the state is merely to arbitrate between the self-serving interests of its citizens, for which only understanding is required of it, certainly not virtue. In the Hobbesian liberal state, the state is the strictly neutral umpire between competing individual goods, simply supplying the legal and constitutional framework within which individuals can pursue their own individual goods. Adam Smith's invisible hand sees to it that these individual goods add up to the common good. However, citizens can no longer pursue individual goals in ignorance of their collective impact for, as Anthony Giddens remarks, in this era of high social reflexivity, how an individual constructs himself or herself is directly related to the larger questions of social and environmental renewal.

The significance of O'Neill's interpretation of the Aristotelian understanding of a flourishing human life... is that it provides an alternative foundation for an ethically good life, a replacement for the ethically limited one inspired by the ideology of the self-regulating market. Having established the necessity of expanding our understanding of the goods of life, the question to be asked is: can sustainable development constitute the basis for a broader and more satisfactory understanding of the goods of life?

Within O'Neill's expanded framework of understanding of the good life, it is the radical interpretation of sustainability, with its recognition that "quality of life" issues are intimately connected to environmental protection, which has the capacity to encourage that flourishing which is constitutive of a good human life. Moreover, participation processes that encourage input into both policy formation and implementation are more likely to be directed at creating just those conditions. As O'Neill concludes, " ... success in our own lives needs to be clearly bound up with those of future generations," and we might add, the flourishing of the rest of nature.

The implications of such an expanded view of the good life for the social, political and economic institutions of modern societies are quite profound. It is sufficient to say here that public policy must provide for the inclusion of the moral relevance of other species and of future generations at a more profound level than the level of rhetoric, while it is encumbent on societies to design the kinds of institutional arrangements which will encourage the cultivation of those submerged aspects of human capability and human flourishing, for they also now assume renewed ethical significance. As I have shown, these matters cannot be left to the spontaneous order of markets as has been largely the case for those governments under the sway of neo-liberal ideology....

V. Individual Autonomy Versus the Freedom of Nature

In modern societies, individual self-seeking sits in constant tension with the pursuit of the wider social good. Freeing humans from the constraints of nature (its vagaries and miseries) and improving their material welfare was a worthwhile objective in the eighteenth century; this goal was achieved by dominating and harnessing natural processes for human ends. It is not surprising that, with such an aggressive objective enjoined to human capacity, the human-nature relationship has assumed a state of imbalance with human liberty now being opposed to the freedom of nature.

Although the reconciliation of individual and common goods has been an age-old objective, the tension between them has been particularly acute in liberal democracies. However, with the novel environmentally limiting conditions now confronting us, this question is being asked anew, this time in the context of sustainability....

Over the last several centuries, the relationship of humanity to nature among certain world views has become unbalanced, taking the form of human domination over the natural world to which humans belong. According to [Denis] Goulet, "Nature and human liberty have come to be perceived as opposing poles in a dichotomy," an oppositional stance which has become paradoxical in that "human beings are not physically compelled to respect nature but they need to do so if they are to survive and preserve the existential ground on which to assert their freedom....

The question to be asked is whether sustainable development as economic strategy can reconcile personal autonomy with the freedom of nature and the interdependence of all life forms. The capacity to achieve this objective varies with the degree to which human embeddedness in nature is acknowledged. Thus, the conservative approach retains the modernist attitude of a society/nature divide, wherein sustainability is achieved by the management of resources through more efficient energy and resource use and new ecologically benign technologies. By contrast, the radical approach, in accepting human embeddedness in nature, recognizes that it is human activities which have to be managed in order to achieve ecosystemic and social viability. In empirical/policy terms, this distinction, between interpreting sustainability as a problem of human activity rather than as a problem of resource management, allows consumption patterns and the values which underpin them as well as the structural rigidities which militate against the assumption of environmentally sustainable practices to come into focus. It then becomes clear that the transition to an ecologically sustainable society requires a more integrated policy approach, in which a range of policy modes is utilized, including standard setting, regulatory intervention, institutional reform, markets, economic instruments, and technological innovation. Important elements of any policy mix will include an educative function and incentives for environmentally sound practices in order to stimulate the assumption of ecologically sympathetic attitudes and values.

In summary, then, human freedom has been conceived as freedom from constraint. However, by reconciling individual goods with the common good and expanding the notion of the "good," as O'Neill has done, the inherent tension of modern life can be relieved, and autonomy can become the positive freedom to unfold and blossom according to one's evolutionary potential. Human freedom then depends not on escaping nature's physical constraints by dominating it but on respecting it. By respecting nature, humans "preserve the existential ground on which to assert their freedom." If humans are to have any possibility of being at least part authors of their lives, of having a range of life choices, uncoerced by others or by nature, and of possessing the necessary capacities and resources for their self-chosen paths . . . , then individual development depends very much on ecologically responsible behavior. It depends upon human development strategies which enhance the mutuality of ecological and social goals.

VI. Reconciling Individual Autonomy and the Common Good

. . . One negative manifestation of our contemporary era of social and environmental disintegration is the breakdown of community and other binding ties and the loss of traditions. The breakdown of community ties has been paralleled by the emergence of an egoistic individualism, grounded in individual rights, but unaccompanied by corresponding responsibilities. . . .

The key normative idea which expresses the interdependence of human autonomy and welfare and community well-being is that of the "individual-in-community" or, . . . as Daly and Cobb's notion of "person-in-community." . . . The ideal of "person-in-community," in which the individual is constituted by his or her relationships, is contrasted with the implicit ideal in the market economy according to which the good of society is identical with the sum of the goods and services accruing to individual society members. . . . The success of a market society is measured by its GNP and by its rate of economic growth —the total aggregation of goods and services and the speed at which they are produced—not in how well it fosters and supports the pattern of personal and communal relationships, which make up the community; nor in how well it cares for the ecosystems on which the community depends for life support. . . .

It is clear that continued economic growth along the current trajectory and the maintenance of ecological viability are mutually exclusive. While governments retain sole reliance on experts and see environmental problems and resource depletion as questions of appropriate management, while they focus only on remediation of environmental damage, environments will continue to degrade and communities to disintegrate. In simply seeking to avoid risk, "top-down" management structures, which limit public participation and view it as having only instrumental value in the implementation of projects and programs initiated by the aforesaid experts, will fail to rebuild and support communal relationships. The radical approach to sustainability, on the other hand, uses the ecological crisis to reflect on the practices, values, knowledges and institutions of industrial society and therefore to rethink social relationships. In privileging the intrinsic value of participation as a valuable learning process, public input into community projects is encouraged, for, in the participative/ consultative process, relevant communities of interest, in negotiating their responsibilities to each other and to environmental protection, undergo processes of social/environmental learning, and thereby build new solidarities and develop new understandings of well-being. Such processes also stimulate the long-term commitment that having ownership of environmental problems produces but which "top-down" management systems are incapable of generating.

VII. Conclusion

Any economic strategy which attempts to match human needs and demands with the demands of nature and to guide human development must foster respect for the nonhuman environment as well as a sense of the mutuality of social and ecological values. Simple economic restructuring in order to contain economic growth within environmental limits is an insufficient attempt at a reformulation of the economic problem.... [W]e now have some idea of what is required of a sustainable system of production and consumption. First, sustainable development must ensure sound human flourishing, by furnishing those goods which ensure human autonomy (survival, opportunities for participation, and a good life); second, it must preserve and foster forms of community well-being, which ensure connection with past and future time perspectives; and third, it must preserve and foster ecosystem viability. Sound human development consistent with ecosystem viability is really only possible with the radical interpretation of sustainable development....

Further, sustainability as ethical ideal challenges the view of the liberal state as neutral umpire between different conceptions of the good. In its ethically strong form the goal of sustainability demands a conception of human flourishing that recognizes the intrinsic value of other nature, whose own flourishing is constitutive of a good human life....

A flourishing human life which furnishes the conditions for autonomy in a highly reflexive society implies reciprocity and mutual obligation in relationships and new bases for solidarity. Since market societies have removed many of the guide posts to sustainable existences, the fostering of new soli-darities which encourage a sense of identity over time is a requirement of any sustainable economic strategy.

The notion that the well-being of a community is constituted by the well-being and virtue of its citizens is recognized in the concepts of person-community, which acknowledge the symbiotic balance between individual and community needs. A sustainable economic order would support the relation-ships of community recognizing their significance for social and ecological integrity....

... Notwithstanding the arguments contained in this discussion for strong sustainability, that goal is not where social transformation should end, for sus-tainability is a principle only for stabilization and survival.... Sustainability may function as the notion which bridges the gap between a resource-greedy modernity and an ecologically enlightened "post-modernity." An essential pre-condition for this transition will be to overcome the structural rigidities of cap-italist market economies, namely to replace the imperative of ever-increasing quantitative growth and individual consumption with an imperative that fur-nishes qualitative social development and improved communal and ecological well-being.

Doomsday Every Day

Sustainable development" was the galvanizing theme of the 1992 Earth Summit in Rio de Janeiro. Based on the work of the Brundtland Commission in 1987, the goal of sustainable development has been enthusiastically promoted by the World Bank, the U.N. Development Fund, the U. N. Environment Programme, and the United Nations agencies promoting "world governance." It inspires President Clinton's Council on Sustainable Development. It has precipitated an avalanche of World Bank publications, such as the fourteen volumes of the *Environmentally Sustainable Development Proceedings* series of the 1990s, transforming untold acreages of forest into official paper. The phrase occurs frequently in the Chinese Communist press, usually in conjunction with news about the progress being made in the family planning program (Hong 1998). The two topics—sustainable development and "family planning"—are linked throughout the literature.

Economists have struggled, without much success, to reconcile the various definitions that have been offered for "sustainable development." Herman Daly, an economist who has been involved since the beginning, says not to worry —lots of good ideas can't be defined (1996,2). Daly, long associated with the World Bank, has written the seminal works in the field and is now joined by a host of authors producing textbooks for the college generation. Instruction in "sustainable economics" suffuses or replaces introductory economics courses at a number of institutions.

Whatever it is, sustainable development promises to transform life on this planet. The Rio conference produced agreements on everything from land-use planning (including "sustainable mountain development") and greenhouse gases to, or course, birth control. There were agreements on "human settlements," "sustainable agriculture," "biodiversity," and on and on in its "Agenda 21" and its Climate Convention and its Convention on Biological Diversity (*Agenda 21* 1992). Though Congress did not adopt the program, the Clinton administration proceeded as if it had, adopting new federal regulations and appointing a President's Council on Sustainable Development, made up of federal officials and prominent environmentalists, to pursue the agenda with vigor.

From Jacqueline R. Kasun, "Doomsday Every Day: Sustainable Economics, Sustainable Tyranny," *The Independent Review*, vol. IV, no. 1 (Summer 1999). Copyright © 1999 by *The Independent Review*. Reprinted by permission. References omitted.

The Clinton Council on Sustainable Development has issued its own version of Agenda 21, declaring that we must "change consumption patterns," "restructure" education, "conduct a high-visibility public awareness campaign... to adopt sustainable practices," "create a network of conservation areas for each bioregion... based on public/private partnerships" (so much for private property), "realign social, economic and market forces... to embrace conservation," "use building codes [to secure]... environmental benefits," have "local... community planning... to develop a common vision," create "a council of... key stakeholders to... achieve sustainable management of forests," and "promote development of compact... neighborhoods" (good-bye, suburbs) (President's Council 1995).

Moreover, it decreed that "population must be stabilized at a level consistent with the capacity of the earth to support its inhabitants," whatever that capacity might be (President's Council 1995). The definitions may be elusive, but the program is uniform throughout the literature. It is to create massive, new bioregional conservation areas; control land use, consumption, and markets; re-educate the masses; and control population.

The Sierra Club announced at the U.N. Population Conference in Cairo in 1994 that "local activists" of the club in the United States were working "in a consensus-based... process to establish... threholds for... population and consumption impact on the local ecoregion.... Addressing local carrying capacities will improve the quality of life for all and help develop sustainable communities" (Sierra Club 1994). The club didn't specify what action those local activists would take if it turns out that local populations exceed carrying capacity, but, as will be shown, other devotees of sustainability have done so.

Since the Rio conference, more than 130 countries have created new bureaucracies to implement Agenda 21 and its requirements for sustainable development, according to the Earth Council, whose head is Maurice Strong, director of the Rio conference and now assistant secretary general of the United Nations (Earth Council 1997). Many local and regional compacts for sustainable development exist in the United States, stretching from Florida through Missouri to Santa Cruz and Humboldt County, California. Henry Lamb of the Environmental Conservation Organization has described some of them, including the statewide plans for Florida and Missouri (1998).

Sustained by foundation money and federal grants, rarely mentioning Agenda 21, salaried environmental activists are convening unsuspecting local citizens to engage in the "visioning" process to plan for the sustainable community in their future. Vice President Gore's Clean Water Initiative and the administration's American Heritage Rivers Initiative are nurturing the process by encouraging local "watershed councils" to make comprehensive plans for their regions.

Herman Daly's Apocalyptic Vision

Probably not many of these souls have read the works of Herman Daly or Maurice Strong, the Rio documents, or the modern college textbooks in sustainable economics. If they had, they might be less eager to help. Daly, an economist, first came to national attention during the 1970s when the Joint Economic Committee of Congress published his plan for reducing births by government licensing. As in China, the government would issue the licenses in the restricted numbers requisite for achieving its population targets, and persons attempting to give birth without licenses would be punished. Unlike the Chinese system, the licenses could be bought or sold, as in the modern schemes for emissions control (Daly 1976).

People of common sense hearing such schemes tend to find them fantastic and amusing. But the World Bank was so enchanted by Daly's notions that it gave him a job as a senior economist in the Environment Department. In 1990 he and a theologian co-author, John B. Cobb, Jr., published their comprehensive plan for the salvation of the world, *For the Common Good: Redirecting the Economy towards Community, the Environment and a Sustainable Future.* Disputing major teachings of economics, the authors called for university "reform" to reduce the influence of economics and increase attention to the "social and global crisis" (357–60). That reform, of course, is now going forward. Like other leaders of mass movements, they argued that logical reasoning is greatly overdone and called for "a conscious shift toward... relativisation" (359). Such a shift also is rapidly occurring. Daly's hostility toward economics is not unique; many aspiring world-changers have seen economics, with its emphasis on logical reasoning based on fact, as the enemy of their plans.

Daly and Cobb called for the conversion of "half or more" of the land area of the United States to unsettled wilderness inhabited by wild animals (255), the abolition of private land ownership (256–59), a giant forced reduction in trade and a change to self-sufficiency at not only the national level but at local levels also (229–35, 269–72), government controls to reduce output to "sustainable biophysical limits" (whatever those might be) (143), and the resettlement of a large portion of the population to rural areas (264, 311)—remember Cambodia and Pol Pot, who has been called "the ultimate deep ecologist."

Moreover, they wanted a prohibition of the movement of private wealth (221, 233)—so much for any escape from the sustainable paradise—the abolition of direct elections, except for local officials who would in turn elect higher officers of the government (177), and, of course, complete population control by means of birth licenses. The intent was to promote the "biospheric vision" in the spirit of "deep ecology," which sees the need for a "substantial decrease in the human population" to promote "the flourishing of nonhuman life" (377). They added that this necessary reduction in the "human niche," a phrase echoed in subsequent United Nations documents, might be achieved either by a fall in population or by a decline in resource consumption (378).

Daly and Cobb understood that these vast changes would require some readjustments in attitudes, to say the least, and saw hope in the "influence of ecological and feminist sensitivities" (377). Not only have those attitude adjustments materialized, but academic economics, identified by Daly as the enemy, has also been remarkably helpful, producing quantities of new books and courses on sustainable development and related topics. Generous grants from government, foundations, and international agencies have encouraged this outpouring.

The justification for these massive changes in human life on the planet lay in what Daly and Cobb called "the wild facts"—that is, the alleged extinction of species, the ozone hole, the greenhouse effect, acid rain, and the imminent exhaustion of oil supplies. The last, of course, has disappeared from the current list of portending calamities; but never mind, we now have deforestation and the methane crisis. In any event, the bottom line was that we suffer from an excessively human-centured point of view, and people should be taught to adopt the "biospheric vision" (376) in recognition of our "community with other living things" in the spirit of "deep ecology."

Daly and Cobb provided no evidence of any of the catastrophes they listed and even acknowledged some uncertainty about the "precise physical effects" (416). Nevertheless, they insisted that the impending crises were "facts" that could not be denied. Scientific disputes over these matters have expanded since then, prompting the True Believers to develop new arguments.

Some of us may wonder whether the work we do makes any difference in the scheme of things. Daly and Cobb need have no such concerns. Their words, phrases, and arguments now appear throughout the United Nations documents on the sustainable society and the literature of sustainable economics. And Daly, now at the University of Maryland, has reiterated his vision in a 1996 book, *Beyond Growth: The Economics of Sustainable Development.* Together with Robert Costanza, Daly now directs the International Society for Ecological Economics, based in Solomons, Maryland.

Steven Hackett's Contribution

The nature of current college instruction in the field can be seen in a new textbook, *Environmental and Natural Resources Economics: Theory, Policy, and the Sustainable Society* (1998), by Steven C. Hackett, who teaches economics at Humboldt State University. As in Daly's case, Hackett's justifications for proposing fundamental social change are the imperiled biosphere and "the continued growth of human population," which causes "loss of biodiversity" and "deteriorating... wilderness areas" (12, 13), and many other ills.

On these points, there is serious debate, as the author admits. He insists nevertheless on "the potential for catastrophic change in the global climate... rising sea levels... inundation of... low-lying areas... desertification of... grain-producing areas... mass hunger... and... rapid loss of biodiversity" (12). These dire forecasts, of course, have been featured on television for a generation and will probably not unduly alarm modern students. Nor will these hardened young consumers of doomsday prophecies be surprised to learn that population growth threatens the "habitats of many of the world's species of animals and plants... the integrity of the world's remaining temperate zone wilderness areas, coral reefs and other marine ecosystems, and tropical rainforests" (12, 13)....

Is the Earth Overpopulated?

Overpopulation, according to Hackett, is a major cause of our doleful condition. Having softened the obviously elitist implications of the diagnosis by professing his concern for injustice, he can get on with the real message. The prolific people of the less developed countries are wreaking havoc on their "fragile environments," engaging in "deforestation... migration to... polluted urban areas... massive environmental degradation" (13), and so forth. Unmentioned are the government policies that create these disasters, such as the destructive taxation of farmers' productivity, the government monopolies that underpay and overcharge the people, the confiscation of traders' stocks and pack animals, the endless wars financed by foreign aid.

Hackett doesn't mention the large current declines in fertility and population growth rates throughout the world. United Nations figures show that seventy-nine countries with 40 percent of the world's population now have fertility rates too low to prevent ultimate population decline in those countries (U.N. Population Division 1996). But this evidence gives little comfort to Hackett, who quotes estimates showing that "2 to 5 hectares of productive land are needed to support... the average person... in an industrialized country [whereas]... the world has only 1.5 hectares per capita of ecologically productive land... and... only 0.3 hectare per capita are suitable for agricultural production" (263). In other words, not only does the less developed world have far too many rapidly multiplying people, but *population in the industrialized countries is several times too large.*

As he does throughout the book, Hackett hedges by saying that we don't really know our "carrying capacity," but the undergraduate reader is going to learn that, whatever that capacity may be, there are already far, far too many people on the earth. In a like vein, Paul Ehrlich, famous for his unblemished record of wrong forecasts, has said the world has "perhaps" five times as many people as it can tolerate (Ehrlich 1989).

Let us not imagine, therefore, that the advocates of the sustainable society are merely talking about cleaning up pollution and giving birth control pills to people in Africa, Asia, and Latin America. Although present State Department and U.N. efforts to restrain the increase of dark-skinned people are very strenuous indeed, they are seen as not nearly enough. Hackett quotes Devall on the desirability of "a substantial decrease of the human population" (20). And he describes the "coercive fertility-control" in China (234) and the proposals of Daly and Cobb and Kenneth Boulding for birth quotas. Spokesmen for the Clinton administration, such as Timothy Wirth, have specified that world population control must include the United States (Wirth 1996). Notice, too, that all of the sustainable society documents call for "population stabilization," without saying whether that is to occur at a population size larger or smaller than the present population.

We hope no guilt-ridden students rush to jump out of our overladen lifeboat before, first, asking why Hackett, Daly, Cobb, and Ehrlich have not done so already and, second, hearing some other information. Again according to Ehrlich and other, more reliable, sources, human beings actually occupy between 1 and 3 percent of the world's land area (Vitousek et al. 1986). The entire world population could be put into the state of Texas, leaving the rest of the world devoid of people. The population density of that giant city of Texas would be about 20,000 persons per square mile, which is somewhat higher than in San Francisco but lower than in Brooklyn (5.9 billion world population divided by 262,000 square miles of land in Texas implies 22,500 persons per square mile, or 1,200 square feet per person).

Farmers use less than half of the world's arable land (Revelle 1984). The world food supply has increased a great deal faster than population since 1950, according to the Food and Agriculture Organization (U.N. Food and Agriculture Organization 1996)....

Market Failure?

Hackett has little hope that existing institutions can steer the earth away from the looming catastrophes. As for markets, they "*reinforce* self-interested behavior" (29). One searches Hackett's book in vain for any sign of understanding Adam Smith's "invisible hand" that leads men to serve one another and to economize in their use of resources as they pursue their own self-interest. There is no sign that Hackett has ever read the great economist John Maurice Clark, who called the market "our main safeguard against exploitation" because it performs "the simple miracle whereby each one increases his gains by increasing his services rather than by reducing them" (1948). He seems unaware of Walter Eucken's perception that markets break up the great concentrations of economic power (1950) or F. A. Hayek's (1948) and Ludwig von Mises's (1949) realization that markets provide otherwise unavailable information about the scarcity of the resources that are the focus of his concerns.

This is not to argue that markets will solve all economic problems. Well-known and much-discussed problems of externalities, public goods, and common pool resources, sometimes arise, as Hackett notes. But the nonmarket economies of this century have provided vivid object lessons in the pitfalls of "communitarian" planning, and the work of James Buchanan, Gordon Tullock, and others has pointed up the perverse incentives that infest the public sector as it goes about trying to correct "market failure."

At times Hackett acknowledges that public ownership and management do not always produce ideal results, but for the most part he sees the market as the villain and concludes that our best hope lies in "cooperative rather than noncooperative decision making" (91). It is a conclusion he draws from game theory, and it leads to his hopes for "sustainable development" through small-group negotiations. On this issue, more later....

Not surprisingly, Hackett finds private property highly suspect: "It is clear that systems centered around private property... can conflict with the common good" (26). After a brief discussion of John Locke and proposals for protecting natural resources by assigning private property rights to them, Hackett points students to a patron saint of the French Revolution: "From Rousseau's perspective... private property rights... alienate people from nature... [and] lead to inequality... and wars." He quotes the great man: "Competition and rivalry...

opposition of interests... and always the hidden desire to profit at the expense of others. All these evils were the first effect of property" (25–26).

Such an indictment demands a response. Private owners did not hunt the buffalo almost to extinction. And it was not a private property system that sent millions to the gulag. When the Ethiopian government socialized the privately owned donkeys, most of them perished (Deressa 1985). I keep the off-road vehicles out of my private forest. And the biblical good shepherd was not the government or the assembly of "stakeholders" in the "sustainable community"; he was the *owner* of the sheep. Where does the common good lie in these decisions? And, most important, Who decides what the common good is? In fairness, also, Hackett might have mentioned the bloodbath that Rousseau's ideas encouraged. Like Devall, Daly, and other environmental utopians of our own time, Rousseau distrusted reason and argued for going "back to Nature." Ever the romantic, he sent his five children to a foundling home (Gauss 1972).

Economists have long noted that voluntary trade must make its participants better off or they wouldn't engage in it, whether they are children trading the contents of their trick-or-treat bags or Mexicans buying used bottles from California to turn them into gravel. Adam Smith and David Ricardo, and even Sir Dudley North before them, saw it as the solution to the uneven distribution of resources. Hackett, however, like Daly and Cobb, whom he quotes at length, lists many objections to trade. It "may... allow rich countries to import pollution-intensive, resource-intensive, and endangered-species products they do not wish to produce themselves and to export their toxics and trash" (225). It "tends to erode livable wages, the bargaining power of unions, and environmental and other standards of communities" (226). It "undermines sustainability" (227) and "has put great pressure on... endangered wildlife" (229).

Nevertheless, Hackett concludes that although there are "important questions" about how much and what kind of trade to allow, "it is neither practical nor desirable to eliminate trade completely" (230). What a relief. Clearly, however, what is left will be a far cry from free trade, just as all other human activity will be far from free in the "sustainable society."

Throughout the world, controllers and would-be controllers have seen, to use Smith's phrase, the human "propensity to truck, barter, and exchange" as a resource to be exploited or suppressed for the benefit of those in power. From mercantilist England, France, and Spain to the recent Soviet Union and modern Ethiopia, governments have sought to channel this propensity, always with the result of impoverishing their subjects. To illuminate the ill effects of trade controls was the main task of Smith's *Wealth of Nations*. That modern proponents of the "sustainable society" should be so eager to revive such controls should give us pause—doubly so because these people *intend* to reduce human consumption, and they understand very well that trade restrictions do impoverish people.

Like Daly, Hackett takes a dark view of what he calls "mainstream economics." Students who have studied economics, according to Hackett, are less altruistic than other students (28). Economics itself, he maintains, tends to reduce everything to a monetary cost-benefit comparison without recognizing "intrinsic" values. In his view, however, not all intrinsic values are equally worthy of recognition. Individual rights are especially suspect. By contrast, the "sustainability ethic holds the interdependent health and well-being of human communities and earth's ecology over time as the basis of value" (209), and is therefore clearly superior to the viewpoint of mainstream economics.

Economics and Ethics

Private property, the market, and economics itself, it would seem, are the bad fruit of a bad tree, the disordered ethical system of contemporary society. Hackett blames the shortcomings of economics on its "teleological ethics"—that is, the end justifies the means—attributing the idea to "religious philosophers" (21). This reference enables him to take a swipe at both religion and economics. Evidently, Hackett either never had catechism or was inattentive when Sister told him the end does not justify the means. His example is "utilitarianism," which he describes as the "normative base" for "much of the traditional economic perspective" (21). His straw man is Jeremy Bentham, a nineteenth-century eccentric who had his body stuffed and put in a glass case after he died so it could be on view for University College, London, undergraduates for all time (Mack 1972).

Bentham's mechanical pleasure-pain calculus has amused students for generations, but other men—Smith, Jean Baptiste Say, Ricardo, Carl Menger, Alfred Marshall, and others—did the serious work of showing how the market reveals and reconciles the varied and conflicting desires of multitudes of individuals, channeling their self-interest to the service of others in their pursuit of individual gain.

These monumental themes receive barely a glance from Hackett, who remains intent on showing the failures of market calculations and the need for more sublime direction by persons imbued with the spirit of the sustainable community and tutored in sustainable economics. To illustrate, Hackett poses the "question of whether an action (for example, policy protecting old-growth forest) is to be judged on its intrinsic rightness or based on the measurable benefits and costs that might result" and "the proper balance between individual self-interest and the common good," again undefined (17–18).

There ensues a discussion of the "fundamentals of ethical systems," beginning with "deontological ethics," which judges an action by "its intrinsic rightness" (19). As an example, Hackett quotes at length from the "ecosophy," or "earth wisdom," of Bill Devall, George Sessions, and Arne Naess:

> The well-being and flourishing of human and non-human life on Earth have value in themselves....
>
> The flourishing of human life and cultures is compatible with a substantial decrease of the human population. The flourishing of non-human life requires such a decrease.
>
> Those who subscribe to the foregoing points have an obligation... to... implement the necessary changes. (20, quoting Devall 1988)

Clearly, this call is not for minor adjustments in lifestyle. A "substantial decrease of the human population" is no small thing. Our "obligation...to... implement the necessary changes" is a profoundly serious matter. This proposal is not a nickel-and-dime deal. True, Hackett is only quoting Devall at this point, but his discussion makes is clear that Devall's insistence on "intrinsic rightness" is a far more beautiful thing than the crass monetary valuations of "utilitarian" economics.

To make the issue perfectly clear, Hackett offers an example. Suppose an endangered species is threatened by development. Guess what will happen in a "society that views the existence of a species as being of intrinsic value" (à la Bill Devall). Then guess what will happen if a monetary cost-benefit comparison determines the outcome. Obviously, all economists, except an enlightened few, should be taken out and shot.

Nowhere in Hackett's discussion of ethics does he refer to the Judeo-Christian tradition of stewardship—the admonition to "keep" the earth (Gen. 2:15), the prescribed days of rest for men and beasts (Deut. 5:14), the prescribed years of rest for the land (Lev. 25:4), the love of nature with its "Leviathan" taking its sport in the sea and its "coneys" among the rocks (Ps. 104), its cedars of Lebanon (Ps. 92), its hills that "rejoice on every side" and its valleys that "laugh and sing" (Ps. 65), and the strict injunctions against the worship of nature and the human sacrifice that often accompanied it (Deut. 17:3, 20:2-6; 2 Kings 17; Job 31:26).

Modern economic reasoning does not destroy these values any more than modern atmospheric science destroys the beauty of a sunset. Certainly, the sin of greed has always beset the race, as has idolatry. Just as certainly, modern economics has its idolaters as well as its Midases, but such corruption is nothing new on earth. Economic reasoning enables us to compare alternatives. It enables us to see that a society following the romanticism of Devall or Daly would probably be no more attractive or healthful than the one we have. One of the greatest tragedies of our time is not that undergraduates study economics but that they study so little of the great civilizing themes of our heritage—our great literature, art, and music, our legacies from the ancient Greeks, our tradition of human rights and our history of the struggle for liberty—and that they know so little about Christianity or Judaism. Thus deprived, they are left vulnerable, not so much to "utilitarianism" as to environmental lunacy.

Worse yet, as John Grobey, professor of economics and a senior colleague of Hackett at Humboldt State University, has noted, the result must be to deprive young people of the traditional birthright of youth—hope for the future. Taught from their earliest years that their own burgeoning humanity is destroying the earth and all of nature, the youth of today face a more depressing prospect than perhaps any previous generation. No wonder the doubling of the suicide rate among children aged ten to fourteen since 1980 (U.S. Bureau of the Census 1997). No wonder the epidemic of school shootings. No wonder the recent case in Humboldt County in which a young man on trial for attempted murder gave as his defense "overpopulation, dwindling resources and the certain doom of the planet" (Parker 1998).

The changes in "basic economic, technological, and ideological structures" called for by Devall obviously threaten traditional views of individual rights to life, liberty, and property. The question that occurs to a mainstream economist at this point is, Just which individuals will be given the awesome responsibility of determining the "common good" and the best interests of the community and the ecology? And what will happen to human beings, stripped of individual rights, who get in the way of the grand march to the sustainable community? Hackett gives hints but no answers. He acknowledges the seminal work of Daly, but without mentioning Daly's call for massive resettlement of populations. The question remains: Is the centuries-long pilgrimage from Magna Carta through *Areopagitica* and the Bill of Rights to Selma to be renounced now in the name of the environment? Will this denouement be the Clinton legacy?

No Price Is Too Great

. . . Hackett makes it sound as if the sustainable society will be brought about by local meetings of "stakeholders" negotiating over local issues. But undergirding these cozy negotiations will be "regulations, taxes, subsidies, and direct finding of clean technology" (277). Of course, the Sierra Club will be there to help.

Here is the rub. To avert a highly problematic future disaster, much disputed by competent scientists, Hackett and his soul-mates in the United Nations and the Clinton Council on Sustainable Development would require human beings to submit to a gigantic present sacrifice of freedom, human dignity, and material welfare in a regime controlled by unelected officials of a global eco-bureaucracy. Have we learned nothing from the utopian horrors devised for us during the past century?

People do love nature. The tremendous expansion of national parks and conservation areas during this century testifies to that love. The environmental movement itself is an expression of our determination not to let the industrial age destroy the oceanic Leviathan and the cedars of Lebanon. The real danger now, however, is not that we stand on the verge of destroying nature but that, stampeded by environmental terrors on every hand, we are plunging over the cliff into totalitarianism.

POSTSCRIPT

Is Sustainable Development Compatible With Human Welfare?

It is a truism that human impact on the environment depends on both human numbers and human activities. Those who assert that sustainability is not an issue often point out that all six billion living humans could be moved into a small area such as Texas, leaving the rest of the planet empty. However, each human requires space on which to grow food and fiber and from which to extract energy and mineral resources. Also, other essential human needs as potable water, clean air, energy resources, etc., must be considered. Those who deny that sustainability is an issue also ignore the fact that the majority of the world's six billion people do not share in the standard of living typical of the developed nations but are trying very hard to change that. If they succeed, the human impact on the environment will increase tremendously. Many projections contend that demand for land, energy, and resources would then greatly exceed supply. The world would be raped to meet the needs of the present generation and nothing would be left for future generations. In "Windows on the Future: Global Scenarios and Sustainability," *Environment* (April 1998), Gilberto C. Gallopin and Paul Raskin assert that because the world is so interconnected today, there can be no separate solutions for the rich and the poor. Gallopin and Raskin state, "There is no question that the contradiction between the modern world's imperative toward growth and the Earth's finite resources will ultimately be resolved in some way. The only question is how that will come about—whether through enlightened management, economic and environmental catastrophe, or some other means." See also Garrett Hardin's *Living Within Limits: Ecology, Economics, and Population Taboos* (Oxford University Press, 1993).

Sustainable economics is addressed in Louis P. Pojman's *Global Environmental Ethics* (Mayfield, 2000). Pojman reviews the difficulties posed by global inequities.

David Malin Roodman, in *The Natural Wealth of Nations: Harnessing the Market for the Environment* (W. W. Norton, 1998), suggests that taxing polluting activities would stimulate corporations and individuals to reduce such activities or to discover nonpolluting alternatives. In "Building a Sustainable Society," *State of the World 1999* (W. W. Norton, 1999), Roodman adds recommendations for citizen participation in decision making, education efforts, and global cooperation; without which we are heading for "a world order [that] almost no one wants."

Davidson asserts that efforts to achieve sustainability cannot by themselves save the world. But such efforts may give us time to achieve new and more suitable values. Those values may need to be expressed in new "myths," say Stanley Krippner, Ann Mortifee, and David Feinstein in "New Myths for the New Millennium," *The Futurist* (March 1998). Many of our environmental problems, they assert, are a result of our belief in the myth of progress. The myth of sustainability is more conducive to human welfare, and it may actually be taking root. Lester R. Brown, in "Crossing the Threshold," *World Watch* (March/April 1999), sees signs that "the world may be approaching the threshold of a sweeping change in the way we respond to environmental threats . . . a paradigm shift in environmental consciousness."

ISSUE 17

Will Voluntary Action by Industry Reduce the Need for Future Environmental Regulation?

YES: Raymond J. Patchak and William R. Smith, from *ISO 14000 Perspective: So Long! Command and Control ... Hello! ISO 14000* (December 1998)

NO: Linda Greer and Christopher van Löben Sels, from "When Pollution Prevention Meets the Bottom Line," *Environmental Science and Technology* (vol. 31, no. 9, 1997)

ISSUE SUMMARY

YES: Certified hazardous materials managers Raymond J. Patchak and William R. Smith describe the voluntary ISO 14000 environmental program developed by the International Organization for Standardization. They assert that this initiative will result in increased environmental protection by permitting industry more flexibility in achieving pollution prevention than current "command and control" regulations do.

NO: Environmental Defense Fund scientist Linda Greer and project analyst Christopher van Löben Sels conclude from a case study of a Dow Chemical facility that not even projected cost savings will ensure that a corporation will adopt a voluntary pollution prevention plan.

Although the actions and lifestyles of individuals make significant contributions to environmental degradation, the major contributors to local, regional, and global pollution of air, land, and water are commercial and industrial activities. Until recently, governmental efforts to protect the environment have principally taken the form of legislated or court-mandated restrictions on such activities. Applicants for the siting of large new facilities of all types must usually satisfy federal, state, and local environmental requirements and undergo extensive reviews to assess the potential for serious adverse environmental impacts. Limits have been set on the emissions of a variety of pollutants. Toxicity testing has been prescribed for suspect industrial and agricultural chemicals. And standards have been set for the safe disposal of hazardous wastes.

Until recently, the response of industry to environmental regulation was usually strident and strictly one-dimensional—adamant opposition. Denials of the severity, or even the existence, of the negative environmental impact of industrial development were the first line of defense. If overwhelming evidence rendered such assertions implausible, then corporate spokespeople would often argue that the costs of regulatory compliance were simply too great and that the public should accept some decrease in air or water quality as the inevitable price of an increased standard of living.

As the environmental movement has matured, the arguments of all parties to the debate have become more sophisticated. Independent of their personal views, industrial leaders have responded to increasing environmental concerns among consumers by adopting strategies designed to portray their corporations as ecologically responsible—a posture that is referred to as "green" in the contemporary vernacular. Although they continue to lobby against most proposed environmental regulations, many corporate decision makers now try to implement and publicize alternative, less damaging means of production that will not lower their profit margins.

A significant manifestation of this change of attitude is the highly publicized Responsible Care program, initiated in the United States in 1988 by the Chemical Manufacturers Association (CMA). As the 10 "ethical principles" that were established by the CMA to guide its members indicate, a major thrust of that program is to persuade the industry to voluntarily change its priorities and practices in ways that would enhance environmental protection. The CMA has openly admitted that one of the prime motivating factors for the Responsible Care initiative is the desire to counter the low esteem accorded to the chemical industry by the public. Recent polls indicate that despite the CMA's reports of many positive achievements resulting from the program, it has not made much progress in the public relations arena.

A more global effort along the same lines is code 14000 of the International Organization for Standardization (ISO). This voluntary code, issued in its final form in 1996, establishes environmental management standards. A professed goal of supporters of this effort is to persuade governmental environmental agencies to accept self-regulation by industries that make a commitment to comply with ISO 14000. Thus far, more companies in Europe and Asia than in the United States have adopted the standards.

Raymond J. Patchak and William R. Smith enthusiastically support the market-driven ISO 14000 code. In the following selection, they argue that ISO 14000 is a more flexible approach to environmental protection and that it will prove to be more effective than the current "command and control" regulatory strategy. An implicit assumption of those who argue in favor of voluntary industrial environmental initiatives is that corporations will adopt more ecologically sound development strategies when it is profitable for them to do so. In the second selection, Linda Greer and Christopher van Löben Sels report on their study of a Dow Chemical facility's response to a money-saving pollution protection plan, which challenges that assumption.

Raymond J. Patchak and
William R. Smith

 **YES**

So Long! Command and Control...
Hello! ISO 14000

The regulatory system currently in use by the United States Environmental Protection Agency (EPA), as well as other U.S. regulatory agencies, can be characterized as one of command and control. Developments in the regulatory approach taken by other industrialized nations and the advent of environmental management systems have ignited a process of critical review of our command and control system. Many new regulatory programs instituted in Europe seek to capitalize on the synergistic effect that can be obtained when you get both the regulated community and regulatory agencies working together to solve environmental problems. This idea is the basis for a new international environmental standard called ISO 14000. Let's take a quick look at both of these environmental systems.

Command and Control

The current system of environmental protection in the United States is, as a whole, the most advanced in the world. Environmental regulations have achieved a great deal in turning back the effects of toxic substances on the environment. Laws such as the Toxic Substances Control Act (TSCA) and the Resource Conservation and Recovery Act (RCRA) have been good control measures for the treatment and disposal of hazardous substances and wastes, while laws such as the Clean Air Act (CAA), the Safe Drinking Water Act (SDWA), and the Clean Water Act (CWA) have been instrumental in responding to and reducing toxic substance releases into the environment. One can easily argue that these regulations have contributed significantly to decreasing the proliferation of disasters like the ones at Love Canal, Times Beach and the burning of the Cuyahoga River in Ohio.

Two shortcomings exist in the current command and control approach to environmental protection. One is that regulations are often too rigid and complicated to allow innovation and common sense to play a role in environmental protection. Second, the current regulatory scheme does not foster a cooperative relationship between regulating agencies and industry. As a result, instead of working together to find common solutions to a problem, industry and the

From Raymond J. Patchak and William R. Smith, *ISO 14000 Perspective: So Long! Command and Control... Hello! ISO 14000* (December 1998). Copyright © 1998 by Competitive Edge: Environmental Management Systems, Inc. Reprinted by permission. Notes omitted.

EPA all too often find themselves battling it out in the court system. Although these shortcomings have been reduced in recent years, the fact that the EPA is involved in some 600 lawsuits at any given time should be evidence enough to show that there is a fundamental conflict between business issues and regulatory constraints. The sad part about this situation is that valuable resources are being wasted in the courts rather than responding to the problem at hand.

Industry has made great progress in its efforts to learn how to protect the environment. They have spent hundreds of billions of dollars to decrease the release of toxic substances into the environment, while also developing technologies to reduce or eliminate hazardous waste generation. Many industry groups such as the Chemical Manufacturers Association have developed initiatives, which utilize pollution prevention programs as the cornerstone for their environmental protection efforts. These types of industry-led initiatives, coupled with advances in technology, are changing the way that many companies view and are responding to their environmental obligations.

The EPA Administrator, Carol M. Browner, has supported industry in furthering their environmental protection efforts by agreeing to look at beneficial ways to modify the current system. Ms. Browner's Common Sense Initiative (CSI) is a fundamentally different approach to environmental protection. In the development phase of CSI, the EPA worked with six pilot industries to look at the regulations that are impacting their businesses and to identify ways to change the ones that are complicated and inconsistent. The CSI program is attempting to promote creativity and encourage the development of innovative technologies by allowing industry more flexibility in meeting stronger environmental objectives. In this fashion, the goal of the CSI program is to develop a comprehensive strategy for environmental protection that will result in a cleaner environment at less cost.

What Is ISO 14000?

In the spirit of the Common Sense Initiative, ISO 14000 signifies a new generation in environmental protection. This standard is directed at establishing a link between business and environmental management for all companies no matter their size or purpose. ISO 14000 provides industry with a system to track, manage, and improve environmental performance without conflicting with the business priorities of an operation. Business considerations, flexibility, continual improvement and a simplistic approach to environmental protection are the main differences between ISO 14000 and the command and control system currently in place.

By design, ISO 14000 is a set of simplified environmental management standards that takes into account business and economic considerations while improving on already established environmental protection programs. In this context it should be pointed out that ISO 14000 standards are not intended to supersede current state and federal regulations. In fact, the ISO 14001 Specifications specifies the incorporation of these regulations as an integral part of a facility's EMS program.

In recent years there has been increasing interest and commitment to improve environmental management practices. This interest is demonstrated in collaborative international events such as the NAFTA Montreal Protocol and the mandates set during the 1992 Earth Summit in Rio de Janeiro. The birthplace of ISO 14000 can be traced back to the environmental goals established during the Earth Summit. At this Summit the United Nations convened representatives from the world's industrialized countries to discuss global environmental issues and to develop a means to meet their basic goal of "sustainable development."

At the conclusion of the Earth Summit, the International Organization for Standardization (ISO) set out to develop a group of international standards that meet the goals of the United Nations Conference. ISO established Technical Committee (TC) 207. Its mission, "to establish management tools and systems that organizations can voluntarily use for their own purposes which may, over time, improve their environmental performance levels." In this mission TC 207 created and is developing the ISO 14000 series of standards.

TC 207 is comprised of representatives from the ISO member nations. The United States's representative to TC 207 is the American National Standards Institute (ANSI). ANSI is responsible for fielding U.S. participation in the development of these international standards. They receive assistance within the U.S. through its Technical Advisory Group (TAG), which is administered by the American Society of Testing and Materials (ASTM). The TAG is further divided into Sub-TAGs, which are responsible for reviewing and commenting on progressive drafts of the individual policy documents that compose ISO 14000. Once completed the draft documents are voted on by the entire international membership of TC 207. Upon approval by the membership the documents become final standards.

ISO 14000 is the name for a family of environmental standards. This family is composed of five major components, each of which has one or more policy documents. The five major components are as follows:

- Environmental Management Systems (EMS)
- Environmental Auditing (EA)
- Environmental Performance Evaluation (EPE)
- Life-Cycle Assessment (LCA)
- Environmental Labeling (EL)

The **Environmental Management System (EMS)**, which incorporates policy documents numbered 14001 & 14004, were released as final standards in September of 1996. The EMS is the building block on which the other four components are incorporated. Through the EMS a company can identify its environmental goals and establish a program for monitoring their progress in reaching these goals.

Although individual EMS programs will differ from one organization to the next, they will all consist of seven core components. These include the following: Identification of the company's environmental policy, its objectives and targets, guidelines for identifying environmental concerns and applicable regulations, implementing procedures for controlling process and activities impacting the environment, a program for internal and external auditing of the system, clear assignment of responsibility and accountability, and a requirement for periodic review of the EMS by top level management.

The **Environmental Auditing (EA)** standards are composed of documents 14010 through 14012, and were released for publication as final standards in October 1996. These documents detail the requirements for the general principles of auditing. They include guidelines for conducting audits of EMS programs, and criteria for evaluating the qualifications of environmental auditors.

The **Environmental Performance Evaluation (EPE)** guidance is scheduled for publication in the second quarter of 1999. This standard includes document ISO 14031 as well as a technical report unofficially designated as TR 14032. EPE is an internal management process that uses indicators to provide management of an organization with reliable and verifiable information comparing the organization's past and present environmental performance with Management's environmental performance goals. ISO 14031 defines EPE as a "process to facilitate management decisions regarding an organization's environmental performance by selecting indicators, collecting and analyzing data, assessing information against environmental performance criteria, reporting and communicating, and periodic review and improvement of this process." These documents contain guidance for a process that identifies and quantifies the impact that a company has on the environment. These measurements are made against baseline levels, and the results are evaluated based on improvements from these levels.

The **Life Cycle Assessment (LCA)** guidance can be found in document numbers 14040 through 14043. Although no date has yet been set for final publication of ISO 14043, the ISO 14040 and ISO 14041 documents were released in late 1997 and 1998 respectively. These documents provide a means for determining what effects the products that are manufactured will have on today's environment, as well as that of future generations. This assessment looks at the impacts that are associated with the entire life of a product from raw material acquisition through production, use and disposal.

The **Environmental Labeling (EL)** standard is contained in documents 14020, ISO 14021, ISO 14024 and 14025. The ISO 14020 standard, published in late 1998, provides general principles of environmental labeling. The draft international standards ISO 14021 and ISO 14024 are expected for final release in early 1999. The overall goal of the environmental labeling standards is to provide manufacturers with a tool to assess and verify the accuracy of product environmental claims and to encourage the demand for those products that cause less stress on the environment. The intent of developing these EL standards is to stimulate the growth of market-driven continuous improvement of environmental performance.

In its development of ISO 14000, TC 207 was influenced by the British Standard BS 7750: "Environmental Management System" and the Global Environmental Management Initiative (GEMI). These systems utilize a consensus approach for determining their operational effectiveness. For example, GEMI is composed of 21 leading companies dedicated to fostering environmental excellence in businesses worldwide. It is intended to promote a worldwide business ethic for environmental management and sustainable development. The standard places an emphasis on individual companies and the incorporation of environmental goals into the company's overall corporate goals. These standards also set up a third party review process, to establish whether a company is working towards its goals.

How Will ISO 14000 Benefit an Organization?

ISO 14000 is not a rigid system of regulations, it is a flexible standard designed to fit any size and type of operation. Through this type of system a company can gain many benefits. These benefits can come in the form of improvements in public relations, improvements in management effectiveness, decreases in non-compliance fines, and improvements in marketing and customer relations.

The implementation of an effective ISO 14001 program can provide assurance to consumers that a company is committed to being not just environmentally friendly, but environmentally protective as well. In the same fashion, by maintaining conformance with the standard a company can make a much stronger case for their commitment to protecting the environment. ISO 14001 also makes regulatory compliance a more integral part of the business operation, and thus the likelihood of a violation can be minimized along with the stiff fines and penalties that normally accompany them.

Part of the ISO 14001 standard requires that a program for continuous improvement be developed and implemented throughout the company. The elements of a well orchestrated continuous improvement program can save even a small organization thousands of dollars per year in compliance and pollution control costs. These savings can come about as a part of improved compliance and in the form of reductions in fines and penalties, waste disposal costs, energy consumption and raw materials costs. For instance the implementation of an effective waste minimization program will not only decrease disposal costs, but can also reduce regulatory burden on the operation.

Other regulatory benefits related to implementing the ISO 14001 standard include reducing the number of audits by regulators. The EPA has already proposed, as part of their environmental auditing policy statement, that companies with internal self auditing programs, like ISO 14001, could be subject to less regulatory scrutiny and compliance audits.

Recent history has shown us that a company's market segment can also benefit by establishing a program like ISO 14000. This is exactly what happened when the International Quality Management Practices standard (ISO 9001) was released. As companies established and certified their ISO 9001/2 programs, they became strong believers in the importance of the program and in turn many required their major suppliers to become certified. This move is already underway with regard to ISO 14000, as many companies and government agencies in the U.S. and Europe are already posturing in this direction. Specific details about these companies are included in the next section.

One of the other business incentives for adopting and implementing ISO 14000 is that continuous improvement is crucial for maintaining prosperity and growth. As mentioned earlier in the article, ISO 14001 was developed in part from the United Kingdom BS 7750 Standard. Those companies that have already implemented the BS 7750 standard have reported improvements in the following areas:

- Productivity
- Waste reduction
- Paperwork declines, and
- Public relations

Industries' Acceptance of ISO 14000

As more and more companies have become familiar with the many benefits of ISO 14000 its acceptance has increased steadily. Companies within the U.S. are already lagging behind their European counterparts in taking the first step towards ISO 14000 certification. Many companies in Europe have established EMS programs and are certified to the draft ISO 14000 standards. The enthusiasm experienced in Europe has started to overflow into the U.S. This fact becomes apparent when you look at the SGS-Thompson Microelectronics facility in Rancho Bernardo, California. SGS-Thompson is the first U.S. manufacturer to have an EMS program certified; this occurred on January 3, 1996. SGS-Thompson, which is a corporation based in France, already has 5 facilities certified to the draft standards. Further, company representatives vow that all 16 SGS-Thompson facilities located in U.S., Europe, Southeast Asia, and North Africa will be certified by the end of 1997. As of December 1998 there have been over 5,000 facilities registered to the ISO 14001 standard worldwide.

According to a company official at SGS-Thompson, many of their major suppliers and contractors may find themselves forced to implement ISO 14000 standards by 1999 to maintain contractual preference. The official explains, "We assign an overall score to our suppliers based on many elements, including quality and environmental performance.... While an EMS program alone does not seem that important, [because of this scoring system] it may be the deciding factor when we decide who gets equipment and manufacturing supplier contracts."

Businesses are not the only ones that foresee the positive changes that can come about as a part of ISO 14000 implementation. Agencies with the United States government are also boarding the band wagon. Two of these agencies include the Department of Energy (DOE) and the Environmental Protection Agency (EPA). The DOE is encouraging all major contractors to implement an EMS program. Sources at the DOE have stated that specific contractors will be required to have ISO 14001 certification. The mood at the EPA can perhaps be best exemplified by the testimony of one of their representatives, who stated that "the old 'command and control' method of environmental regulation, so important in the first 25 years of the environmental movement, will occupy less time and fewer resources as companies start to use ISO 14000 and the next generation of environmental protection tools."

The EPA, which is studying how the ISO 14000 standards will influence and affect their operations, have already come out in support of them. This support has been heard at all levels including Carol M. Browner. Recently this support was testified to during the March 6, 1996 hearing in front of the Senate Environmental Resources & Energy Committee. During this hearing, James M. Seif, Secretary for the Pennsylvania Department of Environmental Protection, testified that ISO 14000 "represents the next generation of tools needed to more effectively achieve our environmental protection goals." Further, Mr. Seif said, "These new tools approach environmental protection in an entirely new way using performance-based environmental objectives, positive incentives to comply, external validation, a flexible approach to implementation and systems which constantly look for new opportunities to prevent pollution and reduce environmental compliance cost."

In the international arena, ISO 14000 is being seriously considered for adoption under NAFTA [North American Free Trade Agreement] and various GATT [General Agreement on Tariffs and Trade] trade agreements, to prevent the development of artificial trade barriers from country-specific environmental requirements. The standard is also viewed in Europe as meeting major components of the European Union's ECO-Management and Auditing Scheme (EMAS) regulations. And much like the proliferation of ISO 9000, it is expected that multi-national companies will require their suppliers to be ISO 14000 certified in order to do business in Europe.

In the U.S. a number of organizations that directly represent environmental and/or technical based members have already come out in support of ISO 14000. A partial list of these organizations includes:

- The American Society for Testing and Materials (ASTM),
- The American Society for Quality (ASQ),
- The Registration and Accreditation Board (RAB),
- The National Registry of Environmental Professionals (NREP),
- The American Forestry and Paper Association,
- The Academy of Certified Hazardous Materials Managers (ACHMM), and
- The American National Standards Institute (ANSI).

ISO 14000: A Business Decision

ISO 14000 is designed as a market driven approach to environmental protection. This system has the potential to be many times more effective in achieving significant environmental improvements than the current command and control approach.

Business success can be attributed partly to growth, strategic planning, and by maintaining a competitive edge in the market place. Organizations that fall in line behind the competition in market place developments often do not achieve a competitive advantage like the leaders in the market. ISO 14000 is seen by many organizations not as a new environmental compliance burden, but as a market place trend, which can provide for a more productive operation. Only those organizations which look at ISO 14000 as a business decision will be able to utilize it to the fullest extent. These organizations will be able to capitalize on the long term cost savings while at the same time broaden their market and improve their customer relations.

**Linda Greer and
Christopher van Löben Sels**

 **NO**

When Pollution Prevention Meets the Bottom Line

W hat if a manufacturer learned that there were untapped opportunities to reduce waste and emissions within a plant that would also significantly cut costs? Conventional wisdom is that the company would seize on such opportunities and implement them.

But the reality is those opportunities are not always taken. A case study completed in 1996 at a Dow Chemical facility showed that certain pollution prevention strategies would save the company more than $1 million a year, approximately 10–20% of the existing environmental expenditures at the plant. Process changes would have eliminated 500,000 pounds (lb) of waste and allowed the company to shut down a hazardous waste incinerator. Surprisingly, these benefits were not enough of an incentive to outweigh other corporate priorities and the potential loss of future business that might have accompanied the incinerator's shutdown.

These findings came out of a collaborative study. In 1993, the pollution prevention pilot program (4P) was begun by the Natural Resources Defense Council (NRDC), an environmental advocacy group, and Dow Chemical, Monsanto, Amoco, and Rayonier Paper. Study participants, who were all interested in pollution prevention in a real-life industrial setting, wanted to know the reason for the lack of widespread reliance on promising pollution prevention techniques. Was it because there was not much to be gained environmentally or economically by using this environmental management technique? Was it because there were government regulations acting at cross purposes, incorporating barriers to implementation? If these factors did not explain the problem, what did?

Pollution prevention is conceptually quite different from pollution control, which relies on capturing emissions generated in processing before their release into the environment. Pollution prevention seeks opportunities to minimize reliance on toxic chemicals, increase efficiency, and decrease waste and emissions. Instead of focusing on changes required for environmental and health reasons, pollution prevention planners also identify opportunities to save money, making the process a potential "win-win" for industry and environmentalists.

From Linda Greer and Christopher van Löben Sels, "When Pollution Prevention Meets the Bottom Line," *Environmental Science and Technology,* vol. 31, no. 9 (1997). Copyright © 1997 by The American Chemical Society. Reprinted by permission.

This approach has failed, however, to take hold in the business world and at EPA and most state agencies. In fact, total waste production reported to the Toxics Release Inventory (TRI) in 1995 is up 6% from 1991, even though industry's TRI emissions have decreased. Some believe that, ironically, EPA'S regulations are responsible for this trend, because its highly prescriptive end-of-the-pipe nature discourages companies from implementing more holistic, innovative ideas at their plants. Others believe the more important obstacles to pollution prevention lie within the companies themselves. They suggest that the companies do not prioritize waste reduction initiatives in their business operations.

Texas Chemical Plant Selected

The study organizers picked a Dow Chemical Company facility in La Porte, Tex., as the primary site to evaluate. Located near the Houston ship channel, this relatively small, well-run chemical manufacturing operation produces methylene diamine diisocyanate (MDI), the major ingredient of foamed and thermoplastic polyurethane. Polyurethane foams appear in a variety of rigid foam products, from automobile parts to insulation in water heaters and picnic coolers. Polyurethane thermoplastic resins are used in tool handles and other clear plastics.

Dow sells most of the MDI from the plant as raw material to companies that combine it with various polyols to create foam and plastic; the rest is combined with polyols on site for some smaller volume Dow product lines. The La Porte facility's gross annual revenues are more than $350 million per year, and its estimated annual environmental expenditures are $5 million to $10 million.

Dow has several voluntary environmental improvement goals: reduction of dioxin emissions; decreased reliance on incinerators throughout the company; and, by 2005, a 50% decrease in the amount of waste generated prior to treatment. The La Porte facility's environmental staff are more interested in pollution prevention than most people in industry, making them good study participants. Because the La Porte facility's manufacturing operations are not especially unusual, study organizers thought that the results of this case study would be broadly applicable.

Dow La Porte's basic manufacturing process first combines formaldehyde and aniline to form methylenedianiline (MDA). The carbon atom from formaldehyde forms the methylene bridge between the two aniline molecules. MDA is then purified, placed in solution in monochlorobenzene (MCB), and reacted with phosgene (produced on site) to form monomeric and polymeric methylenebis(phenylisocyanate) (MDI and PMDI, respectively). In the final process step, MDI and PMDI are purified and then sold.

In 1993, La Porte's TRI releases for this process totaled 506,457 lb, well below the 1989 level of 1,137,300 lb (Table 1). Most of these releases were to air, followed by water. Because these emissions put the Dow La Porte facility in the top 4% of TRI facilities for total releases and transfers, its industrial operations were significant locally and nationally.

Table 1

Toxic Releases and Transfers, in Thousands of Pounds, from a Dow Chemical Facility

| | *Year* | | | | | | |
Category	1987	1988	1989	1990	1991	1992	1993
MCB air releases	712.0	980.0	1036.0	876.0	520.0	462.0	406.0
MDI transfers	620.0	426.0	630.4	181.0	230.0	182.6	227.0
Water releases	89.7	4.8	4.0	3.1	3.2	0.7	0.1
Other air releases	167.9	135.1	97.3	106.7	224.7	184.7	100.3
Other transfers	137.0	226.3	122.0	72.8	125.5	147.8	227.0
TOTALS	1726.6	1772.2	1889.7	1239.6	1103.4	977.8	960.4

Toxic releases from Dow Chemical's facility in La Porte, Tex., have declined steadily since 1989. Much of the decline has come from cutting monochlorobenzene (MCB) releases.

Source: Dow Chemical

At the time of the study, MCB was the leading chemical released annually from La Porte. In 1993, 406,000 lb of MCB (80% of the facility's total releases) were emitted (Table 1). In addition to being a toxic chemical, MCB is a volatile organic compound, emissions of which affect the region's ability to meet its ozone attainment levels. (After this study was completed, MCB emissions were substantially reduced; most are now captured and vented to the hazardous waste incinerator.)

La Porte treated about 1.5 million lb of TRI waste at the site in 1993, nearly three times as much as it released to the environment. Almost all of this treatment occurred in an on-site hazardous waste incinerator, covered by a Resource Conservation and Recovery Act (RCRA) permit. Phosgene and methanol provided the largest quantities of wastes burned, followed by 170,000 lb of MCB (see Table 2). (The amount of MCB burned today is greater than 170,000 lb, because additional quantities are now being captured.)

Assessment of Pollution Prevention

La Porte already had a pollution prevention plan, as required by the state of Texas, and Dow was planning to capture most of the remaining MCB air emissions and incinerate them on site. However, this action was on hold pending a decision to upgrade the plant's section that produced these continuing emissions. All the other chemicals in Dow's existing pollution prevention plan were ozone depleters, required to be phased down under the Montreal Protocol.

Conventional wisdom suggests that good opportunities to reduce waste and emissions have already been identified by large, environmentally sensitive companies. In fact, when this study began, plant personnel said they believed no other "low-hanging fruit" remained at the plant; that is, no other opportunities to reduce wastes and emissions remained that could be readily implemented to the financial or environmental benefit of the facility.

Pollution prevention literature, however, suggests that conventional wisdom might be wrong, and that various barriers within companies (1-3) or in government regulations (4) keep many important opportunities from being identified or implemented. To find out whether this was the case at La Porte, the study team first examined its pollution issues, reviewed existing pollution prevention plans, and assessed further opportunities for prevention. Once the "fact pattern" was established, we identified various barriers to expanded use of pollution prevention and sought agreement on recommendations to further its use.

In the first phase of the project, to understand its environmental impact, the coverage of existing regulations, and the plant's view of opportunities and barriers to additional environmental improvement, we submitted written questions to the facility and obtained an extensive, documented response. We then toured the plant to clarify the written responses to our questions. From this work, we characterized the status of the plant before the project's pollution prevention assessment.

In the second phase of this 18-month project, a third-party pollution prevention assessor, Bill Bilkovich of Environmental Quality Consultants, Tallahassee, Fla., went to the plant to seek pollution prevention opportunities. Bilkovich spent about 300 hours at the plant, talking with the staff and investigating opportunities. Our group met many times with the consultant.

Table 2

TRI Wastes (in Pounds) Incinerated at Dow Chemical Facility in 1993

Ammonia	15,000
Aniline	6800
Chlorine	1400
Monochlorobenzene	170,000
Methanol	630,000
Phosgene	840,000
1,1,1 TCA	2575
TOTAL	**1,665,775**

Pollution prevention assessments begin with an inventory of all wastes generated at a plant before treatment, recycling, or emission. They also require a chemical-use inventory for the site, which includes consumed chemicals that do not contribute to waste and emissions.

The inventory at La Porte required considerable work, as is common in a pollution prevention assessment. Even though major waste streams had been identified and tracked at the facility for pollution control (regulatory) purposes, data on their chemical composition were lacking. Such information is necessary for pollution prevention. For example, to continue the pollution prevention potential for one waste stream, we needed to know what components were present in the waste stream in less than 5% concentration. In another waste stream, a high degree of confidence in the distribution of minor constituents was required. Because this sort of information often is of no regulatory or process significance, it is not gathered. Much of the data could have been easily collected, however, if made a priority at the plant.

Following the waste and chemical-use inventory, we assessed opportunities for reduction through substitution, efficiency improvements in the process, recycling, and other options. Priorities can be set on the basis of financial considerations by working first on those projects that would deliver the highest rates of return. At La Porte, we set priorities primarily according to potential human health and environmental impact, which translated into high interest in MCB emissions to the air and the wastes burned in the hazardous waste incinerator.

For each chemical or waste being assessed for reduction opportunity, the team had to identify the reason the waste was generated. Answering this question required in-depth knowledge of the manufacturing process and basic plant chemistry as well as the conditions under which waste was generated (e.g., was it generated continuously or intermittently, under upset or normal conditions, etc.).

Because these critical pollution prevention issues are considered of little or no relevance in conventional assessments of a plant's environmental issues, it becomes critical to engage the production engineering personnel, especially the process chemistry experts, in pollution prevention planning. At La Porte, meetings with process engineers were key to developing a process flow diagram that showed material flow throughout the plant and indicated where each priority waste was generated.

Pollution Prevention Opportunities

After examining the process flow and waste information, project participants concluded that the site's single largest environmental opportunity would involve capturing and recycling MCB and ending the incineration of 500,000 lb of chlorinated hydrocarbons. If all of the other waste streams to the on-site RCRA incinerator (called a thermal oxidizer, or TOX) could also be reduced, recycled, or otherwise managed, the TOX could be closed altogether, a broader prospect with superb environmental and cost-savings benefits. (The cost saving from eliminating the TOX was determined to be substantial, in light of upcoming re-permitting requirements, soon-to-be-required upgrades in the unit, and operation costs.) Thus, we had found an option that would be good for the environment and save the company a lot of money.

To proceed on the TOX closure opportunity, six major waste streams that entered the TOX had to be examined: methanol contaminated with amines and ammonia, MCB air emissions captured by the pressure swing absorption/ carbon adsorption system, other organic compounds captured on carbon, a phenyl isocyanate/MCB waste, phosgene manufacture vent gases (phosgene and carbon monoxide), and MCB from a groundwater pump-and-treat system that processed historical contamination from a previous owner. A pollution prevention assessment was undertaken for each of these waste streams. What follows is a consideration of options for the first two wastes, methanol and MCB, which offered the most interesting opportunities for reduction. We determined that the others were best addressed by using alternative treatment options.

Dealing With Waste Streams

At La Porte, options for conventional end-of-the-pipe alternative treatment of the methanol waste stream include processing in the wastewater treatment plant or sale as a product. Incineration, however, had been considered the best option because methanol is essentially a clean fuel, and its use reduced the need for TOX operations' supplementary fuels. The pollution prevention assessment started with a different set of questions about methanol, and it presented some interesting options.

First we asked, Where does the methanol originate? Although methanol is a major waste stream generated at the plant, a flow chart of the basic process chemistry does not show an obvious source. Interviews with plant personnel revealed that methanol enters the process in the formaldehyde-water solution (formalin) used to manufacture MDA. Formaldehyde is manufactured from methanol; and some residual methanol, in this case about 0.5%, is kept in the commercial product as a stabilizer. This methanol must be removed by La Porte before the phosgenation step. Because La Porte uses millions of pounds of formaldehyde each year, the amount of methanol waste being burned in the TOX reaches hundreds of thousands of pounds annually

The next obvious question was, Can we substitute formaldehyde with a less toxic chemical that does not generate a waste stream? Formaldehyde is used in the plant to provide a carbon atom to connect two aniline molecules and form the intermediate MDA. Perhaps carbon dioxide (CO_2) could be used to achieve the same end. The assessor researched the use of CO_2 and found that, although there was a patent on the use of CO_2 in the manufacture of toluene di-isocyanate (TDI), the process had never been commercialized and was not applicable to the manufacture of MDI. Other alternative sources of the carbon bridge atom were discussed with Dow research and development personnel, but we found no other good options and stopped this line of inquiry.

The next option was to look for ways to reduce or eliminate the methanol in the formalin. But conversations with a major formaldehyde manufacturer indicated that, under the temperature, humidity, and transit time conditions common in the Gulf Coast, at least 0.3% methanol is necessary to prevent the in-transit polymerization of formaldehyde. Thus, we could not decrease methanol waste to insignificant quantities by using this approach at this plant.

The identification of an alternative stabilizer, one that might be effective at part-per-million (ppm) levels, was not explored for La Porte; the driving force for developing an alternative would have to come more broadly from other formaldehyde users across the country. However, the assessment did raise interesting questions about the treatment and disposal costs incurred nationwide for the management of waste methanol at plants that use formaldehyde as a raw material. Approximately 8 billion lb of formaldehyde are used annually in U.S. manufacturing, and calculations show that 40 million lb of methanol are being handled or disposed of by the plants purchasing this chemical. Full-cost accounting of the cost per pound of formaldehyde purchased as a raw material could reveal that the cost for residual waste methanol management is as high as or higher than the raw material cost—and open a market opportunity for higher priced, methanol-free formaldehyde.

Next we asked, If methanol is needed to stabilize the formaldehyde, might there be a way to remove the methanol as a clean waste stream before the formaldehyde enters production and comes in contact with aniline? Interest in this option was depressed by the low value of recovered methanol and the high cost of constructing and operating separation equipment to process millions of pounds of formaldehyde each year.

Because the trace quantities of aniline and other nitrogenous compounds make the waste methanol difficult to sell, we then asked whether the aniline could be removed from the methanol. Even though trace quantities of aniline could be removed and yield a salable methanol–water mix, the assessment team did not believe that aniline could be eliminated to levels considered safe for unrestricted commercial use of the methanol waste.

Project organizers then asked whether a customer could be found for methanol containing aniline. A cursory review of TRI data revealed no facility close to La Porte that had methanol and aniline emissions, and we did not pursue an in-depth review.

The final option we evaluated was returning the waste methanol to the formalin manufacturer for future processing into the formalin product Dow was interested in this option and would consider a "take back" provision in its purchase contract quite favorably, although this option has not been issued to date.

MCB Waste Stream

Analysis of MCB took the same initial path: identifying the use of MCB in processing (solvent carrier in the phosgenation step of the process) and seeking less toxic–chlorinated alternatives for this purpose. Finding no alternatives, we shifted our focus to recycling MCB instead of incinerating it. Plant personnel reported they had briefly considered this alternative when they decided to capture the MCB air emissions to reduce their air pollution, but they decided to incinerate the solvent because it was convenient and legal to do so, and they did not believe this practice would pose significant risk.

The plant personnel also believed that the presence of water in the used MCB would preclude recycling. The assessor researched this and found that water was not actually the principal barrier to recycling. To the contrary, the virgin MCB purchased by the plant is routinely treated on site to remove water introduction in a molecular sieve bed before introduction into the process. Production staff then raised a more important concern: If the waste MCB contained any impurities, they could build up in the system if MCB were recycled. Impurities were possible, but sufficient information was not available about the time course of their buildup. The uncertainty raised by this issue could be easily resolved by analysis of the actual level and identity of trace contaminants in this stream, however.

At the end, the assessment team's short-term recommendations included recycling the MCB waste stream that is currently incinerated; selling the methanol or burning it at an alternative, off-site incinerator; reducing levels of phosgene sent for treatment; and scrubbing the remaining waste phosgene instead of incinerating it.

If all these waste streams could thus be addressed, and several very minor waste streams were sent off site, Dow La Porte could conceivably shut down its on-site hazardous waste incinerator and avoid the cost of RCRA re-permitting. Dow personnel estimated the rate of return on investment for this project at 20–70%. The investment would pay for itself somewhere between 15 months and 5 years, depending on how the various projects were configured. Estimated savings are $1 million a year, derived from reduced raw material costs, incinerator operating costs, and re-permitting costs with regulatory authorities. Virgin solvent purchases alone might drop by 90%.

Barriers to Implementation

There were no significant regulatory barriers to adopting this plan. The problem lay within the company. Specifically, these opportunities were weak candidates in the capital investment process at Dow. The project was considered for implementation twice by the urethane business group within Dow Chemical and was put off both times because other, more financially attractive business opportunities were given higher priority.

Had EPA required that Dow reduce these waste streams, the 4P projects would have been mandatory, and the rate of return of the project would have been irrelevant to Dow's decision making. However, because these were voluntary opportunities, they were considered in the same way as other business opportunities would be. To succeed, these opportunities needed to do more than reduce waste and save money; they needed to be superior to other options for capital investment.

Although Dow hopes to implement the pollution prevention plan in the future, the La Porte pollution prevention project rests in an odd position: It is not required for the purposes of environmental compliance, and it is not of central interest to production engineers whose main priorities are in capacity building. Nor is the project highly compelling to business line personnel with profit-and-loss authority. They are more concerned with maximizing profit for their business among various Dow plant locations around the world.

Conventional wisdom says that most good opportunities to reduce waste and emissions have already been identified, but that belief was incorrect in this case: The 4P project found very promising opportunities that had not been identified by Dow. More significantly once pollution prevention opportunities were found, corporate business priorities and decision-making structures posed formidable barriers to implementing those opportunities.

Most environmental professionals outside of industry incorrectly assume that a pollution prevention plan that actually saves money and is good for the environment will be quickly seized upon by U.S. businesses. This work shows that at least in one firm, such opportunities may not be sufficiently compelling as a business matter to ensure their voluntary implementation.

References

1. Porter, M. E.; van der Linde, C. *Harvard Business Review* September-October 1995, 120–34.
2. New Jersey Department of Environmental Protection and Energy. *Industrial Pollution Prevention Planning: Meeting Requirements Under the New Jersey Pollution Prevention Act;* Office of Pollution Prevention, State of New Jersey: Trenton, July 1993.
3. Little, A. D. "Hitting the Green Wall," *Perspectives;* Arthur D. Little: Cambridge, MA, 1995.
4. Schmitt, R. E. *Natural Resources and Environment,* 1994, 9, 11–13, 51.

POSTSCRIPT

Will Voluntary Action by Industry Reduce the Need for Future Environmental Regulation?

Most assessments of the environmental achievements of the past three decades agree with Patchak and Smith that regulations have achieved a great deal. At the same time, however, they also note that degradation of the world's ecosystems has only been slowed, not halted or reversed. It is true that governmental environmental agencies have encouraged voluntary efforts like Responsible Care and ISO 14000, but there is little evidence that the idea that these initiatives can replace or reduce the need for "command and control" regulation and enforcement is being broadly embraced. At best, as Paulette L. Stenzel writes in "Can the ISO 14000 Series Environmental Management Standards Provide a Viable Alternative to Government Regulation?" *American Business Law Journal* (Winter 2000), the ISO 14000 standards "provide a useful supplement to environmental regulation. They can facilitate the work of the U.S. Environmental Protection Agency and promote worldwide pursuit of sustainable development."

As Greer and van Löben Sels argue, many environmental advocates assume that all that is necessary to get industry to implement more environmentally appropriate practices is to demonstrate that the changes will have a short-term economic payback. The discovery that this is often untrue has soured several cooperative efforts between corporations and environmental organizations. The problem is that implementing most production changes requires an up-front investment, and management will usually look for the most profitable investment of available revenues. This means that many opportunities that can be proven to have less potential profitability, even if they may appear to be cost-effective, will be rejected. Of course, environmental protection also often requires industries to make changes that are costly and will not improve the bottom line!

For a recent industry assessment of the potential of ISO 14000, see "Environmental Management with ISO 14000," by Steven Voien, *EPRI Journal* (March/April 1998). Ronald Begley reports on the scepticism of U.S. regulators in his article "ISO 14000: A Step Toward Industry Self-Regulation," *Environmental Science and Technology* (July 1996). In a follow-up article entitled "Value of ISO 14000 Management Systems Put to the Test," in the August 1997 issue of the same journal, Begley describes a 10-state research project designed to evaluate the new standards. For an assessment of the accomplishments of the first 10 years of the Responsible Care program, see "Responsible Care: Doing It Right," by Marc S. Reisch, *Chemical and Engineering News* (October 26, 1998).

ISSUE 18

Are Aggressive International Efforts Needed to Slow Global Warming?

YES: Christopher Flavin, from "Last Tango in Buenos Aires," *World Watch* (November/December 1998)

NO: Jerry Taylor, from "Global Warming: The Anatomy of a Debate," *Vital Speeches of the Day* (March 15, 1998)

ISSUE SUMMARY

YES: Worldwatch Institute vice president Christopher Flavin cites evidence that human-induced global warming has begun. He calls for decisive action based on a new approach to reducing greenhouse gas emissions.

NO: Jerry Taylor, the Cato Institute's natural resource studies director, contends that the uncertainties regarding the likely magnitude and consequences of global warming makes the implementation of an expensive agreement that he believes will have little effect on the future climate an unwise gamble.

The physics of the situation is clear. Sunlight warms the Earth, which radiates that warmth back to space as infrared radiation. Molecules of certain atmospheric gases—including carbon dioxide, water vapor, methane, chlorofluorocarbons, and nitrous oxide—absorb the infrared radiation and reemit it. Some of the reemitted infrared radiation heads back toward Earth, where it adds to the warming of the planet.

If there were no such infrared-absorbing "greenhouse gases" in the atmosphere, the Earth would have an average temperature of approximately 63 degrees Fahrenheit colder than its present average of 59 degrees Fahrenheit. If there were more than the present amounts of greenhouse gases in the atmosphere, more heat would be retained, and here is the crux of the problem. Since the dawn of the industrial age, humans have been burning vast quantities of fossil fuels, releasing the carbon the fuels contain as carbon dioxide. Because of this, some estimate that by the year 2050, the amount of carbon dioxide in the air will be double what it was in 1850. By 1982 the increase was apparent.

See Spencer R. Weart, "The Discovery of the Risk of Global Warming," *Physics Today* (January 1997).

How serious is the warming? Climate is by nature variable, but the 1990s provided several years of record-breaking warmth, and 1998 was the warmest year in the last millennium. A recent analysis of oceanographic records finds that the deep layers of the sea are warming faster than anyone had suspected. Spring is arriving earlier, and growing seasons are lengthening. See Douglas Gantenbein, "The Heat Is On," *Popular Science* (August 1999).

It is difficult to say just how warm it will get, but there is no doubt that atmospheric carbon dioxide and other greenhouse gas levels are rising. Most climatologists agree that the warming will continue but that the factors that shape climate are numerous and interact in complex ways. Consequently, there is uncertainty about the eventual outcome and what it will mean for life on Earth. Expected effects include rising sea levels (a serious hazard for low-lying coastal and island nations); changes in rainfall patterns (of obvious concern to global agriculture); increasing numbers of serious storms; shifts in the climatic zones, which define where forests can thrive; and movement of tropical diseases into temperate regions. The disease prospects are discussed by Paul R. Epstein in "Is Global Warming Harmful to Health?" *Scientific American* (August 2000).

In November 1995 the Intergovernmental Panel on Climate Change (IPCC) issued its second major study on global warming, concluding that "the balance of evidence suggests that there is a discernible human influence on climate." This report predicts that global average temperatures will likely increase by 1.8–6.3 degrees Fahrenheit before the end of the twenty-first century. The IPCC study led the United States to reverse its position and to accept the goals for reducing greenhouse gas emissions that were negotiated as part of the Framework Convention on Climate Change at the 1992 Earth Summit in Rio de Janeiro. However, the United States opposed the stronger actions advocated by other nations at the 1997 meeting in Kyoto, Japan, where a protocol was developed for achieving the goals of that convention, and at the 1998 session in Buenos Aires, Argentina, to work out the means for implementing the protocol.

In the following selection, Christopher Flavin argues that despite broad international agreement that global warming is a problem that warrants prompt global action, very little is actually happening. He calls for decisive action based on an entirely new approach to reducing greenhouse gas accumulations. In the second selection, Jerry Taylor asserts that the evidence supporting the need for global actions to combat greenhouse gas–induced climate change is "shockingly weak" and that proposed actions will not be very effective. Therefore, an expensive international effort would be unwise.

Christopher Flavin

Last Tango in Buenos Aires

The world's climate rarely sends clear signals. The interactions of hundreds of variables—of sunlight, ocean currents, precipitation, fire, volcanic eruptions, topography, and the respiration of living things—produce a complex system that scientists are just beginning to understand, and that defies precise forecasts. In any given year, some regions are warmer than normal while others are cooler. Almost any short-term climatic phenomenon, even an extreme one, can be explained as something that falls within the enormous range of natural climatic variability. Until this year.

Even before 1998 comes to a close, it is clear that this year is one for the meteorological record books. Although annual temperature records have become routine recently—all 14 of the warmest years since 1860 have occurred in the past two decades—the record is usually broken by a couple of hundredths of a degree. But the average temperature for January—August 1998 was a full four tenths of a degree warmer than the average for 1997, the previous record-setting year (see figure). In fact, six of the first eight months of 1998 set an all-time temperature record for the month—exceeding the monthly figures recorded in the 139 years that global average temperatures have been tracked.

At first, scientists were inclined to attribute these surprising readings to El Niño, a periodic warming of the eastern Pacific that began in 1997 and extended through the first half of 1998. But as they looked back at the historical trend, it became clear that previous El Niño-related warmings had been far more modest. As month after month of record-breaking data spewed from their computers, the atmospheric scientists expressed growing awe. James Baker, administrator of the U.S. National Oceanic and Atmospheric Administration said, "There is no time in recorded data history that we have seen this sequence of record-setting months."

In earlier years, some scientists' concerns about global warming were assuaged by the fact that satellite-based microwave measurements of temperatures high in the atmosphere since 1979 did not appear to reflect the warming trend from ground-based readings. But this slender straw was swept away in August by a report by scientists Frank Wentz and Matthias Schabel that appeared in the British journal *Nature.* It demonstrated that the widely reported satellite data were skewed by the failure to account for the predictable gravity-induced decay

From Christopher Flavin, "Doomsday Every Day: Sustainable Economics, Sustainable Tyranny," *The Independent Review* (November/December 1998). Copyright © 1998 by The Worldwatch Institute. Reprinted by permission of *World Watch*.

in the orbits of the satellites. Once corrected for, the satellite data demonstrate the same broad warming trend as the ground-level thermometers—including the dramatic spike in 1998.

Figure 1

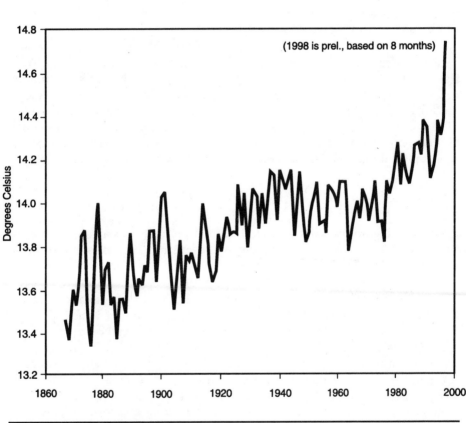

Average Temperature at the Earth's Surface
1866–1998

(1998 is prel., based on 8 months)

Source: Goddard Institute for Space Studies

Scientists have known for some time that the climate is a "non-linear system" that may respond marginally or not at all to initial changes—but then leap suddenly to a new equilibrium, if pushed a little further. Although it is too early to know for sure, the global climate may have just crossed such a threshold. Since the beginning of the twentieth century, human activities have added 925 billion tons of carbon dioxide (CO_2) to the atmosphere, taking concentrations of this heat-trapping gas to the highest levels in 160,000 years. The climate record shows that when CO_2 concentrations reached even close to such levels in the past—during the Eemian interglacial period, for example, beginning 135,000 years ago—they were accompanied by a rapid rise in temperatures.

Though it is impossible to connect any single weather event to global climate change, the past year has been marked by a worldwide pattern of unusually severe weather. China was swept by its worst floods in three decades last summer, with 56 million people reported to be at least temporarily displaced from their homes in the Yangtze basin alone. The $36 billion in estimated damages matches or exceeds the total weather-related losses for the world in every year prior to 1995. Meanwhile, two-thirds of Bangladesh was under water for most of the summer, as torrential monsoon rains cascaded down from the Himalaya and storm surges came up from the sea, covering much of the capital, Dhaka, and destroying the country's rice crop.

At least 54 other countries were hit by severe floods in 1998, and at least 45 were stricken by droughts, many of which led to runaway wildfires. Tropical forests normally do not burn, but unusually harsh droughts contributed to a series of unprecedented fires in southeast Asia starting in late 1997 and in the Amazon through most of 1998. Last spring, much of southern and central Mexico was aflame, leading to air quality alerts in Texas and noticeably smoky air as far north as Chicago. By early summer, scores of fires were sweeping the sub-tropical forests of Florida, leading to the evacuation of an entire country.

꙳

Rarely have the rhythms of the natural world been so out of synch with those of the political world. Even as the climate sent ever-stronger signals of disruption in 1998, efforts to deal with the problem bogged down in glacial and contentious negotiations over the terms of the Kyoto Protocol on climate change.

The effort to build a global climate agreement is in fact already a decade-long saga that began with a major scientific conference on the issue in Toronto in 1988. The scientists there called for a 20 percent cut in carbon dioxide emissions by 2005, which then led to extended efforts on the part of scientists, industrial interest groups, non-governmental organizations, and politicians to forge an international agreement to move in that direction. By the time of the 1992 Earth Summit in Rio de Janeiro, the "Framework" Convention on Climate Change had been forged, but due to the strong objections of the Bush Administration in the United States, still did not include legally binding limits.

After Rio, governments worked for several years to strengthen the climate treaty by adding specific limits on the amounts of greenhouse gases that could be emitted by each industrial country. This process was expected to culminate in the signing of a protocol to the convention that included legally binding emissions limits, in Kyoto, Japan last December. But agreement proved elusive. As the Kyoto conference began, governments were still widely divided on key elements of the agreement, including the overall level to which emissions would be limited. The United States, for example, only wanted to cut emissions back to the 1990 level, while the European Union wanted to cut them to 15 percent below that level.

By the beginning of its final week, the Kyoto conference had become "an emotional roller-coaster for delegates who watched the treaty's fortunes rise, fall, and rise again," according to a *Washington Post* correspondent. Core elements of the treaty remained unresolved, ranging from the level of emission cuts to be mandated to whether planting or protecting trees could be counted against those emission commitments.

With the negotiations bogged down, U. S. Vice President Al Gore, who had devoted much of his 1992 book *Earth in the Balance* to the problem of climate change, was dispatched to Kyoto. Soon after his arrival, the U. S. delegation shifted its position on emission limits and agreed to reduce its emissions 7 percent from 1990 levels—roughly half way between the U. S. and European positions. But on the evening of December 10, as the deadline for concluding the historic conference came and went, other unresolved issues remained—some of which would determine the significance of the numbers that had been agreed to. Raul Estrada Oyuela, the Argentine Chair of the conference, who had been working behind the scenes for months to forge essential compromises, refused to give up. He ordered the "committee of the whole," composed of all 159 national delegations, to re-convene at 1 am, and meet until a conclusion was reached.

Through the wee hours of December 11, Estrada methodically moved the assembled delegates through the remaining passages of disputed text: whether trading of emissions commitments would be permitted among industrial countries, and whether developing countries would be encouraged to adopt voluntary commitments.

As discussions seesawed back and forth, oil producers like Kuwait did their best to derail the agreement, while European and small island countries worked to strengthen it. But the main axis of the battle soon formed around China and the United States, the two largest emitters, who were deeply divided both on trading and on the question of developing country commitments. As positions hardened, hope of an agreement began to fade.

The U. S. delegation, which had brought Under-secretary of State Staurt Eizenstat in from Washington to be its "closer," became so panicky at one point that delegates were standing on their table, waving for Estrada's attention in the huge hall. Given the vice president's close identification with the issue, the Clinton-Gore Administration could not afford to be found holding the noose if the Kyoto agreement was strangled.

As dawn approached, the Kyoto conference hall was beginning to resemble a week-old battlefield. Bleary-eyed reporters and NGO observers wandered the facility searching for remnants of food or coffee, while inside the plenary hall government delegates held their ground on various items, waiting for the other side to back down in the face of mounting sleep deprivation. Many delegates had passed out, one with his head resting in an ashtray. Chinese and Russian speaking interpreters pulled off their headphones and left, and the Japanese conference center staff threatened to cut off the electricity if the conference was not shut down.

But Estrada, an old hand in chairing contentious negotiations, took advantage of the exhaustion. Seizing on a few half-compromises, he began gaveling closed key portions of the agreement. With the spotlight of the world's media upon them, delegates decided they had more to fear from a failed agreement than one with which they only partially agreed, and stood aside as Estrada pushed relentlessly through the text.

At 10:15 am, Estrada called for adoption of the protocol by consensus, and despite remaining reservations, no government was prepared to stand in the way. The deed was done. Hundreds of delegates rushed out to press conferences, declared victory, and headed for the Osaka International Airport.

During the next 24 hours, headlines around the world proclaimed a great success at Kyoto. Chairman Estrada stated that he was "deeply satisfied" with the outcome, and the World Resources Institute called it "an historic step in the history of humanity." Clouds remained on the horizon—particularly the threats of U. S. Senators not to ratify the agreement—but most observers, including this author, were hopeful that the remaining holes could be patched by the Fourth Conference of the Parties in Buenos Aires this November.

Sadly, the past 12 months have turned the Kyoto conference into a kind of high-water mark, from which the climate negotiations have steadily retreated in the past year. Divisions among national governments have only widened since Kyoto, and the holes in the agreement are beginning to seem more substantial than the protocol itself. Indeed, by creatively papering over wide differences between various nations, the Kyoto negotiators may have crafted an agreement that is barely workable in the best of circumstances, and in the current political climate could lead to paralysis.

At its core, the Kyoto Protocol has four major weaknesses that will need to be remedied if it is to be effective in slowing climate change before irreversible damage is done.

1. Weak commitments: Since the 1988 Toronto Conference, the cornerstone of climate negotiations has been the setting of binding limits on the emissions of greenhouse gases by industrial countries—the countries that have accounted for the bulk of the emissions so far. The 1992 Framework Convention includes a voluntary goal of holding those emissions to the 1990 level in 2000. Some European countries are already meeting this goal, thanks mainly to cuts in coal subsidies. But Australia, Canada, the United States, and other industrial nations are not, due in part to their low fuel prices, and to their failure to enact aggressive energy conservation measures. The main goal for the Kyoto agreement was to establish a new legally binding target for the year 2010.

The negotiators in Kyoto settled on nation-by-nation limits that add up to a reduction in greenhouse gas emissions of 5.2 percent below the 1990 level for all industrial nations. Little noticed outside climate policy circles, however, was the curious fact that total CO_2 emissions by industrial countries was—and is—already below 1990 levels, due to steep declines in the former Soviet Union. As a result, the protocols target, were it to cover just CO_2, translates to a mere 2.5 percent cut from the 1997 level.

Within that goal, industrial countries agreed to a range of specific targets —cuts of 8 percent in the European Union, 7 percent in the United States, and 6 percent in Japan—along with an 8 percent increase in Australia. These numbers represent backroom political deals more than they do analyses of the economic potential to reduce emissions in a given country. Australia, for example, has a government dominated by mining interests that wish to boost their export of energy-intensive products to Asian nations—a development that will of course worsen the greenhouse problem.

The anemia of the Kyoto figures can be seen when they are contrasted with what is eventually needed to stabilize CO_2 concentrations. According to the International Panel on Climate Change, the official scientific body that advises the Conference of the Parties, the amount of reduction that eventually will be required is not 5.2 percent, but 60 to 80 percent below the 1990 levels. Yet, when emissions of developing countries are added to those of the industrial countries covered by the protocol, the global total is projected to increase to some 30 percent *above* the 1990 level by 2010.

The most hopeful thing that can be said of the Kyoto Protocol is that it echoes Lao Tse's comment that a journey of a thousand miles begins with single step. The protocol could, perhaps, set the stage for more ambitious agreements later, as has occurred with earlier environmental treaties. But if—as now seems likely—it takes years to ratify the protocol, and years more to enact the national policies needed to achieve its weakened goals, the encounter in Buenos Aires could turn out to be little more than an elaborate tango—a few impressive steps that end up going nowhere.

2. Searching for "flexibility":   As climate negotiations grew tense last year, the Clinton Administration was increasingly desperate to find a way of bridging the huge gulf in emissions goals that separated the United States from the European Union. The key, U. S. officials felt, was to come up with a series of provisions—critics called them loopholes—that would make it less expensive to meet the protocol's goals, and that would avoid the need to take a big bite out of domestic CO_2 emissions. The levels the Europeans were asking for, they believed, would require politically impossible measures that were already being aggressively fought by a multimillion dollar TV and newspaper ad campaign sponsored by the coal, oil, and automobile industries.

Australia, Canada, and New Zealand had similar concerns, and strongly supported the search for "flexibility." European governments were not nearly so worried about tough targets since, unlike the United States, they had not substantially increased their emissions during the 1990s. But even many of their leaders privately welcomed the notion of flexibility that would allow them to delay enacting any new energy taxes or other constraints on politically powerful industries.

During the Kyoto negotiations, the focus turned to a target that would cover a "basket" of six greenhouse gases rather than focusing on each one individually. This "comprehensive" approach seemed logical enough, since it covered all the important greenhouse gases, including methane and HFCs, not just CO_2. But it also happened to be a great convenience to U. S. delegates who were looking for ways to avoid sharp CO_2 reductions that would arouse a hornet's nest of industry outrage. Experts had identified a potential for easy reductions in some of the more minor greenhouse gases, which might offset some of the projected increase in the nation's carbon dioxide emissions.

But this approach seems likely to lead to an accounting nightmare, in part because there is no reliable emissions inventory for some of these gases, and each has a distinct (and in some cases uncertain) lifetime in the atmosphere. As as a result, this "comprehensive" approach is likely to reduce the clarity of the protocol, and could well encourage cheating. (One of the keys to the pioneering 1987 Montreal Protocol on Substances That Deplete the Ozone Layer is that it dealt with each of the offending gases individually and specifically, so that countries knew exactly what was needed—and would be exposed if they did not.)

At the insistence of the United States, as well as Canada and New Zealand, the Kyoto Protocol also allows countries to count carbon absorption by forests (and perhaps later by peat bogs and other carbon "sinks") as offsets against emissions. Under the agreement, carbon flow resulting from both additions to and subtractions from sinks is to be included in national inventories. A coal-burning power company in Ohio, for example, could receive offset credits for financing a tree-planting project in Oregon.

In principle, this idea makes sense—tree planting should be encouraged. But the proposed scheme for doing this is exceedingly complex, combining an accounting maze with uncertain science. Biologists point out that there is not yet enough data on natural carbon cycling to establish full accounting and verification procedures for carbon sinks. And like the provision on the "other" gases, this one complicates monitoring and enforcement, and encourages governments to fiddle with the figures. In response to these concerns, the provision on sinks has been sent back for scientific review, which is to be completed by 2000.

3. Hot air trading: Another form of "flexibility" in the Kyoto Protocol is the concept of emission-allowance trading, an idea pioneered and highly touted by U. S. government regulators, private companies, and even some environmental groups. It is modeled on provisions in the U. S. Clean Air Act that allow power companies to "trade" their sulfur dioxide reduction obligations, in the theory that this will encourage cuts to be made wherever it is least expensive to do so. In the context of global climate change, nations would have the option of buying greenhouse gas emission allowances from other countries that have more than met their own requirements.

The concept has met with considerable skepticism in Europe, as well as in developing countries, which worry that it will dilute the commitments and encourage some governments to avoid difficult domestic policy decisions. Still, a growing number of governments have warmed to the idea in recent months, recognizing that it could improve the economic efficiency of the agreement by channeling capital to economies where it can make the most difference. After a tense standoff, the U. S.-sponsored article on emission allowance trading was accepted, though with obvious reluctance. In a gesture that carried symbolic, if not legal, weight, Estrada cut the trading provision to a few lines and pushed it to the back of the protocol.

It was not until after the weary delegates had arrived home from Kyoto that many of them realized that the United States had pulled a fast one. Under the protocol, Russia and Ukraine must only hold their emissions to the 1990 level, which would allow them to increase emissions 50 and 120 percent respectively from their current depressed levels. Experts do not expect either nation to come close to such increase, even if their economies rebound robustly, so these emission allowances would be available for purchase by countries like the United States, which expect to fall well short of the targets in the protocol. In short, the United States and Russia could make a trade that allowed the United States to take credit for emission reductions that stemmed from the Russian economic collapse of the early 1990s—without reducing future greenhouse gas output by even a molecule.

Although the U. S. government has been vague about its emissions trading intentions, the official plan produced by the White House in July would achieve up to 75 percent of the U. S. reduction requirement by purchasing allowances from the Russians and Ukrainians. While such a deal might result in the United States adding $10-$20 billion a year to Russia's empty treasury, it is hard to see how the climate would benefit. The idea has been widely denounced by everyone from Greenpeace to the U. S. National Coal Association, so it is not clear that it has much of a constituency. Indeed, this kind of trading threatens to undermine more legitimate trading proposals that are tied to specific projects, such as the article on "joint implementation" which is intended to encourage rich industrial countries to invest in climate projects in the former Eastern Bloc. European governments have suggested putting a percentage limit on trading— to encourage adoption of domestic policies—but even this might not be enough to correct a provision that undermines both the effectiveness and legitimacy of the protocol.

4. The ratification trap: While all of these problems are thorny ones, they should in theory be surmountable. But many of the all-important details— including how emissions trading will work—were not included in the Kyoto Protocol, and will have to be added to it if the protocol is to be effective. At negotiating sessions in Bonn, Germany, in June—the last before Buenos Aires —vituperation outweighed progress. In the European and developing country views, the United States has riddled the protocol with sneaky loopholes, while U. S. officials believe that the Europeans are trying to wriggle out of an agreement that was made in good faith in Kyoto.

Complicating the process further is the issue of ratification. The Kyoto Protocol will only go into force if ratified by enough industrial countries to represent at least 55 percent of industrial country emissions. In theory, the protocol could go into force without U. S. assent, as may occur with the land mines treaty negotiated last year. But due to concerns over competitiveness as well as fairness, neither the Europeans nor the Japanese with to move forward with an agreement that excludes the world's largest greenhouse gas emitter.

Leading U. S. Senators, meanwhile, say they will not move forward with ratification without "new specific scheduled commitments to limit or reduce greenhouse gas emissions" by developing countries—a position with which the

Clinton Administration has felt compelled to go along. U. S. officials have been vague as to what such a commitment might consist of, and developing countries are rightly wary of being asked to reduce their emissions, which already average less than one-tenth the U. S. per capita level. A year and a half of arm-twisting has yielded little progress, and the impasse gives the United States an effective veto over the protocol.

As 1998 drags to a close amidst financial crisis and political scandal, the climate negotiations are bogged down by a dangerous combination of impotence and ineptitude. And with the coal, oil, and automobile lobbies again stepping up their climate ad campaigns, it is unclear that the key countries have the political will needed to forge the compromises that are needed. The negotiating process itself seems to have become a kind of diplomatic black hole—sucking in endless quantities of legal, economic, and scientific capital. The thousands of government officials, NGO lobbyists, and observers who follow the process closely continue to circle the globe, attending dozens of meetings on sinks, emissions trading, and other climate issues *du jour.* The climate cognoscenti now speak their own acronym-filled language—interlaced with references to AGBM, QUELRO, SBSTA, and LUCF—and are prone to describing labyrinthine sideways movements as "progress." But the negotiations are increasingly disembodied from the real world threats—from massive floods to rampant disease —that a changing climate represents.

Meanwhile, the process of implementing national policies that will actually reduce emissions—which had gained substantial momentum prior to Kyoto —has stalled since then. President Clinton, for example, who talked extensively about climate change when he was in China this summer, has been unable to persuade the U. S. Congress to adopt even a modest package of new climate policies. Europe has seen a similar lack of progress, as it has bickered over "burden-sharing" its goals and debating emissions trading with the United States. And the new Japanese plan, released in August, is based largely on building 20 nuclear plants—a step that government officials privately acknowledge will never be permitted by the Japanese public.

⚜

One of the ironies of the year since Kyoto is that while the national and international political processes have stagnated, opportunities for economically cutting emissions have blossomed. Motivated in part by the prospect of legally binding emissions limits, companies, cities, and individuals have pursued a host of new approaches. From those emerging possibilities, a less legalistic, more productive approach to the climate problem may emerge.

- British Petroleum President John Browne surprised the oil industry when he announced last year that after extended internal deliberations, his company had concluded that climate change is a serious threat that will inevitably reshape the energy industry. Browne later announced BP's intention to reduce its emissions 10 percent and to step up investment in solar energy. The American Petroleum Institute denounced BP for "leaving the church," but Enron Corp, North America's largest gas company, and Royal Dutch Shell, the world's biggest petroleum firm, have joined BP in acknowledging the severity of the climate problem and beginning to shift their own investment strategies.

- During the very month of the Kyoto Conference, Toyota stunned the auto world with the delivery to its showrooms of the world's first hybrid electric car, the Prius—with twice the fuel economy and half the CO_2 emissions of conventional cars. Marketed as a "green" sedan, the Prius sold so quickly in Japan this year that Toyota had to open a second assembly plant. The shock waves were evident at the massive Detroit Motor Show in January, where each of the U. S. Big Three companies announced plans for new generations of hybrid and fuel cell cars. In a 1998 speech that might be compared to Mao expressing second thoughts about communism, General Motors president John Smith said, "No car company will be able to survive in the 21st century by relying on the internal-combustion engine alone."

- As national governments dither over the Kyoto Protocol, a surprising number of city governments are moving forward with active efforts to reduce their emissions. Over 100 cities, representing 10 percent of global emissions, have joined the Cities for Climate Protection program to reduce those emissions by investing in public transportation, tightening up public buildings, planting trees, and installing solar collectors. Toronto, which was the first city to announce a climate plan—in honor of its role in hosting the first major climate meeting a decade ago —is working to reduce its emissions by 20 percent. And Saarbrucken, a medium-sized city in a coal-mining region of southern Germany, has already cut its emissions by 15 percent, in part via effective energy management and public education campaigns.

- A few national governments are also showing the way. After a decade of effort, Denmark now generates 8 percent of its electricity from wind power, and another fraction from the combustion of agricultural wastes. Already, the Danish wind industry employs 20,000 people, and wind turbines are the country's second largest export. And thanks to Denmark's efforts, wind power, at 25 percent per year, has been the world's fastest growing energy source since 1990.

Taken together, these efforts suggest that it will be easier and less expensive to reduce carbon dioxide emissions than it appeared a few years ago. As has been the case with almost every other environmental problem in the past three decades, once we get serious about slowing climate change, we will likely find a host of innovative and inexpensive ways to do so. Thanks in part to the signal sent by the climate convention, as well as to those coming from the climate itself, that process is under way. The question now is how to speed it up.

꒰◦꒱

tango /tan-go/ *n:* a ballroom dance of Latin-American origin in $^2/_4$ time with a basic pattern of step-step-step-step-close and characterized by long pauses and stylized body positions; also: the music for this dance

After a decade of stylized steps—and long pauses—doubts are growing as to whether this climate dance will ever be successfully completed. The brief glow of optimism that surged from Kyoto last fall has faded. Well-meaning diplomats have become handcuffed by an increasingly complex and unruly process. While it is essential to take the long view with a problem such as climate change, even that perspective provides little comfort today: The commitments agreed to in Kyoto are less clear, and arguably less stringent—once all of the "flexibility mechanisms" are included—than the voluntary emission goals established in Rio in 1992.

The glacial pace of climate negotiations in the past decade can be attributed in part to a powerful combination of forces: the intergenerational scale of the problem; only modest public alarm; broad and well-organized industry opposition; and the complex, multi-faceted nature of the problem being addressed. Together, these four factors make the climate-policy process an order of magnitude more challenging than any others so far. Solving this problem will truly put human institutions and ingenuity to the test.

If they are to get the protocol back on track, the negotiators who meet in Buenos Aires this November would do well to remember a distinction made by environmental negotiations expert David Victor of the Council on Foreign Relations in a 1998 book: when it comes to international environmental treaties, compliance and effectiveness are two different things. Unless it leads to major governmental policy changes that in turn lead to lower emissions of the most important greenhouse gas, carbon dioxide, the Kyoto Protocol will have fallen into the trap of debilitating compromise and complication that has plagued several other environmental agreements.

The climate negotiations have been guided in part by the lessons of one of the most successful of those agreements, the 1987 Montreal Protocol to protect the ozone layer. That agreement, in which many of the Buenos Aires negotiators were involved, led to an 80 percent reduction, within a decade of its adoption, in production of the gases that most damage the ozone layer. It did so by relying on three simple principles: Politics must follow science; environmental goals should be clear and simple; and industry consensus is essential to drive the process forward.

But the challenge of Kyoto is far greater, in large measure because where the ozone problem could be solved with the effective participation of just a few dozen companies, the climate problem is driven by millions of actors, indeed by society as a whole. Although the new business created by the move away from fossil fuels is likely to roughly equal the business lost, the potential losers are far better organized.

As with the Montreal Protocol, the central focus of the Framework Convention on Climate Change is national emissions limits, leaving individual nations to decide how to achieve them. While the principle seems like a solid one, since it allows for national differences and permits flexibility, it has not worked in the climate arena, where carbon dioxide, unlike chlorofluorocarbons, is a virtual currency of modern energy economies that can be reduced only by addressing structural economic issues.

Even among industrial countries, the differences in emissions levels, economic structures, and political philosophies are so wide that no single goal has universal logic. One of the problems in Kyoto was the fact that countries such as the United States, which had substantially increased their emissions since 1990, were panicked by the challenge of meeting goals that seemed reasonable to other countries that had already reduced theirs. But once governments began differentiating the goals in Kyoto, the negotiations became a political free-for-all that undermined the credibility of the entire process. In addition, by bundling together six gases, and adding the highly complicated issues of sinks and trading to the protocol, the negotiators have created an agreement that will be nearly impossible to review or enforce, and that at best sends an ambiguous signal to governments and industries.

The challenge now is to renovate the baroque structure that the Kyoto plan has become—or else scrap it and get ready to start over. The negotiators who have labored so hard over the past decade to get the foundation of the protocol in place deserve one more try in Buenos Aires. But if—as now seems likely—that try produces no serious prospect of ratifying the protocol and implementing it, new approaches may be needed.

David Victor points out that when other environmental treaties have run into similar problems, a leadership group of more committed governments has sometimes formed—adopting a more stringent set of voluntary goals, which they then move immediately to implement. In the 1980s, European negotiations to reduce North Sea pollution and nitrogen oxide emissions each ran aground due to vehement opposition of major governments. But other countries moved ahead with voluntary commitments—complementing more modest, legally binding agreements that were also agreed to. Similarly, the international land-mines treaty of 1997 was spearheaded by NGOs and by a small group of like-minded governments. They formulated an agreement that quickly won the support of most—though not all—governments. Holdouts like the United States are expected to join eventually.

This approach might well work for climate policy, building on the leadership roles of several European countries, and building support outward from there. Taking the idea a step further, it might even be feasible to bring regional and city governments and companies into such an agreement. They would pledge to each other not just to meet certain levels of emissions reductions, but to identify and adopt specific policy changes and investments—such as incentives for purchasing more efficient cars or rejuvenating public transportation —that will expeditiously achieve these reductions. They might also agree to experiment with emissions trading and CO_2 taxes.

The guiding principle of this new initiative would be to make climate stabilization an economic opportunity as well as an environmental necessity. As John Topping of the Climate Institute puts it, "A strategy that works is going to have to be one that has its own very positive economic feedbacks, one that extends opportunity rather than slowing it down." Like the Kyoto Protocol itself, this approach would still require political support—but on a local, regional, or national level.

Such an initiative would start with a relatively small group of committed institutions, drawing in a larger circle of participants over time, and gradually marginalizing those who are so mired in the status quo that they refuse to go along. The psychology of marginalization—and of shame—could turn out to be a powerful spur to action. If history is a guide, it might eventually lead to a second generation protocol—one that really works.

The key to any approach, of course, is strong public support for action on climate—support that is substantial enough to overcome the unavoidable tendency of many industries to fight change. Environmental groups need to do a better job than they have so far in mobilizing public action—but in the end, it may come down to the weather. Catastrophes have been the driving forces behind many previous environmental agreements. Tragically, the probability of such crisis is rising with the temperature.

NO ↩

Global Warming: The Anatomy of a Debate

Delivered to the Johns Hopkins University Applied Physics Laboratory, Baltimore, Maryland, January 16, 1998

The national debate over what to do, if anything, about the increasing concentration of greenhouse gases in the atmosphere has become less a debate about scientific or economic issues than an exercise in political theater. The reason is that the issue of global climate change is pregnant with far-reaching implications for human society and the kind of world our children will live in decades from now.

Introducing nuance and clear-headed reason to this debate is something of a struggle. As Cato Institute chairman William Niskanen has noted, for any international action to merit support, all of the following propositions must be proven true:

1. A continued increase in the emission of greenhouse gases will increase global temperature.
2. An increase in average temperature will generate more costs than benefits.
3. Emissions controls are the most efficient means to prevent an increase in global temperature.
4. Early measures to control emissions are superior to later measures.
5. Emissions controls can be effectively monitored and enforced.
6. Governments of the treaty countries will approve the necessary control measures.
7. Controlling emissions is compatible with a modern economy.

The case for any one of those statements is surprisingly weak. The case for a global warming treaty, which depends on the accuracy of all those statements, is shockingly weak. My talk this afternoon will concentrate on a few of the most important of those propositions.

From Jerry Taylor, "Global Warming: The Anatomy of a Debate," *Vital Speeches of the Day* (March 15, 1998). Copyright © 1998 by Jerry Taylor. Reprinted by permission of City News Publishing Company, Inc.

A Continued Increase in the Emission of Greenhouse Gases Will Increase Global Temperature

First off, this subject is terribly complex; the 2nd Assessment Report of the International Panel on Climate Change [IPCC] is 500 pages long with 75 pages of references. As Ben Santer, author of the key IPCC chapter that summarized climate change science, has noted, there are legions of qualifications in those pages about what we know and what we don't. But, unfortunately, those qualifications get lost in the journalistic and political discourse.

I will dispense with an introductory discussion of the rudimentary elements of greenhouse theory. I'm sure you're all familiar with it. Largely on the basis of computer models, which attempt to reflect what we know, what we assume, and what we can guess, many people believe that continued emissions of anthropogenic greenhouse gasses will increase global temperatures anywhere from 1 to 3.5 degrees Celsius.

At this point, I should note that those estimates have been coming down over time. The 1990 IPCC report predicted a little more than twice this amount of warming, and projections have been declining ever since as better models have been constructed. One wonders, at this rate, whether the models will continue to predict increasingly smaller amounts of warming until even the upper bound forecasts become so moderate as to be unimportant.

What We Know—And What We Don't Know

Here's what the data say, about which there is little debate; ground-based temperatures stations indicate that the planet has warmed somewhere between .3 and .6 degrees Celsius since about 1850, with about half of this warming occurring since WWII. Moreover:

- Most of the warming occurs over land, not over water;
- Most of the warming occurs at night; and
- Most of the warming moderates wintertime low temperatures.

But even here, we have uncertainties. Shorter sets of data collected by far more precise NASA satellites and weather balloons show a slight cooling trend over the past 19 years, the very period during which we supposedly began detecting the greenhouse signal. Those data are generally more reliable because satellite and balloons survey 99% of the earth's surface, whereas land-based data (1) only unevenly cover the three-quarters of the earth's surface covered by oceans and (2) virtually ignore polar regions.

While some of that cooling was undoubtedly a result of Mt. Pinetumbo and the increased strength of the El Nino southern oscillation, those events fail to explain why the cooling occurred both before and after those weather events were played out and why, even correcting for those events, the temperature data show no significant warming during the 19-year period.

While it is true, as critics point out, that the satellite and weather balloons measure temperatures in the atmosphere and not on the ground where ground-based measurements are most reliable—over the North American and European land masses, the correlation coefficient between satellite and surface measurements is 0.95—close to perfect agreement, and the computer models predict at least as much warming in the lower atmosphere as at the surface, so if warming were occurring, it should be detectable by the satellites and weather balloons.

Even assuming that ground-based temperature data are more reflective of true climate patterns, that still leaves us with a mystery. When fed past emissions data, most of the computer models predict a far greater amount of warming by now than has actually occurred (the models that are reasonably capable of replicating known conditions are a tale unto themselves to which I'll return in a moment). Notes the IPCC, "When increases in greenhouse gases only are taken into account... most climate models produce a greater mean warming than has been observed to date, unless a lower climate sensitivity is used." Indeed, the most intensive scientific research is being done on why the amount of warming that has occurred so far is so low. After all, a .3-.6 degree Celsius warming trend over the last 150 years all but disappears within the statistical noise of natural climate variability.

There are three possibilities:

- something's wrong with the temperature data;
- something's masking the warming that would otherwise be observed; or
- the atmosphere is not as sensitive to anthropogenic greenhouse gases as the models assume.

Indirect Evidence of Global Temperature

Scientists who argue the first possibility cite the largely incompatible, imprecise, and incomplete nature of even recent land-based temperature records. Those observations, of course, are absolutely correct. Instead, these scientists concentrate on indirect evidence suggesting that the planet has been warming and has been warming significantly over the relatively recent past. They typically point to precipitation trends, glacial movement, sea level increases, and increased extreme temperature variability as suggestive of a significant warming trend. Let's take each of these issues in turn.

Precipitation trends According to the IPCC, global rainfall has increased about 1% during the 20th century, although the distribution of this change is not uniform either geographically or over time. Evidence gleaned from global snowfall is definitely mixed. Still, measuring either rain or snowfall is even more difficult than measuring simple temperature. As the IPCC notes, "Our ability to determine the current state of the global hydropologic cycle, let alone changes in it, is hampered by inadequate spatial coverage, incomplete records, poor data quality, and short record lengths."

Recent evidence from climatologist Tom Karl that the incidence of 2-inch rainfalls has increased in the U.S. received sensational coverage but even according to Karl amounts to "no smoking gun." Why? Because he found only one additional day of such rainfall every two years—well within statistical noise —and that most of those days occurred between 1925 and 1945, a time period that does not coincide with major increases in emissions of anthropogenic greenhouse gases.

Glacial movement   The data here are contradictory. Glaciers are expanding in some parts of the world and contracting in others. Moreover, glacial expansion/ contraction is a long-running phenomenon and trends in movement do not appear to have changed over the past century.

Sea level   While there is some evidence that sea levels have risen 18 cm over the past 100 years (with an uncertainty range of 10–25 cm), there is little evidence that the rate of sea level rise has actually increased during the time that, theoretically, warming has been accelerating. Says the IPCC, "The current estimates of changes in surface water and ground water storage are very uncertain and speculative. There is no compelling recent evidence to alter the conclusion of IPCC (1990) that the most likely net contribution during the past 100 years has been near zero or perhaps slightly positive."

Concerning both ice and sea level trends, the IPCC reports that "in total, based on models and observations, the combined range of uncertainty regarding the contributions of thermal expansion, glaciers, ice sheets, and land water storage to past sea level change is about 19 cm to +37 cm."

Extreme weather variability   Again, the data here are mixed. Reports the IPCC, " . . . overall, there is no evidence that extreme weather events, or climate variability, has increased, in a global sense, through the 20th century, although data and analyses are poor and not comprehensive. On regional scales, there is clear evidence of changes in some extremes and climate variability indicators. Some of these changes have been toward greater variability; some have been toward lower variability."

The Masking Theory

The second theory is more widely credited. The most likely masking culprit according to the IPCC are anthropogenic aerosols, primarily sulfates, that reflect some of the sun's rays back into space and thus have a cooling effect on the climate. That aerosols have this effect is widely understood. But as ambient concentrations of anthropogenic aerosols continue to decline (yes, global pollution is on the decline, not on the rise), the argument is that this artificial cooling effect will be eliminated and the full force of anthropogenic greenhouse gas loading will be felt in short order.

This theory becomes particularly attractive when the details of temperature variability are considered. The warming, as noted a moment ago, is largely a nighttime, winter phenomenon; patterns, which suggest increased cloud cover, might have something to do with the temperature records.

The best evidence marshaled thus far in support of the masking theory was published in *Nature* in the summer of 1996. The study, by Santer et al., used weather balloon temperature data from 1963 to 1987 to determine temperature trends in the middle of the Southern Hemisphere, where virtually no sulfates exist to counter greenhouse warming. The article, which caused a sensation in the scientific world, showed marked warming and seemed to confirm the argument that, when sulfates were absent, warming was clearly evident. The article was featured prominently in the 1995 IPCC report as strong evidence that artificial sulfate masking was behind the dearth of surface warming.

Yet it turns out that, if one examines a fuller set of data from the Southern Hemisphere (1958–95, 13 years' worth of data that Santer et al. did not use), no warming trend is apparent. Moreover, if we carefully examine the land-based temperature records, we discover that it is the regions most heavily covered by sulfates—the midlatitude land areas of the Northern Hemisphere—that have experienced the greatest amount of warming. That, of course, is the exact opposite of what we should discover if the masking hypothesis were correct.

Climate Sensitivity

As I noted a few moments ago, a few of the climate models come reasonably close to replicating past and present climatic conditions when historical data are entered. Those models, interestingly enough, predict the least amount of future warming based on present trends. The two most prominent of those models, those of the National Center for Atmospheric Research and the U.K. Meteorological Organization, predict warming of only 1.2 degrees Celsius and 1.3 degrees Celsius over the next 50 years; the lower-bound estimates reported by the IPCC.

The argument for moderate climate sensitivity to anthropogenic greenhouse gas emissions largely rests on three observations:

First, there appear to be carbon sinks that continue to absorb more carbon dioxide than can be explained. While most models assume that those sinks are presently or nearly beyond their carrying capacity, we have no way of knowing that.

Second, 98% of all greenhouse gases are water vapor, and many atmospheric physicists, most notably Richard Lindzen of MIT, doubt that a doubling of anthropogenic greenhouse gases would have much climate effect absent a significant change in the concentration of atmospheric water vapor.

Finally, a warming planet would probably lead to increased cloud cover, which in turn would have uncertain effects on climate. Concedes the IPCC, "The single largest uncertainty in determining the climate sensitivity to either natural or anthropogenic changes are clouds and their effects on radiation and their role in the hydrological cycle... at the present time, weaknesses in the parameterization of cloud formation and dissipation are probably the main impediment to improvements in the simulation of cloud effects on climate."

The Anatomy of the "Consensus"

Despite all the uncertainty, we are constantly told that there is a "consensus" of scientific opinion that human-induced climate changes are occurring and that they are a matter of serious concern. That belief is largely due to the weight given the IPCC report, where this consensus is supposedly reflected. Here is the talismanic sentence of that report, inserted by a small, politically appointed committee after the large-scale peer review was completed: "the balance of the evidence suggests a discernible human influence on global climate." Now, compare that statement with this, which appears on p. 439 of the report:

> Finally, we come to the difficult question of when the detection and attribution of human-induced climate change is likely to occur. The answer to this question must be subjective, particularly in the light of the large signal and noise uncertainties discussed in this chapter. Some scientists maintain that these uncertainties currently preclude any answer to the question posed above. Other scientists would and have claimed, on the basis of the statistical results presented in Section 8.4, that confident detection of significant anthropogenic climate change has already occurred.

On p. 411, the statement is even clearer:

> Although these global mean results suggest that there is some anthropogenic component in the observed temperature record, they cannot be considered as compelling evidence of clear cause-and-effect link between anthropogenic forcing and changes in the Earth's surface temperature.

Counterbalancing IPCC's note of cautious concern are other, far harsher judgements about the scientific evidence for global climate change:

4,000+ scientists (70 of whom are Nobel Prize winners) have signed the so-called Heidelberg Appeal, which warns the industrialized world that no compelling evidence exists to justify controls of anthropogenic greenhouse gas emissions.

A recent survey of state climatologists reveals that a majority of respondents have serious doubts about whether anthropogenic emissions of greenhouse gases present a serious threat to climate stability.

Of all the academic specialists, climatologists (only about 60 of whom hold Ph.D.'s in the entire U.S.) and atmospheric physicists are those most qualified to examine evidence of climate change. It is those professions that are most heavily populated by the so-called "skeptics."

A recent joint statement signed by 2,600 scientists under the auspices of the environmental group Ozone Action is less than compelling. A survey of those signatories by Citizens for a Sound Economy concludes that fewer than 10% of them had any expertise at all in any scientific discipline related to climate science.

An Increase in Average Temperature Will Generate More Costs Than Benefits

How costly might global warming prove to be 100 years hence? Well, that largely depends on the distribution of warming through time and space. It also depends on how much warming occurs; will it be the upper bound or lower bound estimate that comes to pass?

Benign Warming Patterns

For what it's worth, I tend to agree with the IPCC's summary statement that the "balance of the evidence suggests" that anthropogenic greenhouse gas emissions explain some of the detected warming observed thus far over the past 100 years. But as noted earlier, that warming has been extremely moderate, has been largely confined to the northern latitudes during winter nights, and has exhibited no real detrimental effects thus far. I expect those trends to continue and that's the main reason why I doubt that the costs of warming will be particularly consequential.

The present observed warming pattern is certainly consistent with our understanding both of atmospheric physics, which indicates the following:

The driest air-masses will warm faster and more intensely than moister air-masses. The driest air-masses are the coldest; i.e., those in the northern latitudes during the night.

Increased warming will increase the amount of water evaporation, which will in turn result in greater cloud cover. Cloud cover during the daytime has a cooling effect; during the nighttime, a warming effect.

Virginia state climatologist Pat Michaels concludes that:

> If warming takes place primarily at night, the negative vision of future climate change is wrong. Evaporation rate increases, which are a primary cause of projected increases in drought frequency, are minimized with nighttime, as opposed to daytime, warming. The growing season is also longer because that period is primarily determined by night low temperatures. Further, many plants, including some agriculturally important species, will show enhanced growth with increased moisture efficiency because of the well-known "fertilizer" effect of CO2. Finally, terrestrial environments with small daily temperature ranges, such as tropical forests, tend to have more bio-mass than those with large ones (i.e., deserts and high latitude communities) so we should expect a greener planet.

Nighttime warming also minimizes polar melting because mean temperatures are so far below freezing during winter that the enhanced greenhouse effect is sufficient to induce melting.

Indeed, this warming scenario predicts benign, not deleterious, effects on both the environment and the economy.

But what if the warming turns out to be more serious than this? What if the median estimate reported by the climate models comes to pass: a 2.5 degree Celsius warming over the next 100 years?

There have been six particularly comprehensive or prominent serious studies undertaken to estimate the macro-economic consequences of such a warming. None of them gives us much reason for alarm. The main reason is that most modern industries are relatively immune to weather. Climate affects principally agriculture, forestry, and fishing, which together constitute less than 2 percent of U.S. gross domestic product (GDP). Manufacturing, most service industries, and nearly all extractive industries remain unaffected by climate shifts. A few services, such as tourism, may be susceptible to temperature or precipitation alterations: a warmer climate would be likely to shift the nature and location of pleasure trips.

1974 Department of Transportation Study

Back when the world was more concerned with global cooling than global warming, the DOT brought together the most distinguished group of academics ever assembled before or after to examine the economic implications of both cooling and warming. In 1990 dollars, the DOT study concluded that a .9 degree Fahrenheit warming would save the economy $8 billion a year. Only increases in electricity demand appeared on the "cost" side of the warming ledger. Gains in wages, reduced fossil fuel consumption, lower housing and clothing expenses, and a slight savings in public expenditures appeared on the "benefit" side. The amount of warming examined by DOT is roughly equivalent to what the ground-based monitors suggest the planet has experienced over the last 100 years.

1986 EPA Study

Crafted mostly by internal staff (not one of whom had any economics training), the EPA produced few figures, and no quantitative estimates of costs or benefits, failed to even refer to the DOT study of only 12 years earlier, and was littered with qualifications like "could" and "might." While conceding that global warming would reduce mortality slightly, the report nonetheless concluded impressionistically that warming would probably cost the economy.

1991 Nordhaus Study

Perhaps the most prominent academic study of the economic consequences of warming was produced by Yale economist William Nordhaus, an informal adviser to the Clinton administration. Nordhaus calculates that a doubling of atmospheric carbon dioxide concentrations would cost the economy approximately $14.4 billion in 1990 dollars, or about 0.26% of national income. On the "cost" side, Nordhaus places increased electricity demand, loss of land due to flooding, coastal erosion, and the forced protection of various threatened seaboard properties. On the "benefits" side, Nordhaus places reductions in demand for non-electric heat. He concludes that agricultural implications are too uncertain to calculate but estimates that losses could be as great as $15 billion annually while gains could reach $14 billion annually. Finally, Nordhaus assumes that unmeasured impacts of warming could dwarf his calculations, so he arbitrarily quadruples his cost estimates to produce an estimate of warming costs somewhere around 1% of GDP.

1992 Cline Study

One of the most extensive treatments of the economic consequences of climate change and climate change abatement was produced by economist William Cline of the Institute for International Economics. Instead of assuming a median—4.5 degree Fahrenheit—estimate of warming a century hence (as all other studies tend to do), he assumes 18 degree Fahrenheit warming by 2300 and works back from there. Moreover, Cline includes an extremely low "social" discount rate to calculate the value of future investment. Despite all this, his preliminary calculations reveal that, for every $3 of benefits to be gained by emission restrictions, $4 of costs is incurred. Only by applying arbitrary adjustments after his initial calculations are performed does he find that the benefits of control exceed their cost; but that won't occur, even according to Cline, for at least a century.

Even more controversial are Cline's allocations of costs and benefits of warming. He finds no benefits whatsoever. Costs are found not only in the traditional places (sea level rise, species loss, and moderately increased hurricane activity) but also in areas where most economists have found benefits: agricultural productivity, forest yields, overall energy demand, and water demand. His net estimate is that, spread out over 300 years, the costs of warming will be approximately $62 billion annually.

Unfortunately, it is the Cline study that receives the lion's share of attention from the IPCC. The existence of contrary studies is often simply ignored in the document.

1997 Mendelsohn Study

Robert Mendelsohn of the Yale School of Forestry and Environmental Studies calculated late last year that a temperature hike of 2.5 degrees Celsius would lead to a net benefit of $37 billion for the U.S. economy. Farming, timber, and commercial energy sectors all benefit, with agriculture enjoying "a vast increase in supply from carbon fertilization."

1998 Moore Study

Economist Thomas Gale Moore of Stanford University might be termed the "anti-Cline." Whereas Cline has reported the steepest potential costs of warming, Moore's review of the literature this year, in addition to his own investigation, pegs net annual benefits of the median warming scenario at $105 billion. While Moore too, finds costs in species loss, sea level rise, increased hurricane activity, and increased tropospheric ozone pollution, he finds moderate benefits in agricultural productivity, forest yields, marine resource availability, and transportation. Moreover, he argues that major benefits will accrue from reduced energy demand, improved human morbidity, an increase from miscellaneous amenity benefits, lower construction costs, greater opportunities for leisure activities, and increased water supplies.

Historical Evidence

There is some historical precedent for optimism regarding the consequences of the median computer model warming scenario. The period 850 AD–1350 AD experienced a sharp and pronounced warming approximately equivalent to that predicted by the median warming scenario; 2.5 degrees Celsius. That period is known to climate historians as the Little Climate Optimum. While there were some climatic dislocations such as coastal flooding, there were marked increases in agricultural productivity, trade, human amenities, and measurable improvements in human morbidity and mortality.

Only when the climate cooled off at the end of the Little Climate Optimum did trade drop off, harvests fail, and morbidity and mortality rates jump largely due to an increase in diseases, particularly the plague.

The reason for optimism here is that human civilization was far more weather dependent a millennia ago than it is today. And even our more primitive, weather dependent ancestors appeared to do fairly well during their episodic warming.

Early Measures to Control Emissions Are Superior to Later Measures

Assuming even the worst about the consequences of unabated anthropogenic greenhouse gas emissions and their economic consequences does not necessarily imply that emissions controls today make more sense than emissions controls tomorrow.

There is no compelling need to act now. According to a recent study by Wigley et al. in *Nature*, waiting more than 20 years before taking action to limit anthropogenic greenhouse gas emissions would result in only about a .2 degree Celsius temperature increase spread out over a 100-year period.

Why might we want to wait a couple of decades before acting? First, we might profitably "look before we leap." There are a tremendous number of uncertainties that still need to be settled before we can be reasonably sure that action is warranted. Second, we can't anticipate what sorts of technological advances might occur in the intervening period that might allow far more efficient and less costly control or mitigation strategies than those before us today. Given the low cost of waiting, it would seem only prudent to continue to try to answer the open questions about climate change before making major changes to Western civilization.

Controlling Emissions Is Compatible With a Modern Industrialized Economy

The restrictions on greenhouse gas emissions agreed to in Kyoto are not in any way minor or insubstantial. Reducing U.S. emissions 7% below what they were in 1990 by the year 2012 means reducing emissions almost 40% below what they would be absent the agreement. Adjusted for expected population growth, this means a 50% reduction per capita in greenhouse gas emissions. Virtually everyone agrees that these targets can be met only by reducing fossil fuel consumption, the main source of virtually all anthropogenic emissions.

Environmentalists argue that such reductions can occur relatively painlessly, that we can cut the amount of fuel we use by 50% and actually produce even more economic growth as a result. Virtually no mainstream academic economist shares that opinion. The two most prominent and well respected academic specialists—Robert Stavins of the John F. Kennedy School of Government at Harvard and William Nordhaus of Yale—maintain that only the functional equivalent of a $150 per ton carbon tax can accomplish this, which they calculate would reduce GDP by 3%, or, as Stavins puts it, "approximately the cost of complying with all other environmental regulations combined." A recent survey in *Forbes* summarizing the recent macro-economic modeling that's been done on the subject broadly agrees with Stavins's and Nordhaus's estimates.

Then there is the matter of whether the emissions cuts presently on the table are even worth the bother. According to the best computer model from the National Center for Atmospheric Research, the Kyoto agreement, even if signed by all the nations of the world, would reduce global warming by an infinitesimal .18 degrees Celsius over the next 50 years. That's not much bang for the global warming buck.

The reason is that, according to all observers, actually stopping any further global warming from occurring (assuming the median predictions of present climate models) would require a 70% reduction of present emissions, roughly the equivalent of completely abandoning the use of fossil fuels. This, according to Jerry Mahlman, director of the Geophysical Fluid Dynamics Laboratory at Princeton, "might take another 30 Kyotos over the next century." Indeed, environmentalists are frequently quoted as saying that, ultimately, we will need to completely restructure society around the objective of energy efficiency and sustainability, the economic and political costs of which we can only imagine.

Unless we're prepared to see that journey to its completion, there's little point in even bothering to sign the Kyoto agreement because, in and of itself, it will make virtually no difference to our planetary climate.

Conclusion: A Matter of Perspective

Let me wind up my comments on a provocative note. We are constantly urged to act because "we shouldn't be gambling with our children's future." In fact, our kids are marshaled endlessly to shame us into planning for the worst... for their sake. But even assuming the absolute worst case about future planetary climate change and the most extreme estimates about what that climate change will ultimately cost society, conservative estimates are that our grandchildren 100 years hence will not be 4.4 times wealthier than we are—as they would be absent global warming—but will instead be only 3.9 times wealthier than we are at present.

I ask you, would you have been comfortable had your grandmother impoverished herself so that you could be 4.4 times wealthier than she rather than 3.9 times wealthier than she? Remember also that increased energy costs are borne most directly by the poor, who spend a greater portion of their income on energy than do the wealthy. Moreover, the poor who will pay the highest price of greenhouse gas abatement will be those in the developing world who will be denied the opportunity to better their lifestyle and standard of living. They will be "saved" from the fate of industrialization and experiencing even the most rudimentary comforts of Western consumer societies.

We're not really gambling with the lives of our grandchildren. They'll be just fine regardless of how the climate plays itself out. We're gambling with the lives of today's poor, who stand to lose the most if we act rashly. Thank you very much.

POSTSCRIPT

Are Aggressive International Efforts Needed to Slow Global Warming?

Taylor makes use of a common debating tactic. He begins by enumerating a set of propositions that he contends must be true to justify international action on climate warming. He then devotes most of the essay to demonstrating that little evidence exists to support these propositions. In addition to taking issue with Taylor's evidence and conclusions, most supporters of the need for a concerted effort to reduce greenhouse gas emissions would dispute the premise that proof of all of Taylor's propositions is required to justify their case. For example, since promoting the use of renewable energy sources as a means of reducing greenhouse gases would be environmentally beneficial, regardless of whether or not it reduced global warming, support for such actions does not require conclusive proof of any of Taylor's propositions.

Scientific acceptance of a phenomenon like global warming, which emerges from a background of normal climatic fluctuation, is a gradual process. If the predictions of greenhouse gas theorists are correct, then belief in the theory will grow as more evidence accumulates. In fact, during 1997 and 1998, continued above-normal worldwide temperatures and other phenomena, such as unusually severe storms, thinning glaciers, and lengthening of growing seasons, have increased scientific support for the position that the consequences of greenhouse gas warming are beginning to be observed. The current consensus is well documented in the March–April 1999 issue of *The Ecologist*, which was devoted to the global warming issue. For a discussion of several schemes for removing carbon dioxide from the atmosphere, see Howard Herzog, Baldur Eliasson, and Olav Kaarstad, "Capturing Greenhouse Gases," *Scientific American* (February 2000).

S. Fred Singer is a member of the group of scientists who dismiss the seriousness of greenhouse gas–induced global warming. Singer presents a succinct summary of his arguments in "Warming Theories Need Warning Labels," *The Bulletin of the Atomic Scientists* (June 1992). Some scientists who oppose efforts to reduce greenhouse gas emissions do so not because they dismiss the climate change threat but because they advocate the alternative of using one of several proposed "technological fixes." Physicist Gregory Benford makes such an argument in his article "Climate Controls," *Reason* (November 1997).

The 1997 international conference in Kyoto that developed a protocol for reducing greenhouse gas emissions stimulated the writing of numerous articles about all aspects of this issue. See, for example, "Kyoto and Beyond," by Robert M. White, and "Implementing the Kyoto Protocol," by Rob Coppock, both in the Spring 1998 issue of *Issues in Science and Technology*.

Are Major Changes Needed to Avert a Global Environmental Crisis?

YES: Chris Bright, from "Anticipating Environmental 'Surprise'," in Lester R. Brown et al., *State of the World 2000: A Worldwatch Institute Report on Progress Toward a Sustainable Society* (W. W. Norton, 2000)

NO: Julian L. Simon, from "More People, Greater Wealth, More Resources, Healthier Environment," *Economic Affairs* (April 1994)

ISSUE SUMMARY

YES: Chris Bright argues that human impacts on the environment are so extensive that we face an era of catastrophic surprises unless we learn to think of the world as a complex system and behave accordingly.

NO: The late professor of economics and business administration Julian L. Simon predicts that over the long term, the brainpower of more people coupled with the market forces of a free economy will lead to improved standards of living and a healthier environment.

In 1972 the results of a study by a Massachusetts Institute of Technology computer modeling team triggered an avalanche of controversy about the future course of worldwide economic growth. The results appeared in a book entitled *The Limits to Growth* (Universe Books, 1972). The book's authors—Donella Meadows, Dennis Meadows, Jorgen Randers, and William Behrens—predicted that exponential growth in population and capital, accompanied by increasing pollution, would culminate in sudden resource depletion and economic collapse before the middle of the next century. The sponsors of the study, a group of rich European and American industrialists called the Club of Rome, popularized its conclusions by distributing 12,000 copies of the book to prominent government, business, and labor leaders. In 1992 the Meadows team published *Beyond the Limits: Confronting Global Collapse, Envisioning a Sustainable Future* (Chelsea Green), a sequel to the earlier report that was based on much-improved computer models. This book presents an even more pessimistic picture of the future.

Critiques of the study emerged from all sectors of the political spectrum. Conservatives rejected the implication that international controls on industrial development were necessary to prevent disaster. Liberals asserted that no-growth policies would hurt the poor more than the affluent. Radicals contended that the results were only applicable to the type of profit-motivated growth that occurs under capitalism. Among the universal criticisms of the study were the simplicity of the computer models used and the questionable practice of making long-term extrapolations based on present trends.

Although the debate about the specific catastrophic predictions of Meadows et al. has died down, the questions raised during that controversy continue to receive attention. In 1980 a three-volume publication entitled *The Global 2000 Report to the President* was released by the U.S. government. This report, which has sold over 500,000 copies, is the result of a study of trends in population growth, natural resource development, and environmental quality through the end of the twentieth century. The projections of this study include increased environmental degradation, continued abuse of natural resources, and a widening of the gap between the rich and the poor. As we enter the new millennium, most environmental analysts would agree that these predictions were accurate.

The *Global 2000 Report* has contributed to the widely held view that present patterns and rates of worldwide industrial growth are likely to cause intolerable environmental stress. This issue, the potential for conflict between the need for development and the need for environmental protection, was the central focus of the 1992 UN Earth Summit in Rio de Janeiro, which, in turn, was organized in response to a recommendation of the World Commission on Environment and Development, established by the UN in 1983 to produce a "global agenda for change." The concept of sustainable development, which requires a fundamental change in the technologies used by the world's economies in order to meet their energy, transportation, agricultural, and industrial production needs, has received increasing attention in the aftermath of the Rio meeting. Recently, several prominent ecological economists have proposed that continued economic growth and sustainable development are incompatible in a world with finite resources. They suggest that true sustainability requires development without net growth. Niles Eldredge, in *Life in the Balance: Humanity and the Biodiversity Crisis* (Princeton University Press, 1998), warns that without dramatic changes in industrial and agricultural activities, population growth, and economics, environmental problems will threaten the future of humanity.

In the first selection, Chris Bright states that simple straight-line extrapolation of past trends is not to be trusted, for the world is a complexly interwoven structure, human activities can have unforeseen and severe effects, and such effects are already upon us. Change in human activity and attitude is essential. In the second selection, Julian L. Simon asserts that there is no evidence that environmental degradation, health problems, or world hunger are increasing. He predicts that the world can look forward to an improved standard of living and a cleaner environment unless governments restrict the free-market trends.

Chris Bright

Anticipating Environmental "Surprise"

If there is any comfort to be found in the prospect of environmental decline, it lies in the idea that the process is gradual and predictable. All sorts of soothing cliches follow from this notion: Even if we have not turned the trends around, our children will rise to the challenge. There's time. We're constantly learning; you can see plenty of progress already.

But this way of thinking is sleepwalking. To understand why, you have to look at decline close up. Here, for instance, is how it has happened in one small country, with big implications. Honduras, in the early 1970s, was caught up in a drive to build agricultural exports. Landowners in the south increased their production of cattle, sugarcane, and cotton. This more intensive farming reduced the soil's water absorbency, so more and more rain ran off the fields and less remained to evaporate back into the air. The drier air reduced cloud cover and rainfall. The region grew warmer—a lot warmer. The local weather station recorded an increase in the median annual temperature of 7.5 degrees Celsius between 1972 and 1990, by which time it had exceeded 30 degrees.

The hotter, drier landscape was poor habitat for the kind of mosquitoes that carry malaria, so the mosquitoes largely died off and malaria infection declined. But of course the land was also becoming less productive, so people began to leave. Many found work on big plantations that were being carved out of the rainforests to the north. The plantations were growing export crops too, primarily bananas, melons, and pineapples. But it is difficult to mass-produce big, succulent fruits in a rainforest—even a badly fragmented rainforest—because there are so many insects and fungi around to eat them. So the plantations came to rely heavily on pesticides. From 1989 to 1991, Honduran pesticide imports increased more than fivefold, to about 8,000 tons.

This steaming, ragged forest was perfect habitat for malaria mosquitoes. Around the plantations, the insecticide drizzle suppressed them for a time, but they eventually acquired resistance to a whole spectrum of chemicals, and that basically released them from human control. When their populations bounced back, they encountered a landscape stocked with their favorite prey: people. And since these people were from an area where malaria infection had become rare, their immunity to the disease was low. Malaria rapidly reasserted itself: from 1987 to 1993, the number of cases in Honduras jumped from 20,000 to 90,000.

From Chris Bright, "Anticipating Environmental 'Surprise'," in Lester R. Brown et al., *State of the World 2000: A Worldwatch Institute Report on Progress Toward a Sustainable Society* (W. W. Norton, 2000). Copyright © 2000 by The Worldwatch Institute. Reprinted by permission of W. W. Norton & Company, Inc. Notes omitted.

The situation was brought to light in 1993 by a group of researchers concerned about the public health implications of environmental decline. But their primary interest was not in what had already happened—it was in what might happen next. Some very nasty surprises might be tangled up somewhere in this web of pressures. They argued, for example, that deforestation and changing patterns of disease had made the country "especially vulnerable to climatic change and climate instability."

They were right. In October 1998, Hurricane Mitch slammed into the Gulf coast of Central America and stalled there for four days. Nightmarish mudslides obliterated entire villages; half the population of Honduras was displaced and the country lost 95 percent of its agricultural production. Mitch was the fourth strongest hurricane to enter the Caribbean this century, but much of the damage was caused by deforestation: had forests been gripping the soil on those hills, fewer villages would have been buried in mudslides. And in the chaos and filth of Mitch's wake there followed tens of thousands of additional cases of malaria, cholera, and dengue fever.

It is hard to shake the feeling that "normal change"—even change for the worse—should not happen this way. In the first place, too many trends in this scenario are spiking. Instead of gradual change, the picture is full of discontinuities—very rapid shifts that are much harder to anticipate. There is a rapid warming in the south, then an abrupt expansion in deforestation in the north, as plantations are developed. Then malaria infections jump. Then those mudslides, in addition to killing thousands of people, cause a huge increase in the rate of topsoil loss.

There also seem to be too many overlapping pressures—too many synergisms. The mudslides were not the work of Mitch alone; they were caused by Mitch plus the social conditions that encouraged the farming of upland forests. The malaria emerged not just from the mosquitoes, but from the movement of a low-immunity population into a mosquito-infested area, and from heavy pesticide use.

Such discontinuities and synergisms frequently catch us by surprise. They tend to subvert our sense of the world because we so often assume that a trend can be understood in isolation. It is tempting, for example, to believe that a smooth line on a graph can be used to see into the future: all you have to do is extend the line. But the future of a trend—any trend—depends on the behavior of the entire system in which it is embedded. When we isolate a phenomenon in order to study it, we may actually be preventing ourselves from knowing the most important things about it.

This fragmented form of inquiry is becoming increasingly dangerous—and not just because we might miss problems in small, poor countries like Honduras. After all, there is nothing special about the pressures in the Honduran predicament. Deforestation, climate change, chemical contamination—these and many other forms of environmental corrosion are at work on a global scale. Each has engendered its own minor research industry. But even as the publications pile up, we may actually be missing the biggest problem of all: what might the inevitable convergence of these forces do?

"When one problem combines with another problem, the outcome may be not a double problem, but a super-problem." That is the assessment of Norman Myers, an Oxford-based ecologist who is one of the most active pioneers in the field of environmental surprise. We have hardly begun to identify those potential super-problems, but in the planet's increasingly stressed natural systems, the possibility of rapid, unexpected change is pervasive and growing. . . .

Tropical Rainforests: The Inferno Beneath the Canopy

Eight thousand years ago, before people began to clear land on a broad scale, more than 6 billion hectares, or around 40 percent of the planet's land surface, were covered with forest. Today, Earth's tattered cloak of natural forests (as opposed to tree plantations) amounts to 3.6 billion hectares at most. Every year, at least another 14 million hectares are lost—and maybe considerably more than that. This is an enormous evolutionary tragedy. Among the many thousands of species that are believed to go extinct every year, the overwhelming majority are forest creatures, primarily tropical insects, who have been denied their habitat. That, anyway, is the best estimate, but the forests are vanishing far more rapidly than they can be studied. We really don't even know what we are losing.

Currently, well over 90 percent of forest loss is occurring in the tropics— on a scale so vast that it might appear to have exceeded its capacity to surprise us. In 1997 and 1998, fires set to clear land in Amazonia claimed more than 5.2 million hectares of Brazilian forest, brush, and savanna—an area nearly 1.5 times the size of Taiwan. In Indonesia, some 2 million hectares of forest were torched during 1997 and 1998. All this is certainly news, but if you are interested in conservation, it is the kind of dreadful news you have come to expect.

And yet our expectations may not be an adequate guide to the sequel, assuming the destruction continues at its current pace. A substantial portion of the damage is "hidden"—it does not show up in the conventional analysis. But once you take the full extent of the damage into account, you can begin to make out some of the surprises it is likely to trigger.

Consider, for example, the destruction of Amazonia. Over half the world's remaining tropical rainforest lies within the Amazon basin, where more forest is being lost than anywhere else on Earth. Deforestation statistics for the area are intended primarily to track the conversion of forest into farms and ranches. Typically, the process begins with the construction of a road, which opens up a new tract of forest to settlement. In June 1997, for instance, some 6 million hectares of forest were officially released for settlement along a major new highway, BR-174, which runs from Manaus, in central Amazonia, over 1,000 kilometers north to Venezuela. Ranchers and subsistence farmers clearcut patches of forest along the road and burn the slash during the July-November dry season. (The farmers generally have few other options: Brazil has large numbers of poor, land-hungry people, and the plots they cut from the forest usually lose their fertility rapidly, so there is a constant demand for fresh soil.)

But the damage to the forest generally extends much farther than the areas that are "deforested" in this conventional sense, because of the way fire works in

Amazonia. In the past, major fires have not been a frequent enough occurrence to promote any kind of adaptive "fire proofing" in the region's dominant tree species. Some temperate-zone and northern trees, by contrast, are "fire-adapted" in one way or another—they may have especially thick bark, for example, or the ability to resprout after burning. The lack of such adaptations in Amazonian trees means that even a small fire can begin to unravel the forest.

During the burning season, the flames often escape the cuts and sneak into neighboring forest. Even in intact forest, there will be patches of forest floor that are dry enough at that time of year to allow a small "surface fire" to feed on the dead leaves. Surface fires do not climb trees and become crown fires. They just crackle along the forest floor, here and there, as little patches of flame, going out at night, when the temperature drops, and rekindling the next day. They will not kill the really big trees, and they do not cover every bit of ground in a burned patch. But they are fatal to most of the smaller trees they touch. Overall, an initial surface fire may kill perhaps 10 percent of the living forest biomass.

The damage may not look all that dramatic, but another tract of forest may already be doomed by an incipient positive feedback loop of fire and drying. After a surface fire, the amount of shade is reduced from about 90 percent to around 60 percent, and the dead and injured trees rain debris down on the floor. So a year or two later, the next fire in that spot finds more tinder, and a warmer, drier floor. Some 40 percent of forest biomass may die in the second fire. At this point, the forest's integrity is seriously damaged; grasses and vines invade and contribute to the accumulation of combustible material. The next dry season may eliminate the forest entirely. Once the original forest is gone, the scrubby second growth or pasture that replaces it will almost certainly burn too frequently to allow the forest to restore itself.

New roads admit not just settlers and ranchers but loggers as well. Commercial logging involves a form of damage that is in some ways similar to the surface fires. Unlike its temperate-zone counterparts, the Amazonian timber industry does not generally clearcut. Most of its operations are what foresters call "high grading"—the best specimens of the most desirable species are cut and hauled out. The result is not outright deforestation, but the forest loses its largest trees and suffers extensive collateral damage from the felling and hauling. An intricate network of logging roads undermines the canopy—a human termite's nest through the wood. An Amazonian timber operation typically kills 10–40 percent of the forest biomass, thereby reducing canopy coverage by up to 50 percent. The forest floor grows warmer and drier; the forest becomes increasingly flammable.

As the main mechanisms of deforestation, the logging and burning are hardly surprising per se, but a great deal of the resulting damage is still obscure. Deforestation estimates are derived primarily from satellite photos, and while the photos show a great deal of detail, the mapping process is generally designed to give a picture of gross canopy destruction. Damage that leaves most of the canopy intact, or that has been masked by second growth, it usually not factored in to the estimates.

In a recent survey, researchers cross-checked satellite maps with field observations and concluded that conventional deforestation estimates for Brazilian Amazonia were missing some 1–1.5 million hectares of severe forest damage done by logging every year. Surface fire damage is harder to quantify, but the same researchers did a fire survey and found that the amount of standing forest that had suffered a surface fire in 1994 and 1995 was 1.5 times the area fully deforested in those years. Overall, they suggested, the area of Amazonian forest attacked by surface fire every year may be roughly equivalent to the area deforested outright. And in some parts of the basin, the extent of this cryptic damage is so great that the conventional measurements may no longer be all that useful. In one region, around Paragominas in eastern Amazonia, the researchers found that although 62 percent of the land was classified as forested, only about one tenth of this consisted of undisturbed forest.

Apart from these direct losses, the logging and burning are likely to trigger various forms of second-order damage through fragmentation. Cutting the forest up into smaller and smaller pieces renders the surviving tracts increasingly vulnerable to "edge effects." Near the edge of a major clearing, competing vegetation often invades forest, choking out saplings. Higher winds dry out the soil and sometimes topple trees. In Amazonia, these effects may extend a kilometer from the edge itself.

As the Amazonian forest dwindles, a more surprising second-order effect may emerge as the hydrological cycle changes. Because trees exhale so much water vapor, a forest to some degree creates its own climate. Much of this water vapor condenses out below the canopy and drips back into the soil. Some of it rises higher before falling back in as rain—researchers estimate that most of the Amazon's rainfall comes from water vapor exhaled by the forest. Widespread deforestation will therefore tend to make the region substantially drier, and that will accelerate the feedback loop created by the fires.

Some degree of deforestation-induced drought already appears to have affected other parts of the humid tropics—parts of Central America, for instance, Côte d'Ivoire, and peninsular Malaysia, where the drying has been severe enough to force the abandonment of some 20,000 hectares of rice paddy. It may not be possible to define the point at which such a drought takes hold in Amazonia, but about 13 percent of the Brazilian Amazon has now been deforested outright. If you add to that the tracts of forest that have been seriously degraded—by logging, surface fires, and fragmentation—that fraction could rise to more than a third.

The fire feedback loop is also likely to gain momentum from forces outside the region. Over the past two decades, Amazonia has seen several unusually intense dry seasons, during which the burning was far worse than normal. These periods correspond with recent El Niño weather events (in 1982–83, 1992–93, and 1997–98). El Niño events appear to be growing longer, more intense, and more frequent. Many climatologists regard this trend as a likely effect of climate change—the change in the behavior of Earth's climate system caused by increasing atmospheric concentrations of carbon dioxide (CO_2) and other greenhouse gases. Climate change, in other words, may be accelerating the Amazonian fire cycle. By burning large amounts of coal and oil, the United States, China, and other major carbon-emitting countries may in effect be burning the Amazon.

Other kinds of surprises are lurking in tropical forests as well. As developing countries industrialize, some forest maladies better known in the industrial world are likely to appear in these countries too. Acid rain, for example, is already reported to be affecting the forests of southern China. In parts of South Asia, Indonesia, South America, and West Africa, this form of pollution is bound to increase as industrialization proceeds and cities enlarge. The soil in these areas tends to be fairly acidic already, which would make it incapable of buffering large doses of additional acid. At least in some of these places, acid-induced decline may therefore be much more abrupt than in the temperate zone....

The Atmosphere: An Invisible Confluence of Poisons

The delicate membrane of gases that makes up our atmosphere is as thin, comparatively speaking, as the skin of an onion. The atmosphere's outer border is very diffuse—a faint scattering of gas molecules extends into space for hundreds of kilometers—but 90 percent of those molecules lie within 16 kilometers of sea level. Every ecosystem on Earth is linked to the chemistry of the membrane that separates us from outer space, and that chemistry is changing in many ways. Levels of some "trace gases," such as sulfur dioxide, nitrogen oxides, and carbon dioxide, are increasing. Many novel compounds have entered the brew as well—for instance, chlorofluorocarbons, the ozone-destroying chemicals used as refrigerants. From this immense potential for change, look at just three basic phenomena: acid raid, nitrogen pollution, and increasing levels of CO_2.

Fossil fuel combustion is the source of acid rain (which falls not just as rain but also as dry particles). Acid rain is composed in large measure of sulfuric acid, which derives from the sulfur dioxide released by coal-burning power plants and metal smelters. (Sulfer is a common contaminant of coal and metal ores.) Smoke stack "scrubbers" and a growing preference for low-sulfur coal and natural gas have helped reduce sulfur dioxide emissions in North America and Western Europe, although high sulfur emissions are still common elsewhere. The other primary constituent of acid rain is nitric acid, which is generated from the nitrogen oxides in fossil fuel emissions. Unfortunately, nitric acid is likely to be more difficult to control than sulfuric acid, since a substantial share of the nitrogen oxides comes from gasoline burned in the world's expanding fleets of cars.

Acid rain can travel downwind for hundreds of kilometers—then fall on forests and farmland, where the idea of air pollution may seem quite incongruous. The acid can damage plant tissues directly, but its worst effects come as a series of discontinuities that are much harder to see. As the acid drips into the soil, decade after decade, it tends to leach out the stock of calcium and magnesium, both essential plant nutrients. Depending upon how nutrient-rich a soil is to begin with, this process may or may not be an immediate concern, but if it persists, the nutrient decline will eventually cross a threshold of scarcity: it will begin to cripple plant growth.

A second discontinuity will occur once soil calcium has grown scarce. Without calcium to neutralize it, the incoming acid will just build up in the soil —soil acidification will increase abruptly, even if the amount of incoming acid remains constant. The growing acidity will work another change, by releasing aluminum from its mineral matrix. Aluminum is a common soil constituent; when bonded to other minerals, it is biologically inert, but free aluminum in acid conditions is toxic to both plants and animals. In plants, acidity plus aluminum damages the fine roots. That could affect water uptake, and thereby increase susceptibility to drought. Root damage will also cripple a plant's ability to absorb whatever nutrients remain in the soil.

Acid lowers the calcium level, which allows the acid to build up, which releases aluminum, which interferes with calcium uptake. There is a cascade of chemical effects here, reinforcing the nutrient starvation. Other kinds of second-order effects may emerge as well. In the United States, for example, recent research over a swath of the Midwest from southern Illinois to southern Ohio has uncovered a correlation between increasing acidity in forest soils and a decline of soil organisms—earthworms, beetles, and so on. As biological activity in these soils has dropped off, decay of woody debris appears to have slowed radically, so the calcium "locked up" in the dead wood is not being released back into the soil. The nutrient cycle has apparently been constricted to some degree.

Because they are chronic, these changes in soil chemistry also create many opportunities for synergisms. Aluminum, for example, is not the only metal that acid tends to "mobilize." Toxic heavy metals such as cadmium, lead, and mercury may also be present in the soil, or they may arrive in trace amounts, on the same winds that brought in the acid. (Like sulfur, heavy metals are common contaminants of coal and ores, although usually in much smaller amounts.) Increasing acidity will tend to make these metals more soluble and toxic as well.

Even though it involves discontinuities, acid-induced decline may still unfold for decades as a hidden process that largely escapes casual notice. Does a tract of forest have fewer large trees than it once did, or fewer species that need more alkaline soil? Even if it does, it may still be perfectly green. It may still show vigorous growth, but the growth may be concentrated in younger plants, and in acid-tolerant species. It may be on its way to becoming a kind of "acid thicket."

In the eastern United States, over a large portion of Appalachia, the death rate of oaks appears to have doubled and that of hickories to have nearly tripled from 1960 to 1990; a recent review found a strong correlation between these declines and exposure to acid rain and ozone pollution. In the mountains of New Hampshire, at the Hubbard Brook Experimental Forest, acid-induced leaching of minerals has been identified as the main reason vegetation there has shown virtually no net growth since the mid-1980s. Here and there throughout the U.S. Northeast, acidity may be a factor in the failure of the sugar maple, one of the region's dominant tree species, to generate new seedlings. Acid rain is a threat elsewhere in the country as well—in the southern Appalachians, for example, and in the mountains of Colorado.

The worst acid rain, however, is in Asia, particularly in China, which gets 73 percent of its energy from burning coal. The vast quantities of sulfur dioxide released in the process are reportedly now affecting some 270 million hectares of land—more than a quarter of the country's land mass. The acid is reaching Japan and the Koreas as well; Japan, for instance, currently receives more than a third of its sulfur deposition from China. In a recent study, scientists built a computer model of China's energy development and concluded that if the country does not curb its appetite for coal, then over the next two decades acid rain could overwhelm many of the region's soils.

The study is not wholly pessimistic: under its best case scenario, state-of-the-art pollution controls are installed at all of China's coal-burning power plants and factories, causing sulfur dioxide emissions to fall to 31 percent of their 1990 value by 2020. But under a more realistic scenario—with state-of-the-art pollution control installed only in new power plants and some fuel-switching to cleaner energy—sulfur dioxide emissions actually increase by 40 percent over the same time frame. China's coal use is therefore an invitation to widespread discontinuity—not just on an ecosystem level, but also, because of its potential for poisoning cropland, on a social level as well.

Acid rain overlaps with another, broader form of global change: the alteration of the planet's "nitrogen cycle." Nitrogen is an essential plant nutrient and the main constituent of the atmosphere: 78 percent of the air is nitrogen gas. But plants cannot metabolize this pure, elemental nitrogen directly. The nitrogen must be "fixed" into compounds with hydrogen or oxygen before it can become part of the biological cycle. In nature this process is accomplished by certain types of algae and bacteria, as well as by lightning strikes, which fuse atmospheric oxygen and nitrogen into nitrogen oxides.

Humans have radically amplified this process. Farmers boost the nitrogen level of their land through fertilizers and the planting of nitrogen-fixing crops (actually, symbiotic microbes do the fixing). The burning of forests and the draining of wetlands releases additional quantities of fixed nitrogen that had been stored in vegetation and organic debris. And fossil fuel combustion produces still more fixed nitrogen, in the form of nitrogen oxides. Natural processes probably incorporate around 140 million tons of nitrogen into the terrestrial nitrogen cycle every year. (The ocean cycle is largely a mystery.) Thus far, human activity has at least doubled that amount.

Fixed nitrogen is often a "limiting nutrient" in terrestrial ecosystems: it is in high demand and relatively short supply, so its availability determines the amount of plant growth. If you add more, you get more growth—at least in some species. That is why fertilizer is mostly nitrogen. But if you keep adding more, you run into trouble. The excess nitrogen becomes a kind of poison that may interact synergistically with acid rain and other pressures. (Of course, since nitrogen compounds often contribute to the acidity, the processes are not entirely distinct.)

In forests, for example, excess nitrogen tends to inhibit fine root growth, just as the acid-aluminum combination does. Acidity plus nitrogen pollution could therefore deal a double blow to trees' ability to withstand drought and to take in calcium and magnesium. Above ground, excess nitrogen may stimulate extra growth, but it is likely to be produced faster than the tree can absorb those mineral nutrients. The new growth will therefore tend to be weak —essentially, malnourished. This effect also can be exacerbated by acid rain, since acid leaches out those minerals in the first place.

The weakness of the new growth is not just physical—there can be chemical weaknesses too. Since nitrogen may also be a limiting nutrient for insects and other small organisms that feed on trees, nitrogen-rich foliage is likely to be very attractive to pests. And the low mineral concentrations within the tree's tissues can interfere with its ability to produce the chemicals that make up its "immune system"—compounds that, for example, inhibit infection or make foliage less palatable to insects. Other physiological effects are probably at work as well; excess nitrogen, for instance, appears to lower a tree's ability to cope with cold weather. Combinations of various effects like these may eventually produce substantial discontinuities. In one monitoring experiment in the U.S. Northeast, researchers found that by increasing the nitrogen in pine and spruce-fir stands, they induced declines in growth and increases in tree mortality over a period of just six years.

Nor is it just forests that are at risk from excess nitrogen. In prairies and heaths, too much nitrogen can favor the terrestrial equivalent of algae—whatever fast-growing, weedy species happen to be present—at the expense of slower-growing species that do not have the adaptations necessary to use the extra fertilizer. Several field experiments have shown that this process can have a dramatic homogenizing effect. In one such study, a highly diverse prairie in Minnesota dissolved into a luxuriant patch of fast-growing, aggressive grass. The main beneficiaries of this process are often likely to be non-native "exotic" species, since the ability to grow fast (and therefore to capitalize on any extra nitrogen) will tend to make a plant a good invader in the first place. In this way, nitrogen pollution could converge synergistically with bioinvasion, the spread of exotic species.

Now factor in climate change. Although the processes of climate change are too complex to permit accurate prediction of local effects, the higher latitudes are generally expected to warm much more than the tropics. In the north, the warming is likely to proceed faster than forests can respond by "migrating" further north (where the soil and water conditions permit such movement). Unless carbon emissions are reduced, the result is likely to be substantial forest decline. The immediate causes of decline will likely vary from place to place, but will often involve drying and changes in the freeze-thaw regime during winter and early spring. (Such changes can cause trees to start growing too early in the season.) Both types of change invite an overlap with acid rain and nitrogen pollution, which can make trees less drought- and frost-tolerant. The biggest potential for such an overlap may be in Siberia, where air pollution has degraded vast areas and destroyed some 1 million hectares of forest outright.

In northern forests, unusually warm years often provoke massive defoliation from insects. Recent warming in Alaska, for example, apparently underlies a spruce budworm attack that had chewed up some 20 million hectares of forest by the end of 1998. In parts of northern Europe, southeastern Canada, and the

U.S. Northeast, any such climate-driven insect response could combine with the tendency of nitrogen pollution to promote insect damage.

The warming may trigger less direct stresses as well. In the north, climate change is likely to weaken the stratospheric ozone layer, thereby allowing more harmful ultraviolet radiation to reach Earth's surface. (Greenhouse gases are keeping more heat in the lower atmosphere, so the stratosphere is cooling, and air currents are likely to exacerbate that effect in the north. Stratospheric cooling affects the ozone layer because ozone-destroying substances are more effective at lower temperatures.) The extra ultraviolet radiation will damage the foliage of many trees—another overlap with air pollution stress. Chronic damage to foliage slows growth and tends to increase susceptibility to other pressures, such as drought and pests.

Drought stress, pollution, insect attack, ultraviolet light—the critical issue here is not whether any particular synergism will occur, it is the increase in the aggregate risk of a major surprise. As the pressures build, so does the chance of triggering some unanticipated "super-problem."

An Agenda for the Unexpected

Human pressures on Earth's natural systems have reached a point at which they are more and more likely to engender problems that we are less and less likely to anticipate. Dealing with this predicament is obviously going to require more than simply reacting to problems as they appear. We need to forge a new ethic for managing our relationship with nature—one that emphasizes minimal interference in the lives of wild beings and in the broad natural processes that sustain all living things. Such an ethic might begin with three basic principles.

First, nature is a system of unfathomable complexity. Our predominant response to that complexity has been specialization, in both the sciences and public policy. Learning a lot about a little is a form of progress, but it comes at a cost. Expertise is seductive: it is easy for specialists to get into the habit of thinking that they understand all the consequences of a plan. But in a complex, highly stressed system, the biggest consequences may not emerge where the experts are in the habit of looking. This inherent unpredictability condemns us to some degree of error, so it is important to err on the side of minimal disruption whenever possible.

Second, nature gives away nothing for free. You cannot get an appreciable quantity of anything out of nature without sacrificing something in the process. Even sustainable resources management is a trade-off—it's simply one we regard as acceptable. In our dealings with nature, as with any other sort of transaction, we need to know the full cost of the goods before deciding whether they are worth the price, or whether there is a better way to pay for them.

Third, nature has no reset button. Environmental corrosion is not just killing off individual species—it is setting off system-level changes that are, for all practical purposes, irreversible. Even if, for example, all the world's coral reef species were miraculously to survive the impending bout of rapid climate change, that does not mean that our descendents will be able to reconstruct reef communities. The near impossibility of restoring complex systems to some previous state is another strong argument for minimal disruption.

These are basic features of the natural world: we will never understand it completely, it will not do our bidding for free, and we cannot put it back the way it was....

Solutions are almost never permanent, so plan to keep on planning. In the 1950s, organochlorine pesticides were hailed as a permanent "fix" for insect pest problems; given the pervasive ecological damage that these chemicals are now known to cause, the idea of a permanent chemical solution to anything may seem rather naive today. But because our relationship with nature is in a constant state of flux, even realistic fixes will need regular revision. The Montreal Protocol is not a permanent patch for the ozone layer, in part because climate change will probably exacerbate ozone loss. The Green Revolution is not a permanent answer to world hunger, in part because conventional agriculture is overtaxing aquifers. The growing strain on Earth's natural systems will probably force an increase in the tempo of policy revision—so it makes sense to take full advantage of the powerful new information and communications technologies. Because of their ability to bring together enormous quantities of data from different areas and disciplines, such technologies could help counter the blinkering effects of specialization.

None of us may find the answer alone, but together we probably can. In social as well as natural systems, there is a potent class of properties that exists only on the system level—properties that cannot be directly attributed to any particular component. In a political system, for example, institutional pluralism can create a public space that no single institution could have created alone. One of the most important policy activities may therefore be to encourage innovation outside policy institutions. Policy may need to become increasingly a matter of creating not so much solutions per se as the conditions from which solutions can arise. In the face of the unexpected, our best hopes may lie in our collective imagination.

NO ↩

Julian L. Simon

More People, Greater Wealth, More Resources, Healthier Environment

This is the economic history of humanity in a nutshell. From 2 million or 200,000 or 20,000 or 2,000 years ago until the 18th century there was slow growth in population, almost no increase in health or decrease in mortality, slow growth in the availability of natural resources (but not increased scarcity), increase in wealth for a few, and mixed effects on the environment. Since then there has been rapid growth in population due to spectacular decreases in the death rate, rapid growth in resources, widespread increases in wealth, and an unprecedently clean and beautiful living environment in many parts of the world along with a degraded environment in the poor and socialist parts of the world.

That is, more people and more wealth have correlated with more (rather than less) resources and a cleaner environment—just the opposite of what Malthusian theory leads one to believe. The task before us is to make sense of these mind-boggling happy trends.

The current gloom-and-doom about a 'crisis' of our environment is wrong on the scientific facts. Even the US Environmental Protection Agency acknowledges that US air and water have been getting cleaner rather than dirtier in the past few decades. Every agricultural economist knows that the world's population has been eating ever-better since the Second World War.

Every resource economist knows that all natural resources have been getting more available not more scarce, as shown by their falling prices over the decades and centuries. And every demographer knows that the death rate has been falling all over the world—life expectancy almost tripling in the rich countries in the past two centuries, and almost doubling in the poor countries in only the past four decades.

Population Growth and Economic Development

The picture is now also clear that population growth does not hinder economic development. In the 1980s there was a complete reversal in the consensus of thinking of population economists about the effects of more people. In 1986,

From Julian L. Simon, "More People, Greater Wealth, More Resources, Healthier Environment," *Economic Affairs* (April 1994). Copyright © 1994 by Julian L. Simon. Reprinted by permission.

the National Research Council and the National Academy of Sciences completely overturned its 'official' view away from the earlier worried view expressed in 1971. It noted the absence of any statistical evidence of a negative connection between population increase and economic growth. And it said that 'The scarcity of exhaustible resources is at most a minor restraint on economic growth'. This U-turn by the scientific consensus of experts on the subject has gone unacknowledged by the press, the anti-natalist [anti-birth] environmental organisations, and the agencies that foster population control abroad.

Long-Run Trends Positive

Here is my central assertion: Almost every economic and social change or trend points in a positive direction, as long as we view the matter over a reasonably long period of time.

For a proper understanding of the important aspects of an economy we should look at the long-run trends. But the short-run comparisons—between the sexes, age groups, races, political groups, which are usually purely relative —make more news. To repeat, just about every important long-run measure of human welfare shows improvement over the decades and centuries, in the United States as well as in the rest of the world. And there is no persuasive reason to believe that these trends will not continue indefinitely.

Would I bet on it? For sure. I'll bet a week's or month's pay—anything I win goes to pay for more research—that just about any trend pertaining to material human welfare will improve rather than get worse. You pick the comparison and the year.

Let me quickly review a few data on how human life has been doing, beginning with the all-important issue, life itself.

The Conquest of Too-Early Death

The most important and amazing demographic fact—the greatest human achievement in history, in my view—is the decrease in the world's death rate. Figure 1 portrays the history of human life expectancy at birth. It took thousands of years to increase life expectancy at birth from just over 20 years to the high twenties in about 1750. Then life expectancy in the richest countries suddenly took off and tripled in about two centuries. In just the past two centuries, the length of life you could expect for your baby or yourself in the advanced countries jumped from less than 30 years to perhaps 75 years. What greater event has humanity witnessed than this conquest of premature death in the rich countries? It is this decrease in the death rate that is the cause of there being a larger world population nowadays than in former times.

Then starting well after the Second World War, the length of life you could expect in the poor countries has leaped upwards by perhaps 15 or even 20 years since the 1950s, caused by advances in agriculture, sanitation, and medicine (Figure 2).

Let me put it differently. In the 19th century the planet Earth could sustain only 1 billion people. Ten thousand years ago, only 4 million could keep

Figure 1

History of Human Life Expectancy at Birth (3000BCE–2000CE)

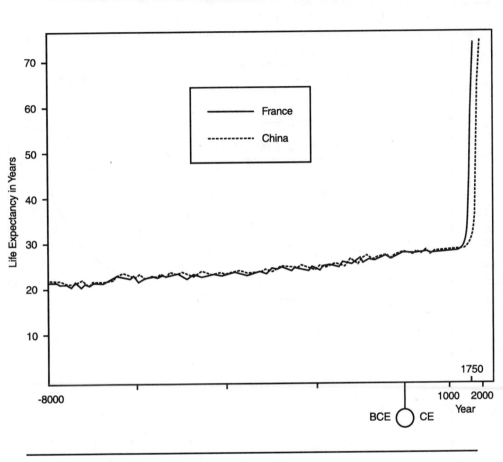

BCE: Before the Christian Era.

themselves alive. Now 5 billion people are on average living longer and more healthily than ever before. The increase in the world's population represents our victory over death.

Here arises a crucial issue of interpretation: One would expect lovers of humanity to jump with joy at this triumph of human mind and organisation over the raw killing forces of nature. Instead, many lament that there are so many people alive to enjoy the gift of life. And it is this worry that leads them to approve the Indonesian, Chinese and other inhumane programmes of coercion and denial of personal liberty in one of the most precious choices a family can make—the number of children that it wishes to bear and raise.

Figure 2

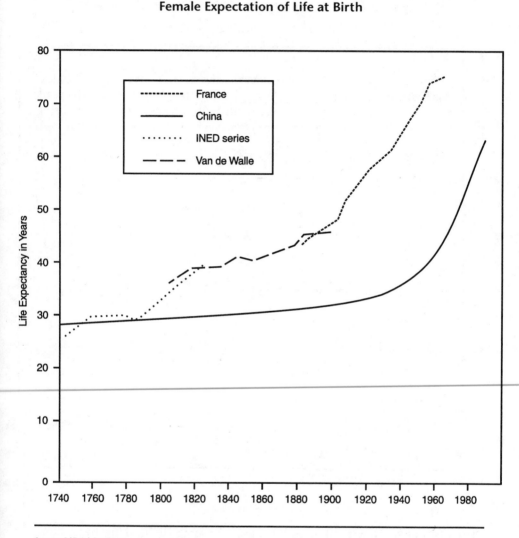

Female Expectation of Life at Birth

Legend: France, China, INED series, Van de Walle

Source: Official Statistics

The Decreasing Scarcity of Natural Resources

Throughout history, the supply of natural resources has worried people. Yet the data clearly show that natural resource scarcity—as measured by the economically-meaningful indicator of cost or price—has been decreasing rather than increasing in the long run for all raw materials, with only temporary exceptions from time to time: that is, availability has been increasing. Consider copper, which is representative of all the metals. In Figure 3 we see the price relative to wages since 1801. The cost of a ton is only about a tenth now of what it was two hundred years ago.

This trend of falling prices of copper has been going on for a very long time. In the 18th century BCE in Babylonia under Hammurabi—almost 4,000 years ago—the price of copper was about a thousand times its price in the USA now relative to wages. At the time of the Roman Empire the price was about a hundred times the present price.

In Figure 4 we see the price of copper relative to the consumer price index. Everything we buy—pens, shirts, tyres—has been getting cheaper over the years because we have learned how to make them more cheaply, especially during the past 200 years. Even so, the extraordinary fact is that natural resources have been getting cheaper even faster than consumer goods.

So, by any measure, natural resources have been getting more available rather than more scarce.

Figure 3

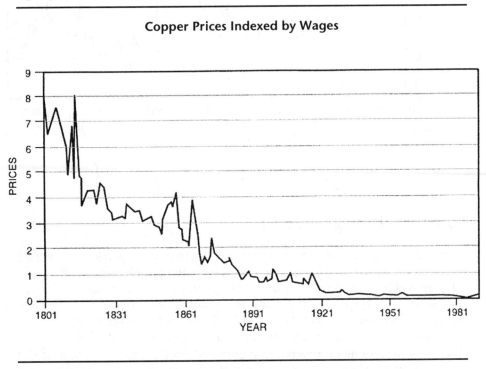

Copper Prices Indexed by Wages

In the case of oil, the shocking price rises during the 1970s and 1980s were not caused by growing scarcity in the world supply. And indeed, the price of petroleum in inflation-adjusted dollars has returned to levels about where they were before the politically-induced increases, and the price of gasoline is about at the historic low and still falling. Taking energy in general, there is no reason to believe that the supply of energy is finite, or that the price of energy will not continue its long-run decrease indefinitely. . . .

Figure 4

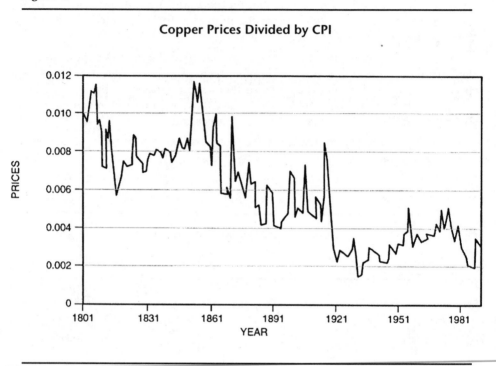

Copper Prices Divided by CPI

Food—'A Benign Trend'

Food is an especially important resource. The evidence is particularly strong for food that we are on a benign trend despite rising population. The long-run price of food relative to wages is now perhaps only a tenth as much as it was in 1800 in the USA. Even relative to consumer products, the price of grain is down because of increased productivity, as with all other primary products.

Famine deaths due to insufficient food supply have decreased even in absolute terms, let alone relative to population, in the past century, a matter which pertains particularly to the poor countries. Per-person food consumption is up over the last 30 years. And there are no data showing that the bottom of the income scale is faring worse, or even has failed to share in the general improvement, as the average has improved.

Africa's food production per person is down, but by 1994 almost no-one any longer claims that Africa's suffering results from a shortage of land or water or sun. The cause of hunger in Africa is a combination of civil wars and collectivisation of agriculture, which periodic droughts have made more murderous.

Consider agricultural land as an example of all natural resources. Although many people consider land to be a special kind of resource, it is subject to the same processes of human creation as other natural resources. The most important fact about agricultural land is that less and less of it is needed as the decades pass. This idea is utterly counter-intuitive. It seems entirely obvious that a growing world population would need larger amounts of farmland. But the title of a remarkably prescient article by Theodore Schultz in 1951 tells the story: 'The Declining Economic Importance of Land'.

The increase in actual and potential productivity per unit of land has grown much faster than population, and there is sound reason to expect this trend to continue. Therefore, there is less and less reason to worry about the supply of land. Though the stock of usable land seems fixed at any moment, it is constantly being increased—at a rapid rate in many cases—by the clearing of new land or reclamation of wasteland. Land also is constantly being enhanced by increasing the number of crops grown per year on each unit of land and by increasing the yield per crop with better farming methods and with chemical fertiliser. Last but not least, land is created anew where there was no land.

The One Scarce Factor

There is only one important resource which has shown a trend of increasing scarcity rather than increasing abundance. That resource is the most important of all—human beings. Yes, there are more people on earth now than ever before. But if we measure the scarcity of people the same way that we measure the scarcity of other economic goods—by how much we must pay to obtain their services—we see that wages and salaries have been going up all over the world, in poor countries as well as rich. The amount that you must pay to obtain the services of a barber or cook—or economist—has risen in the United States over the decades. This increase in the price of people's services is a clear indication that people are becoming more scarce even though there are more of us.

Surveys show that the public believes that our air and water have been getting more polluted in recent years. The evidence with respect to air indicates that pollutants have been declining, especially the main pollutant, particulates. With respect to water, the proportion of monitoring sites in the USA with water of good drinkability has increased since the data began in 1961.

Every forecast of the doomsayers has turned out flat wrong. Metals, foods, and other natural resources have become more available rather than more scarce throughout the centuries. The famous Famine 1975 forecast by the Paddock brothers—that we would see millions of famine deaths in the US on television in the 1970s—was followed instead by gluts in agricultural markets. Paul Ehrlich's primal scream about 'What will we do when the [gasoline] pumps run dry?' was followed by gasoline cheaper than since the 1930s. The Great Lakes are not dead; instead they offer better sport fishing than ever. The main pollutants, especially the particulates which have killed people for years, have lessened in our cities. Socialist countries are a different and tragic environmental story, however!

... But nothing has reduced the doomsayers' credibility with the press or their command over the funding resources of the federal government....

With respect to population growth: A dozen competent statistical studies, starting in 1967 with an analysis by Nobel prizewinner Simon Kuznets, agree that there is no negative statistical relationship between economic growth and population growth. There is strong reason to believe that more people have a positive effect in the long run.

Population growth does not lower the standard of living—all the evidence agrees. And the evidence supports the view that population growth raises it in the long run.

Incidentally, it was those statistical studies that converted me in about 1968 from working in favour of population control to the point of view that I hold today. I certainly did not come to my current view for any political or religious or ideological reason.

The basic method is to gather data on each country's rate of population growth and its rate of economic growth, and then to examine whether —looking at all the data in the sample together—countries with high population growth rates have economic growth rates lower than average, and countries with low population growth rates have economic growth rates higher than average. All the studies agree in concluding that this is not so; there is no correlation between economic growth and population growth in the intermediate run.

Of course one can adduce cases of countries that seemingly are exceptions to the pattern. It is the genius of statistical inference, however, to enable us to draw valid generalisations from samples that contain such wide variations in behaviour. The exceptions can be useful in alerting us to possible avenues for further analysis, but as long as they are only exceptions, they do not prove that the generalisation is not meaningful or useful.

Population Density Favours Economic Growth

The research-wise person may wonder whether population density is a more meaningful variable than population growth. And, indeed, such studies have been done. And again, the statistical evidence directly contradicts the common-sense conventional wisdom. If you make a chart with population density on the horizontal axis and either the income level or the rate of change of income on the vertical axis, you will see that higher density is associated with better rather than poorer economic results....

The most important benefit of population size and growth is the increase it brings to the stock of useful knowledge. Minds matter economically as much as, or more than, hands or mouths. Progress is limited largely by the availability of trained workers. The more people who enter our population by birth or immigration, the faster will be the rate of progress of our material and cultural civilisation.

Here we require a qualification that tends to be overlooked: I do not say that all is well everywhere, and I do not predict that all will be rosy in the future.

Children are hungry and sick; people live out lives of physical or intellectual poverty, and lack of opportunity; war or some new pollution may finish us off. What I am saying is that for most relevant economic matters I have checked, the aggregate trends are improving rather than deteriorating.

Also, I do not say that a better future happens automatically or without effort. It will happen because women and men will struggle with problems with muscle and mind, and will probably overcome, as people have overcome in the past—*if the social and economic system gives them the opportunity to do so.*

The Explanation of These Amazing Trends

Now we need some theory to explain how it can be that economic welfare grows along with population, rather than humanity being reduced to misery and poverty as population grows.

The Malthusian theory of increasing scarcity, based on supposedly fixed resources (the theory that the doomsayers rely upon), runs exactly contrary to the data over the long sweep of history. It makes sense therefore to prefer another theory.

The theory that fits the facts very well is this: More people, and increased income, cause problems in the short run. Short-run scarcity raises prices. This presents opportunity, and prompts the search for solutions. In a free society, solutions are eventually found. And in the long run the new developments leave us better off than if the problems had not arisen.

To put it differently, in the short run more consumers mean less of the fixed available stock of goods to be divided among more people. And more workers labouring with the same fixed current stock of capital means that there will be less output per worker. The latter effect, known as 'the law of diminishing returns', is the essence of Malthus's theory as he first set it out.

But if the resources with which people work are not fixed over the period being analysed, the Malthusian logic of diminishing returns does not apply. And the plain fact is that, given some time to adjust to shortages, the resource base does not remain fixed. People create more resources of all kinds.

When we take a long-run view, the picture is different, and considerably more complex, than the simple short-run view of more people implying lower average income. In the very long run, more people almost surely imply more available resources and a higher income for everyone.

I suggest you test this idea against your own knowledge: Do you think that our standard of living would be as high as it is now if the population had never grown from about 4 million human beings perhaps 10,000 years ago? I do not think we would now have electric light or gas heat or cars or penicillin or travel to the moon or our present life expectancy of over 70 years at birth in rich countries, in comparison to the life expectancy of 20 to 25 years at birth in earlier eras, if population had not grown to its present numbers....

The Role of Economic Freedom

Here we must address another crucial element in the economics of resources and population—the extent to which the political-social-economic system provides personal freedom from government coercion. Skilled people require an appropriate social and economic framework that provides incentives for working hard and taking risks, enabling their talents to flower and come to fruition. The key elements of such a framework are economic liberty, respect for property, and fair and sensible rules of the market that are enforced equally for all.

The world's problem is not too many people, but lack of political and economic freedom. Powerful evidence comes from an extraordinary natural experiment that occurred starting in the 1940s with three pairs of countries that have the same culture and history, and had much the same standard of living when they split apart after the Second World War—East and West Germany, North and South Korea, Taiwan and China. In each case the centrally planned communist country began with less population 'pressure', as measured by density per square kilometre, than did the market-directed economy. And the communist and non-communist countries also started with much the same birth rates.

The market-directed economies have performed much better economically than the centrally-planned economies. The economic-political system clearly was the dominant force in the results of the three comparisons. This powerful explanation of economic development cuts the ground from under population growth as a likely explanation of the speed of nations' economic development.

The Astounding Shift in the Scholarly Consensus

So far I have been discussing the factual evidence. But in 1994 there is an important new element not present 20 years ago. The scientific community of scholars who study population economics now agrees with almost all of what is written above. The statements made above do not represent a single lone voice, but rather the current scientific consensus.

The conclusions offered earlier about agriculture and resources and demographic trends have always represented the consensus of economists in those fields. And the consensus of population economists also is now not far from what is written here.

In 1986, the US National Research Council and the US National Academy of Sciences published a book on population growth and economic development prepared by a prestigious scholarly group. This 'official' report reversed almost completely the frightening conclusions of the 1971 NAS report. 'Population growth [is] at most a minor factor. . . .' As cited earlier in this paper, it found benefits of additional people as well as costs.

A host of review articles by distinguished economic demographers in the past decade has confirmed that this 'revisionist' view is indeed consistent with the scientific evidence, though not all the writers would go as far as I do in pointing out the positive long-run effects of population growth. The consensus is more towards a 'neutral' judgement. But this is a huge change from the earlier judgement that population growth is economically detrimental.

By 1994, anyone who confidently asserts that population growth damages the economy must turn a blind eye to the scientific evidence.

Summary and Conclusion

In the short run, all resources are limited. An example of such a finite resource is the amount of space allotted to me. The longer run, however, is a different story. The standard of living has risen along with the size of the world's population since the beginning of recorded time. There is no convincing economic reason why these trends towards a better life should not continue indefinitely.

The key theoretical idea is this: The growth of population and of income create actual and expected shortages, and hence lead to price rises. A price increase represents an opportunity that attracts profit-minded entrepreneurs to seek new ways to satisfy the shortages. Some fail, at cost to themselves. A few succeed, and the final result is that we end up better off than if the original shortage problems had never arisen. That is, we need our problems though this does not imply that we should purposely create additional problems for ourselves.

I hope that you will now agree that the long-run outlook is for a more abundant material life rather than for increased scarcity, in the United States and in the world as a whole. Of course, such progress does not come automatically. And my message certainly is not one of complacency. In this I agree with the doomsayers—that our world needs the best efforts of all humanity to improve our lot. I part company with them in that they expect us to come to a bad end despite the efforts we make, whereas I expect a continuation of humanity's history of successful efforts. And I believe that their message is self-fulfilling, because if you expect your efforts to fail because of inexorable natural limits, then you are likely to feel resigned; and therefore literally to resign. But if you recognise the possibility—in fact the probability—of success, you can tap large reservoirs of energy and enthusiasm.

Adding more people causes problems, but people are also the means to solve these problems. The main fuel to speed the world's progress is our stock of knowledge, and the brakes are (a) our lack of imagination, and (b) unsound social regulation of these activities.

The ultimate resource is people—especially skilled, spirited, and hopeful young people endowed with liberty—who will exert their wills and imaginations for their own benefit, and so inevitably benefit not only themselves but the rest of us as well.

POSTSCRIPT

Are Major Changes Needed to Avert a Global Environmental Crisis?

It is tempting to accept Simon's rosy predictions for the future and his faith in the ability of human beings to solve whatever problems they confront. However, he undermines his argument by falsely suggesting that environmental degradation is a problem only in poor and socialist parts of the world. He states that the Environmental Protection Agency (EPA) acknowledges that air and water in America have generally been getting cleaner. In actuality, however, while citing some significant local improvements, the EPA and most other analysts report continued deterioration of air and water quality in many parts of the United States as well as in the rest of the highly populated regions of the world.

Simon's view that increasing world population is positive rather than problematic is fully explicated in his book *The Ultimate Resource* (Princeton University Press, 1981). In the March/April 1997 issue of *The Futurist*, the topic "The Global Environment: Megaproblem or Not?" is debated by Simon and three other participants. A very pointed rebuttal to Simon and his perspective is Robert Cohen's guest opinion "Cornucopians, Global Resources and Technological Fixes," in the September 6, 1998, issue of the *Boulder Daily Camera* of Boulder, Colorado.

People who are concerned about a future environmental crisis usually promote the need for "sustainable development." This concept was popularized by the World Commission on Environment and Development in a much-publicized report entitled *Our Common Future* in 1987. Commission chairperson Gro Harlem Brundtland, then the prime minister of Norway, has actively publicized its findings and recommendations. Her keynote address at the 1989 Forum on Global Change, "Global Change and Our Common Future," was published in *Environment* (June 1989). For a detailed discussion of sustainable development from the perspective of Herman Daly and other environmental economists who believe that it requires development without growth, see the four weekly issues of *Rachel's Environment and Health Weekly* beginning November 12, 1998.

The limits on economic development and population growth based on the Earth's "carrying capacity" appear to be rejected by Simon. For more insight into this controversial concept, see the Worldwatch Institute's *State of the World* series of annual books (published by W. W. Norton). The need for a restructured, sustainable economy is a recurring theme. The 2000 volume's opening chapter, Lester R. Brown's "Challenges of the New Century," states, "If we cannot stabilize climate and we cannot stabilize population, there is not an ecosystem on Earth that we can save.... There is no middle path. The challenge is either to build an economy that is sustainable or to stay with our unsustainable economy until it declines. It is not a goal that can be compromised. One way or another, the choice will be made by our generation, but it will affect life on Earth for all generations to come."

Contributors to This Volume

EDITOR

THEODORE D. GOLDFARB is a professor of chemistry at the State University of New York at Stony Brook. He earned a B.A. from Cornell University and a Ph.D. from the University of California, Berkeley. He is the author of over 35 research papers and articles on molecular structure, environmental chemistry, and science policy, as well as the book *A Search for Order in the Physical Universe* (W. H. Freeman, 1974). He is also the editor of *Sources: Notable Selections in Environmental Studies* (Dushkin/McGraw-Hill). Dr. Goldfarb is a recipient of the State University of New York's Chancellor's Award for Excellence in Teaching. In addition to teaching undergraduate and graduate courses in environmental and physical chemistry, he has taught summer institutes and special seminars for college and secondary school teachers and for undergraduate and graduate research students on a variety of topics, including energy policy, integrated waste management strategies, sustainable development, and ethics in science. He is presently directing a program funded by the National Science Foundation that is designed to promote the incorporation of ethics and values issues in the teaching of secondary school science. Dr. Goldfarb has served as consultant and adviser to citizens' groups, town and city governments, and federal and state agencies on environmental matters. He is an active member of several professional organizations, including the American Chemical Society, the American Association for the Advancement of Science, and the New York Academy of Sciences.

STAFF

Theodore Knight List Manager
David Brackley Senior Developmental Editor
Juliana Gribbins Developmental Editor
Rose Gleich Administrative Assistant
Brenda S. Filley Director of Production/Design
Juliana Arbo Typesetting Supervisor
Diane Barker Proofreader
Richard Tietjen Publishing Systems Manager
Larry Killian Copier Coordinator

392

AUTHORS

JANET N. ABRAMOVITZ is a senior researcher at the Worldwatch Institute.

DENNIS T. AVERY is director of the Hudson Institute's Center for Global Food Issues. For nearly a decade, he was the senior agricultural analyst in the U.S. Department of State. He has also done policy analysis for the U.S. Department of Agriculture and President Lyndon Johnson's National Advisory Commission on Food and Fiber. He is the author of *Saving the Planet With Pesticides and Plastics* (Hudson Institute, 1996).

RONALD BAILEY is the science correspondent for *Reason* magazine.

RICK BASS, a nature writer, studied wildlife science and geology at Utah State University. After graduation, he worked as a geologist in Mississippi, wild-catting for oil. His many books of short stories and nonfiction include *The Book of Yaak* (Houghton Mifflin, 1997), *The Deer Pasture: Essays,* rev. ed. (W. W. Norton, 1996), and *The Lost Grizzlies: A Search for Survivors in the Wilderness of Colorado* (Houghton Mifflin, 1995).

CHRIS BRIGHT is a research associate at the Worldwatch Institute, senior editor of *World Watch Magazine,* and the author of *Life Out of Bounds: Bioinvasion in a Borderless World* (W. W. Norton, 1998).

CAROL M. BROWNER has been the administrator for the Environmental Protection Agency since 1993. She has also served as director of Florida's Department of Environmental Regulation, as legislative director for then-senator Al Gore, and as counsel for the U.S. Senate Committee on Energy and Natural Resources. She earned her bachelor's degree and her law degree from the University of Florida in 1977 and 1979, respectively.

LUTHER J. CARTER is an independent, Washington, D.C.-based journalist. He is the author of *Nuclear Imperatives and the Public Trust: Dealing With Radioactive Waste* (Resources for the Future, 1987), which received the Special Forum Award from the U.S. Council on Energy Awareness in 1988. He is also a former writer for *Science,* a journal of the American Association for the Advancement of Science, specializing in energy and the environment.

THEO COLBORN is a zoologist and senior scientist with the World Wildlife Fund in Washington, D.C. He is coauthor, with Dianne Dumanoski and John Peterson Myers, of *Our Stolen Future: Are We Threatening Our Fertility, Intelligence, and Survival? A Scientific Detective Story* (Penguin Group, 1997).

JULIE L. DAVIDSON studies sustainability of liberal democracies and ecological citizenship at the Centre for Environmental Studies at the University of Tasmania.

RICHARD A. DENISON is a senior scientist with the Environmental Defense Fund (EDF) in Washington, D.C., and an expert on waste management and incineration. He headed the EDF's delegation on the EDF-McDonald's Waste Reduction Task Force, beginning in 1990. He earned his Ph.D. in molecular biophysics and biochemistry from Yale University. He is coeditor, with John Ruston, of *Recycling and Incineration: Evaluating the Choices* (Island Press, 1990).

DIANNE DUMANOSKI is an award-winning environmental journalist and a Knight Fellow in Science Journalism at the Massachusetts Institute of Technology. She has been covering environmental issues since Earth Day 1970, and she reported on the Earth Summit in Rio de Janeiro in 1992. During her time as an environmental reporter for the *Boston Globe,* she pioneered in reporting on the new generation of global environmental issues, such as global warming, ozone depletion, human population, and threats to biodiversity. She is coauthor, with Theo Colborn and John Peterson Myers, of *Our Stolen Future: Are We Threatening Our Fertility, Intelligence, and Survival? A Scientific Detective Story* (Penguin Group, 1997).

DAVID E. ERVIN is director of policy studies at the Henry A. Wallace Institute for Alternative Agriculture in Greenbelt, Maryland.

CHRISTOPHER FLAVIN is senior vice president of the Worldwatch Institute.

DAVID FRIEDMAN is a writer, an international consultant, and fellow in the MIT Japan program.

JAN MARIE FRITZ a certified clinical sociologist, is associate professor of planning and health policy at the University of Cincinnati. She is an executive board member of the International Sociological Association and the American Health Planning Association.

PATTI GOLDMAN is managing attorney in the Seattle office of the Earthjustice Legal Defense Fund, an environmental law firm affiliated with the Sierra Club.

LINDA GREER, a senior scientist with the Natural Resources Defense Council, has worked for more than 15 years on the regulation of toxic chemicals and hazardous waste.

BRIAN HALWEIL is a staff researcher at the Worldwatch Institute.

DOUG HARBRECHT is a correspondent for *Business Week* in Washington, D.C.

PAUL HARRISON is the author of *The Third Revolution* (Penguin Books, 1993), which won a Population Institute Global Media Award.

BETSY HARTMANN is director of the Hampshire College Population and Development Program in Amherst, Massachusetts, and she has been a fellow of the Institute for Food and Development Policy. She is the author of *Reproductive Rights and Wrongs: The Global Politics of Population Control,* rev. ed. (South End Press, 1995).

CHRIS HENDRICKSON is a professor of civil engineering in and associate dean of the College of Engineering at Carnegie Mellon University in Pittsburgh, Pennsylvania.

INTERNATIONAL FOOD INFORMATION COUNCIL is a nonprofit organization based in Washington, D.C., whose purpose is to provide sound, scientific information on food safety and nutrition to journalists, health professionals, educators, government officials, and consumers.

JACQUELINE R. KASUN is a professor emeritus of economics at Humboldt State University and editorial director of the Center for Economic Education in Bayside, California.

DAVID LANGHORST is an executive board member of the Idaho Wildlife Federation.

LESTER LAVE is the James H. Higgins Professor of Economics at Carnegie Mellon University in Pittsburgh, Pennsylvania, with appointments in the Graduate School of Industrial Administration, the School of Urban and Public Affairs, and the Department of Engineering and Public Policy. He received a Ph.D. in economics from Harvard University and was a senior fellow at the Brookings Institution from 1978 to 1982.

FRANCIS McMICHAEL is the Blenko Professor of Environmental Engineering at Carnegie Mellon University in Pittsburgh, Pennsylvania.

DANIEL B. MENZEL is a professor in and chair of the Department of Community and Environmental Medicine at the University of California, Irvine. His current research interests are the effects of drinking water contaminants and the effects of air pollution on acute and chronic disease. He is coeditor, with Fred J. Miller, of *Fundamentals of Extrapolation Modeling of Inhaled Toxicants: Ozone and Nitrogen Dioxide* (Hemisphere, 1984).

JOHN PETERSON MYERS is director of the W. Alton Jones Foundation in Rio de Janeiro, Brazil. He is coauthor, with Theo Colborn and Dianne Dumanoski, of *Our Stolen Future: Are We Threatening Our Fertility, Intelligence, and Survival? A Scientific Detective Story* (Penguin Group, 1997).

D. WARNER NORTH is senior vice president of NorthWorks, Inc., in Mountain View, California. A former president of the Society for Risk Analysis, he is currently a member of the Board on Radioactive Waste of the National Research Council of the National Academy of Sciences. In 1997 he received the Frank P. Ramsey medal for distinguished contributions to the field of decision analysis from the Decision Analysis Society of the Institute for Operations Research and the Management Sciences. He earned his B.S. in physics and mathematics from Yale University and his Ph.D. in operations research from Stanford University.

RAYMOND J. PATCHAK, a certified hazardous materials manager, has over 10 years of environmental regulatory compliance experience, during which time he has been responsible for auditing and developing compliance programs at many different types of manufacturing and waste-handling facilities. An active member of the Academy of Certified Hazardous Materials Managers (ACHMM), he is a founding member of the ACHMM's ISO 14000 Committee as well as president of the Michigan chapter of the ACHMM.

THOMAS H. PIGFORD is a professor of nuclear engineering at the University of California at Berkeley and an internationally prominent adviser on radioactive waste and other nuclear issues. He is coauthor, with Manson Benedict and Hans Wolfgang Levi, of *Nuclear Chemical Engineering* (McGraw-Hill, 1981).

MARK L. PLUMMER is a senior fellow at the Discovery Institute, which is a nonpartisan center for national and international affairs. He specializes in environmental issues, and he is coauthor, with Charles C. Mann, of *The*

Aspirin Wars: Money, Medicine, and 100 Years of Rampant Competition (Alfred A. Knopf, 1991).

BERNARD J. REILLY is corporate counsel at DuPont, where he has been managing the legal aspects of the company's Superfund program since 1986.

JOHN F. RUSTON is an economic analyst with the Environmental Defense Fund in New York City. He earned his bachelor's degree in environmental policy and planning from the University of California at Davis and his master's degree in city planning from the Massachusetts Institute of Technology. He is coeditor, with Richard A. Denison, of *Recycling and Incineration: Evaluating the Choices* (Island Press, 1990).

STEPHEN H. SAFE is in the Department of Veterinary Physiology and Pharmacology at Texas A&M University in College Station, Texas. He is the founding principal of Wellington Environmental Consultants, now Wellington Laboratories, and he is a member of the American Association for Cancer Research and the American College of Toxicology.

MARK SAGOFF is director of the Institute for Philosophy and Public Policy in the School of Public Affairs at the University of Maryland. His books include *Ecology and Law: Science's Dilemma in the Courtroom* (University of Maryland, Sea Grant Program, 1991) and *The Economy of the Earth: Philosophy, Law and the Environment* (Cambridge University Press, 1990).

JULIAN L. SIMON is a professor of economics and business administration in the College of Business and Management at the University of Maryland at College Park. His research interests focus on population economics, and his publications include *Population Matters: People, Resources, Environment, and Immigration* (Transaction Publishers, 1990), *The Ultimate Resource,* 2d ed. (Princeton University Press, 1994), and *The Economics of Population: Key Modern Writings* (Edward Elgar, 1997).

WILLIAM R. SMITH, a certified hazardous materials manager and an ISO 14001 Environmental Management Systems (EMS) auditor, is a principal at Competitive Edge Environmental Management Systems, Inc. He has served on the Registrar Accreditation Board's Auditor Certification Board, which governs the ISO 14000 EMS auditor certification program in the United States, and he is a board member of the Academy of Certified Hazardous Materials Managers. He is also the founding chairman of the ISO 14000 Committee.

BYRON SWIFT is a senior attorney and director of the Technology Center at the Environmental Law Institute in Washington, D.C. He is coauthor, with Kathryn S. Fuller, of *Latin American Wildlife Trade Laws*, 2d ed. (World Wildlife Fund, 1985).

JERRY TAYLOR is director of natural resources studies at the Cato Institute in Washington D.C. He is a former editor of *The Environmental Monitor*, an adjunct scholar at the Institute for Energy Research, and an associate editor of *Regulation* magazine.

BRIAN TOKAR is an associate faculty member at Goddard College in Plainfield, Vermont. A regular correspondent for *Z* magazine, he has been an activist for over 20 years in the peace, antinuclear, environmental, and green politics movements. He is the author of *The Green Alternative: Creating an Ecological Future,* 2d. ed. (R & E Miles, 1987).

WILLIAM TUCKER, a writer and social critic, is a staff writer for *Forbes* magazine. His publications include *The Excluded Americans: Homelessness and Housing Policies* (Regnery Gateway, 1989), which won the 1991 Mencken Award for best nonfiction, and *Zoning, Rent Control, and Affordable Housing* (Cato Institute, 1991).

CHRISTOPHER VAN LÖBEN SELS is a senior project analyst for the Natural Resources Defense Council in San Francisco, California.

J. MARTIN WAGNER is director of international programs for the Earthjustice Legal Defense Fund, an environmental law firm affiliated with the Sierra Club.

TED WILLIAMS has been a regular contributor to *Audubon* for over 15 years, during which he has covered a variety of topics, including gold mining in Alaska and the Northern Forest of the Northeast.

BRUCE YANDLE is an alumni professor and senior associate in the Center for Policy and Legal Studies of the College of Business and Public Affairs at Clemson University in Clemson, South Carolina. He has authored or edited a number of books, including *Common Sense and Common Law for the Environment: Creating Wealth in Hummingbird Economies* (Rowman & Littlefield, 1997) and *Land Rights: The 1990s' Property Rights Rebellion* (Rowman & Littlefield, 1995).

Index